Eckart Voland · Karl Grammer (Eds.)
Evolutionary Aesthetics

Springer

*Berlin
Heidelberg
New York
Hong Kong
London
Milan
Paris
Tokyo*

Eckart Voland · Karl Grammer (Eds.)

Evolutionary Aesthetics

With 57 Figures, 11 in Colour, and 7 Tables

 Springer

Professor Dr. ECKART VOLAND
Center for Philosophy and Foundations of Science
Justus-Liebig-University Giessen
Otto-Behaghel-Strasse 10 C
35394 Giessen
Germany

Professor Dr. KARL GRAMMER
Ludwig-Boltzmann-Institute for Urban Ethology
University of Vienna
Althanstrasse 14
1090 Vienna
Austria

ISBN 3-540-43670-7 Springer-Verlag Berlin Heidelberg New York

Library of Congress Cataloging-in-Publication Data
Evolutionary aesthetics / Eckart Voland, Karl Grammer, eds.
p. cm.
"With 57 figures, 10 in color, and 7 tables".
Includes bibliographical references and index.
ISBN 3-540-43670-7 (alk. paper)
1. Aesthetics–Psychological aspects. 2. Evolutionary psychology. I. Voland, Eckart,
1949- II. Grammer, Karl, 1950-
BH301.P78E96 2003
111'.85-dc21

This work is subject to copyright. All rights reserved, whether the whole or part of the material is concerned, specifically the rights of translation, reprinting, reuse of illustrations, recitation, broadcasting, reproduction on microfilm or in any other way, and storage in data banks. Duplication of this publication or parts thereof is permitted only under the provisions of the German Copyright Law of September 9, 1965, in its current version, and permission for use must always be obtained from Springer-Verlag. Violations are liable for prosecution under the German Copyright Law.

Springer-Verlag Berlin Heidelberg New York
a member of BertelsmannSpringer Science+Business Media GmbH
http://www.springer.de

© Springer-Verlag Berlin Heidelberg 2003
Printed in Germany

The use of general descriptive names, registered names, trademarks, etc. in this publication does not imply, even in the absence of a specific statement, that such names are exempt from the relevant protective laws and regulations and therefore free for general use.

Cover Design: Design & Production, Heidelberg
Cover photograph: Bill Lorenz
Typesetting: Mitterweger & Partner, Plankstadt
31/3150W1 - 5 4 3 2 1 0 – Printed on acid-free paper

Preface

This book was initiated by a roundtable meeting in April 2001 which was organized by the Konrad Lorenz Institute for Evolution and Cognition Research (KLI) in Altenberg, Austria. The symposium brought together scientists from various fields who presented their ideas on the evolution of aesthetics. Aesthetics, as a discipline of philosophy, truly has a rich intellectual history. One therefore might be inclined to capitulate to its philosophical complexity.

Recent developments in evolutionary theory, however, have helped to successfully naturalize the specific Humana, above all epistemology and ethics. This observation encouraged us, also trying to bring together aesthetics with the Darwinian paradigm. The present volume, therefore, can be understood as a kind of a "Darwinian sonde" in a field traditionally occupied by non-Darwinian lines of arguments.

We thank the staff of the KLI for their professional assistance with the preparation of the meeting. We are grateful to Thomas Weise and Kai Willführ, who showed both commitment and care when helping out during the various phases of the production of this book. We are especially grateful to Dr. Adolf Heschl (KLI), who, full of sympathy and enthusiasm accompanied and supported the enterprise from the very beginning up to the making of this volume.

Giessen and Vienna, March 2003   Eckart Voland, Karl Grammer

Contents

Introduction 1
KARL GRAMMER and ECKART VOLAND

I The Problem

Darwinian Aesthetics Informs Traditional Aesthetics 9
RANDY THORNHILL

II From Is to Beauty – From Perception to Cognition

The Beauties and the Beautiful – Some Considerations
from the Perspective of Neuronal Aesthetics 39
OLAF BREIDBACH

The Role of Evolved Perceptual Biases in Art and Design 69
RICHARD G. COSS

From Sign and Schema to Iconic Representation.
Evolutionary Aesthetics of Pictorial Art 131
CHRISTA SÜTTERLIN

III The Fitness of Beauty – Aesthetics and Adaptation

Beauty and Sex Appeal: Sexual Selection of Aesthetic Preferences .. 173
UTA SKAMEL

Beyond Nature Versus Culture: A Multiple Fitness Analysis
of Variations in Grooming 201
MICHAEL R. CUNNINGHAM and STEPHEN R. SHAMBLEN

Aesthetic Preferences in the World of Artifacts –
Adaptations for the Evaluation of Honest Signals? 239
ECKART VOLAND

Handaxes: The First Aesthetic Artefacts 261
Steven Mithen

IV Modular Aesthetics

Human Habitat Preferences: A Generative Territory for Evolutionary Aesthetics Research 279
Bernhart Ruso, LeeAnn Renninger and Klaus Atzwanger

Bodies in Motion: A Window to the Soul 295
Karl Grammer, Viktoria Keki, Beate Striebel,
Michaela Atzmüller and Bernhard Fink

Perfumes 325
Manfred Milinski

Do Women Have Evolved Adaptation for Extra-Pair Copulation? ... 341
Randy Thornhill and Steven W. Gangestad

Subject Index 369

List of Contributors

MICHAELA ATZMÜLLER
Ludwig-Boltzmann-Institute for Urban Ethology, University of Vienna,
Althanstrasse 14, 1090 Vienna, Austria

KLAUS ATZWANGER
Institute for Anthropology, University of Vienna, Althanstrasse 14,
1090 Vienna, Austria

OLAF BREIDBACH
Ernst Haeckel Haus, University of Jena, Berggasse 7, 07745 Jena,
Germany

RICHARD G. COSS
Department of Psychology, University of California at Davis, Davis,
CA 95616, USA

MICHAEL R. CUNNINGHAM
Department of Psychological and Brain Sciences, University of Louisville,
Louisville, KY 40292, USA

BERNHARD FINK
Ludwig-Boltzmann-Institute for Urban Ethology, University of Vienna,
Althanstrasse 14, 1090 Vienna, Austria

STEVEN W. GANGESTAD
Department of Psychology, University of New Mexico, Albuquerque,
NM 87131, USA

KARL GRAMMER
Ludwig-Boltzmann-Institute for Urban Ethology, University of Vienna,
Althanstrasse 14, 1090 Vienna, Austria

VIKTORIA KEKI
Ludwig-Boltzmann-Institute for Urban Ethology, University of Vienna,
Althanstrasse 14, 1090 Vienna, Austria

MANFRED MILINSKI
Department of Evolutionary Ecology, Max-Planck-Institute of
Limnology, August-Thienemann-Strasse 2, 24306 Ploen, Germany

STEVEN MITHEN
Department of Archaeology, Faculty of Letters and Social Sciences,
University of Reading, Whiteknights, P.O. Box 218, Reading RG6 6AA,
UK

LEEANN RENNINGER
Ludwig-Boltzmann-Institute for Urban Ethology, University of Vienna,
Althanstrasse 14, 1090 Vienna, Austria

BERNHART RUSO
Department of Retailing and Marketing, University of Economics and
Business Administration, Augasse 2-6, 1090 Vienna, Austria

STEPHEN R. SHAMBLEN
Pacific Institute of Research and Evaluation, 1300 South 4th Street,
Louisville, KY 40208, USA

UTA SKAMEL
Center for Philosophy and Foundations of Science, Justus-Liebig-
University Giessen, Otto-Behaghel-Strasse 10C, 35394 Giessen, Germany

BEATE STRIEBEL
Ludwig-Boltzmann-Institute for Urban Ethology, University of Vienna,
Althanstrasse 14, 1090 Vienna, Austria

CHRISTA SÜTTERLIN
Film Archive of Human Ethology of the Max-Planck-Society and
Human Studies Center at the Ludwig-Maximilian University of Munich,
Von-der-Tann-Strasse 3-5, 82346 Andechs, Germany

RANDY THORNHILL
Department of Biology, University of New Mexico, Albuquerque,
NM 87131, USA

ECKART VOLAND
Center for Philosophy and Foundations of Science, Justus-Liebig-
University Giessen, Otto-Behaghel-Strasse 10C, 35394 Giessen, Germany

Introduction

KARL GRAMMER and ECKART VOLAND

From Darwin's Thoughts on the Sense of Beauty to Evolutionary Aesthetics

Humans tend to judge and sort parts of their social and nonsocial environment permanently into a few basic categories: those parts they like and those parts they do not. Indeed, we have developed aesthetic preferences for those things and people we are exposed to. Moreover, needless to say, these preferences shape our behavioral choices – our tendency to seek out or avoid what the world has to offer to us. For example, humans and other animals have evolved preferences for food and habitats, for naturally occurring sensations like smells and sounds, as well as for the broad array of culturally created artifacts. Last, but not least, humans have also evolved aesthetic preferences for their sexual and social companions. Everyone knows that people may be treated differently according to their physical appearance. This differential treatment by others starts early in life. Three-month-old children gaze longer at attractive faces than at unattractive faces. From these results Langlois et al. (1990) concluded that beauty standards are not learned and that there is an innate beauty detector. Attractive children receive less punishment than unattractive children for the same types of misbehavior. Differential treatment occurs throughout all school levels, from elementary to university (Baugh and Parry 1991). In this part of our lives, attractiveness is coupled with academic achievements. Attractive students receive better grades. However, the differentiation does not stop there. Even when we apply for jobs, appearance may dominate qualification (Collins and Zebrowitz 1995). Differential treatment according to physical appearance extends even as far as our legal systems, where high attractiveness can lead to better treatment and fewer/less harsh convictions. However, this is only the case if attractiveness did not play a role in the crime (Hatfield and Sprecher 1986).

Obviously, we believe that attractive people are better. This mental coupling of "beautiful" and "morally good" has had a long tradition in Western cultural history. Albertus Magnus' sentence "Das Gute ist dem Schönen inhärent, weil das Schöne dasselbe Substrat hat wie das Gute" (loosely translated: "Well is inherent in the beautiful, as they share the same substrate") can be regarded as exemplary for a quite monotonous history of thought on this topic. The tendency to amalgamate aesthetic and moral judgments is obviously deeply seated in our emotional and cognitive mechanisms (Henss 1992).

The question then becomes: where does our obsessive preoccupation with beauty and attractiveness come from? Darwin himself promoted the idea that human aesthetic preferences might be the outcome of the human evolutionary history. In regard to a general sense of beauty, Darwin writes in the *Descent of Man* (p. 99): "...*This sense has been declared to be peculiar to man. I refer here only to the pleasure given by certain colours, forms, and sounds, and which may fairly be called a sense of the beautiful; with cultivated men such sensations are, however, intimately associated with complex ideas and trains of thought.*" and ..."*Why certain bright colours should excite pleasure cannot, I presume, be explained, any more than why certain flavours and scents are agreeable; but habit has something to do with the result, for that which is at first unpleasant to our senses, ultimately becomes pleasant, and habits are inherited. With respect to sounds, Helmholtz has explained to a certain extent on physiological principles, why harmonies and certain cadences are agreeable. But besides this, sounds frequently recurring at irregular intervals are highly disagreeable, as every one will admit who has listened at night to the irregular flapping of a rope on board ship.*" And further on "*The same principle seems to come into play with vision, as the eye prefers symmetry or figures with some regular recurrence. Patterns of this kind are employed by even the lowest savages as ornaments; and they have been developed through sexual selection for the adornment of some male animals.*" and "*Whether we can or not give any reason for the pleasure thus derived from vision and hearing, yet man and many of the lower animals are alike pleased by the same colours, graceful shading and forms, and the same sounds.*"

Darwin, even then, was already tracking the important aesthetic perception components that are used as explanation principles today. Although Darwin had grasped the basic idea, he had no explanation for the sense of beauty in animals and man, and when reading the *Descent of Man* the reader finds Darwin wavering between the idea that evolution must have worked on aesthetic preferences in both animals and humans, and the idea that a general cultural variation for aesthetic preferences exists. After reviewing the evidence he had at this time, he came to the conclusion: (p. 581) "*It is certainly not true that there is in the mind of man any universal standard of beauty with respect to the human body. It is, however, possible that certain tastes may in the course of time become inherited, though there is no evidence in favour of this belief: and if so, each race would possess its own innate ideal standard of beauty. It has been argued that ugliness consists in an approach to the structure of the lower animals, and no doubt this is partly true with the more civilised nations, in which intellect is highly appreciated; but this explanation will hardly apply to all forms of ugliness. The men of each race prefer what they are accustomed to; they cannot endure any great change; but they like variety, and admire each characteristic carried to a moderate extreme. Men accustomed to a nearly oval face, to straight and regular features, and to bright colours, admire, as we Europeans know, these points when strongly developed. On the other hand, men accustomed to a broad face, with high cheek-bones, a depressed nose, and a black skin, admire these peculiarities when strongly marked. No doubt characters of all kinds may be too much developed for beauty. Hence a perfect beauty, which implies many characters modified in a particular manner, will be in every race a prodigy. As the great anatomist Bichat long ago said, if every one were cast in the same mould, there would be no such thing as beauty. If all our women were to become as beautiful as the Venus de Medici, we should for a time be charmed; but we should soon wish for variety; and as soon as we had obtained variety, we should wish to see certain characters a little exaggerated beyond the then existing common standard."*

Today, the generality of attractiveness within cultures is readily accepted. Several rating studies, especially those by Iliffe (1960), have shown that people of an ethnic group share common attractiveness standards. In this standard, beauty and sexual attractiveness seem to be essentially the same, and ratings of pictures show a high congruence over social class, age, and sex. Moreover, recent studies (Cunningham et al. 1995) suggest that the constituents of beauty are neither arbitrary nor culture bound. The consensus on which female is considered to be good-looking, or not good-looking, is quite high across cultures. Thus, Darwin was not completely right in his doubts about a possible trans-culturally uniform beauty standard with respect to the human body.

Nevertheless, with the information available to him, Darwin's conclusion was rather suggestive. He had asked missionaries and ethnographers to describe the beauty standards of different ethnic groups. Darwin's conclusion – that an innate standard for beauty was unlikely to exist – stemmed from the massive variety of answers he received from around the world. Even today, when we only consider the wide diversity of human appearances, we might accept Darwin's conclusions. For many researchers, it still seems obvious that beauty standards are culturally determined. For example, Sarah Grogan (1999) pointed out that beauty standards vary highly over time and history and between (and even within) societies. This high degree of cultural relativism culminates in the following statement: "Evolutionary psychologists have failed to demonstrate convincingly that preferences for particular body shapes are biologically based... Current data suggest that body satisfaction is largely determined by social factors, and is intimately tied to sexuality." Basically, this line of argumentation means that no scientific progress in understanding human beauty has occurred since the times of Darwin, and leaves one to only shake one's head in wonder: should not something like human sexual attraction be linked to human reproductive consequences, and thus be prone to evolutionary processes? Although we observe that aesthetic judgments may be rather variable across people, cultures, and time, the underlying algorithms and constraints should manifest themselves as human universals of a common evolutionary origin.

Darwin's inability to explain the origins of aesthetic preferences was followed by modern evolutionary theories. It was only in the 1960s of the last century that the general theoretical foundation for the explanation of our aesthetic preferences emerged. Konrad Lorenz was one of the first researchers to lay this foundation, with his emphasis on the idea that the human cognitive apparatus has evolved to respond to physical structures that are actually present, and that human thinking and reasoning, just as much as human body form, are also a result of evolution. This means that the cognitive structures we currently possess are actually adaptations to problem solving in our phylogenetic past. In our past, those brains (or cognitive algorithms) who were able to process environmental information more efficiently were those who were selected for, generation after generation. However, Lorenz went further, he argued that in the course of evolution certain environmental stimuli which promoted reproductive success were connected to positive emotional reactions (Lorenz 1973).

On the basis of Lorenz' adapted mind theory, Cosmides and Tooby (1992) proposed that our brains have domain-specific cognitive structures which have evolved in order to process everyday information efficiently and to check this information in their impact on reproductive success. One of the first hints in favor of the idea that we have domain-specific adaptations in aesthetic conduct comes from Aharon et al. (2001). By showing photographs of attractive and unattractive people to males, these authors found out that in all males the same regions of the brain were active when seeing attractive faces. Retrospectively, it can be noted that, beginning with Darwin's thought on the sense of beauty and continuing on to the discovery of neurobiology, the scientific focus on aesthetics has undergone an increasing naturalization. Naturalization can be defined as the attempt to use the methodological and theoretical means of natural sciences to explain certain phenomena that have traditionally been the subject of studies in philosophy and the humanities. Among these scientific approaches, the attempts of naturalization from an evolutionary perspective have a particular status.

Evolutionary epistemology and evolutionary ethics are well-known, successful results of this approach. As far as aesthetics are concerned, we think that comparable attempts, namely, importing naturalistic explanations and by this means empirically corroborating or falsifying the positions of traditional philosophy with the methods of evolutionary anthropology and psychology, are still in the fledgling stages. The biological grounding has not yet been as successfully realized as in evolutionary epistemology and ethics – the "philosophical siblings" of aesthetics. Evolutionary aesthetics, as the importation of aesthetics into natural sciences, and especially its integration into the heuristics of Darwin's evolutionary theory, despite some historical attempts, still seems to be more a vision than a productive project.

There must be certain reasons for the shortcomings of aesthetics research, and it would certainly be very interesting to shed more light on them. One of these reasons might be the fact that there is a fundamental contradiction between conventional philosophy and a Darwinian approach to the aspect of usefulness of aesthetics. The idea that also aesthetic competences have proven themselves to be advantageous in processes of biological selection and therefore can be regarded as functional, is, without doubt, a tremendous challenge to traditional philosophical aesthetics, which have frequently tried to define their subject by the absence of any profitableness. Although only rarely in the history of conventional philosophical ethics and evolutionary epistemology, has there been a link established to biological determination, profitableness, and evolutionary adaptiveness, the hiatus between biology and philosophy on aesthetics is incomparably larger. However, this doesn't have to remain as it is, and maybe the articles of this volume are apt to finally breach the gap between biology and philosophy, by showing it in a new light.

References

Aharon I, Etcoff N, Ariely D, Chabris CF, O'Connor E, Breiter HC (2001) Beautiful faces have variable reward value: fMRI and behavioral evidence. Neuron 32:537–551
Baugh S, Parry L (1991) The relationship between physical attractiveness and grade point average among college women. J Soc Behav Personality 6:219–228
Collins MA, Zebrowitz LA (1995) The contributions of appearance to occupational outcomes in civilian and military settings. J Appl Soc Psychol 25:129–163
Cosmides L, Tooby J (1992) Cognitive adaptations for social exchange. In: Barkow JH, Cosmides L, Tooby J (eds) The adapted mind. Evolutionary psychology and the generation of culture. Oxford University Press. Oxford, p 163–228
Cunningham MR, Roberts AR, Barbee AP, Druen PB, Wu CH (1995) "Their ideas of beauty are, on the whole, the same as ours": Consistency and variability in the cross-cultural perception of female physical attractiveness. J Personality Social Psychol 68:261–279
Darwin C (1871) The descent of man and selection in relation to sex. John Murray, London
Grogan S (1999) Body image. Understanding body dissatisfaction in men, women and children. Routledge, London
Hatfield E, Sprecher S (1986) Mirror, mirror..: The importance of looks in everyday life. State University of New York Press, Albany
Henss R (1992) "Spieglein, Spieglein an der Wand..." – Geschlecht, Alter und psychische Attraktivität. Psychologie Verlags Union, Weinheim
Iliffe A (1960) A study of preferences in feminine beauty. Br J Psychol 51:267–273
Langlois JH, Roggman LA, Reiser-Danner LA (1990) Infant's differential social responses to attractive and unattractive faces. Dev Psychol 26:153–159
Lorenz K (1973) Die Rückseite des Spiegels. Piper, Munich

I The Problem

Darwinian Aesthetics Informs Traditional Aesthetics

Randy Thornhill

Introduction

This paper treats the topics that have been of long interest to aestheticians. Traditional aesthetics, i.e., aesthetics in philosophy, is broad and diverse, including such topics as the beauty of ideas as well as the beauty of body form, natural landscapes, scents, ideas and so on. Some colleagues have suggested that I provide a succinct definition of aesthetics. It is not possible, however, to provide an objective definition based on Darwinian theory. As D. Symons (pers. comm.) put it: "... [T]he whole notion of 'aesthetics,' as a 'natural' domain, i.e., as a domain that carves nature at a joint, is misguided.... All adaptations are aesthetic adaptations, because all adaptations interact in some way with the environment, external or internal, and prefer certain states to others. An adaptation that instantiates the rule, 'prefer productive habitats', is no more or less aesthetic than an adaptation that instantiates the rule, 'prefer a particular blood pressure'." Although there is no way to objectively define the aesthetic domain, there is value, I believe, in treating the various topics of traditional aesthetics in a modern, Darwinian/adaptationist framework.

The Darwinian theory of brain design, whether human or nonhuman, is that of many functionally specific psychological adaptations. Just as human blood pressure regulation and habitat selection are guided by different, functionally specific psychological adaptations, the traditional topics of aesthetics each arise from fundamentally different psychological adaptations. It is the many psychological adaptations that underlie the diversity of aesthetic experiences of interest to aestheticians that I address. Darwinian aesthetics has great promise for elucidating the design of the psychological adaptations involved in these experiences.

The starting point for the Darwinian theory of aesthetics I offer is as follows. Beauty experiences are unconsciously realized avenues to high fitness in human evolutionary history. Ugliness defines just the reverse. Greenough (1958), in reference to architectural structures, defined beauty as the promise of function. The Darwinian theory of human aesthetic value is that beauty is a promise of function in the environments in which humans evolved, i.e., of high likelihood of survival and reproductive success in the environments of human evolutionary history. Ugliness is the promise of low survival and reproductive failure. Human aesthetic value is a scale of reproductive success and failure in human evolutionary history, i.e., over the last few million years.

First, I briefly list the experiential domain of interest to traditional academic aestheticians. I then discuss the adaptationist program and how it applies to these experiences in a general way. Next, I resolve some dilemmas in traditional aesthetics using the adaptationist perspective. Finally, I give a taxonomy of the psychological adaptations underlying the diverse experiences of interest to aestheticians.

Aesthetics: The Topics of Interest

The study of beauty is a major endeavor in academia. Intellectual beauty is the beauty scholars in all academic disciplines find in scholarship. Intellectual beauty is the most noble goal of academic pursuit, and is a sublime reward of the pursuit. In addition, the academic discipline of aesthetics, a part of philosophy, is concerned with the rhetorical meaning of beauty and ugliness. Aesthetics apparently first became a distinct discipline within philosophy with G. Baumgarten's *Aesthetica*, published in 1750, but as documented by the historical aesthetician Kovach (1974), speculation about aesthetics by scholars has been going on in the Western World at least since the sixth century B.C. in Greece. In addition, the arts, as well as other areas of the humanities, are fundamentally concerned with beauty. The humanities are focused on competition in heightening sensation in general, and in particular in generating beauty experiences in the minds of people. Beauty may be generated in the mind by poetry, literature, paintings, dance, oral or written rhetoric, etc. When humanists contemplate beauty and analyze its meaning, they present their views rhetorically, again striving to generate the effect of beauty.

Whereas the arts and humanities compete in creating the effect of beauty in human minds, scientific aestheticians use the scientific method to understand how the effect arises and why it exists. Scientific aestheticians, then, are concerned with proximate causes (the how; physiology, development, cues or stimuli, and information processing) and/or ultimate causes (the why; evolutionary history). Scientific aesthetics is a diverse discipline, which includes the study of aesthetic valuations by non-human animals.

Since Darwin and Wallace, biologists have studied natural beauty's meaning in terms of the evolved signal content of striking phenotypic features such as showy flowers, the peacock's tail, and elaborate courtship behavior. Hypotheses to explain the existence of these kinds of extravagant features, in ultimate or evolutionary functional terms, have ranged from species or sexual identity to advertisement of phenotypic quality, i.e., an organism's advertisement of its ability to deal effectively with adverse environmental agents that impact survival and reproduction. This form of biological aesthetics has become a major research area in the last 25 years as a result of theoretical treatments validating the idea that elaborate features can evolve to honestly signal phenotypic quality, as first suggested for secondary sexual traits by A.R. Wallace (see Cronin 1991) and explicitly treated first by Zahavi (1975).

For extravagant sexual traits such as the peacock's tail, some biologists ally with Darwin's and R. Fisher's formulation about the evolution of beauty and aesthetic preference. This was that the beauty of the peacock's tail merely signals sexual attractiveness to females, not viability or phenotypic and underlying genotypic quality, and the preference for the elaborate tail evolved because it yields attractive and thus, sexually preferred sons, not offspring with high viability (for discussion of recent theory, see Cronin 1991; Ridley 1993). There is total agreement that sexual selection is the biological process that has made extravagant features such as the peacock's tail. Many sexual-selection aestheticians now are engaged in efforts to determine the exact nature of the process of sexual selection responsible for the features, i.e., advertisement of and preference for phenotypic quality, or for sexual attractiveness only. The general consensus is that mate choice is focused on phenotypic (and often genotypic) quality and sexual advertisement is basically about displaying phenotypic quality (for useful reviews, see Cronin 1991; Ridley 1993). This same conclusion seems to best fit the human data of mate choice and sexual display involving physical bodily features (reviewed in Thornhill and Gangestad 1993, 1999a, b; Symons 1995; Thornhill and Grammer 1999). Miller (2001) has persuasively argued that art production is analogous to the honest signal of quality conveyed by the peacock's tail. Accordingly, the psychology that motivates art production is sexually selected and art is then a signal to potential mates of the artist's general fitness. The scientific study of human sexual attraction and attractiveness is carried out by human sexual selection aestheticians.

Biophilia is another branch of scientific aesthetics. Wilson (1984) coined the term "biophilia" to describe the innately emotional affiliation of human beings to natural phenomena such as animals, plants and habitats. Wilson (1984) emphasized that aesthetic judgments are central to biophilia. Considerable recent and current research is clarifying human aesthetic feelings toward habitats, animals and plants (see review in Kellert and Wilson 1993).

The final branch of scientific aesthetics has been called experimental aesthetics by those in this field. Experimental aesthetics is a branch of experimental psychology dating to about the mid-nineteenth century. It focuses on the aesthetic value of shapes and patterns (for reviews, see Berlyne 1971; Solso 1994).

When a scientific aesthetician discovers the scientific meaning of beauty, he or she may generate beauty in the same way that the artist or other humanist does. Any significant scientific discovery may generate the effect of beauty in the mind of the discoverer and the minds of scientists in general.

Adaptationism

In this paper, and in theoretical (evolutionary) biology in general, adaptation refers to goal-directed, i.e., functionally designed, phenotypic features (e.g., Thornhill 1990, 1997; Symons 1992; Williams 1992). As Williams (1992) put it, an adaptation is the material effect of response to selection. Four natural processes are known to cause evolution or changes in gene frequencies of populations, but selection is the only one that can create an adaptation. The other three – mutation, drift, and gene flow – lack the necessary creativity because their action is random relative to individuals' environmental problems. Selection is not a random process. It consists of differential survival and reproduction by individuals because of differences in their phenotypic design for the environment. Thus, selection is defined concisely as nonrandom differential reproduction of individuals. The nonrandomness of selection is a result of individual differences in fitness, i.e., ability to cope with environmental problems. This nonrandomness has differential reproductive consequences. When design differences among individuals reflect genetic differences, selection accumulates genes in subsequent generations that (through the gene-environment interactions that constitute ontogeny) lead to increased fit of individuals to their environment. An adaptation, then, is a phenotypic solution to a past environmental problem that persistently impinged on individuals for long periods of evolutionary time and thereby caused cumulative directional selection, which, in turn, caused cumulative directional change in the gene pool. Evolution by selection is not a purposeful process, but gradually incorporates ever more refined and focused purpose into its products (adaptations) by persistent effects.

Each adaptation is the archive of data of the selection that made it. These data are in the functional design of the adaptation. Said differently, to discover the purposeful design of an adaptation is to discover the kind of selection that made the adaptation. This is why the study of adaptation is fundamental to understanding the evolution of life. Adaptations are our *sole* source of information about the forces of selection that actually were effective in designing phenotypes over the course of evolutionary history.

Note that the data encoded in the design of an adaptation concern the environmental features that caused differences in individual reproductive success during the evolution of the adaptation. This is just another way of putting the fact that adaptations have stamped in their functional designs the selective forces that made them. Thus, *the* way to discover the environmental forces, i.e., the selective forces that made *human* adaptations, is by deciphering the functional design of *human* adaptations. Contrary to the belief of some, the selective forces that designed humans cannot be elucidated directly by the study of nonhuman primates (see Thornhill 1997).

The aim of the adaptationist program is to identify adaptations and characterize their functional designs. An adaptation is a phenotypic feature that is so precisely organized for some apparent purpose that chance cannot be the explanation for the feature's existence. Thus, the feature cannot merely be the chance by-product or incidental effect of an adaptation or the product of random genetic drift. The feature has to be the product of long-term evolution by directional selection. Once a true adaptation is recognized and its apparent purpose perceived, the next step is to examine in detail and fully characterize the functional design of the adaptation to determine the adaptation's actual evolutionary purpose. To discover an adaptation's evolutionary purpose entails describing specifically how the adaptation contributed to reproduction of individuals during its evolution or, put differently, detailing the precise relationship between a phenotypic trait and selection during the trait's evolution into an adaptation (for further discussion of the study of adaptation, see Thornhill and Gangestad, this Vol.).

Adaptationist studies do not typically include analysis of the evolutionary origin of adaptations – that is, the phenotypic precursors that were modified by directional selection over long-term evolution into complex features with identifiable purposeful designs. Instead, the focus is on the understanding of phenotypic design and thus the forces of selection responsible for adaptation. Both the origin and selection history of an adaptation are useful to understand, but origin and selection history are different questions; the two types of questions deal with different historical causes. Thus, the evolutionary purpose/function of an adaptation can be studied productively without any reference to or understanding of the adaptation's origin.

Evolutionary psychology is the discipline of applying the adaptationist program to discover psychological design of animals. Psychological adaptation causally underlies *all* human feelings, emotion, arousal, creativity, learning and behavior (Cosmides and Tooby 1987; Symons 1987); this is indisputable. Behavior and psychological change are each the product of the processing of environmental information by psychological adaptation. Thus, aesthetic judgments are manifestations of psychological adaptation. As Symons (1995, p. 80) put it, "[P]leasure, like all experiences, is the product of brain mechanisms, and brain mechanisms are the products of evolution by .... selection. The question is not whether this view of pleasure is *correct* – as there is no known or suspected scientific alternative – but whether it is useful." Symons meant *scientifically* useful. That is, can the view guide researchers to empirical understanding of aesthetic judgments? An unequivocal "yes" was his answer, and I am in total agreement.

Human psychological adaptations are information-processing mechanisms providing solutions to information-processing problems that influenced the survival and reproductive performance of individuals during our species' evolutionary history. Human psychological adaptations are engineered to process environmental information and to guide feelings, emotions, learning and behavior toward ends that were maximally reproductive in human evolutionary history. Psychological adaptations are characterized in functional evolutionary terms, i.e., by the kinds of information they are designed by selection to process rather than in terms of neurophysiology or neuroanatomy (also see Thornhill and Gangestad, this Vol.; Cosmides and Tooby 1987; Tooby and Cosmides 1989).

Knowledge of human habitat aesthetics illustrates the purposeful design of psychological adaptation. Heerwagen, Kaplan and Orians in a series of papers (see Kaplan 1992; Heerwagen and Orians 1993 for reviews) provide considerable evidence for the existence of a number of *ancestral* cues to productive habitats in human evolutionary history. These are the cues processed by human habitat aesthetics adaptations. In human evolution, individuals with the psychological ability to use these specific cues solved the problem of choosing a habitat that was safe and fruitful. For example, humans like savanna features and specifically those savanna features that provide evidence of productive and safe habitats. People even distinguish the branching patterns of trees and leaf shapes and prefer branching patterns and leaf shapes that connote rich savanna habitats over impoverished savanna habitats. Moreover, people prefer tree-trunk branching patterns that would have allowed climbing and escaping from terrestrial predators. These preferences have been documented cross-culturally in people who have never been in a savanna.

As is the case with all human aesthetic judgments, habitat aesthetic judgments are associated with feelings. The feelings arise from the processing of information, specifically ancestral cues to habitable and unhabitable environments, by the human habitat aesthetics adaptation. These feelings guided our ancestors toward productive and safe habitats and away from unproductive, hostile habitats. In aesthetic adaptation resides not only the data of the environmental features that generated the selection that made the adaptation, but also the data of the feelings of our ancestors about those features.

We can conclude with great confidence that beauty and ugliness were important feelings in the lives of the evolutionary ancestors of humans – i.e., those individuals who out-reproduced others in human evolutionary history. The existence of human aesthetic adaptation demonstrates that human feelings about aesthetic value distinguished our evolutionary ancestors from the other individuals present in human evolutionary history who failed to reproduce or reproduced less. A beautiful idea of evolutionary psychology is that the discipline allows discovery of how human ancestors felt about various aspects of their environments; the discipline allows discovery of our emotional roots.

Characteristics of Adaptations

In addition to possessing purposeful design, which is the evidence of the form of selection that created the design, adaptations have the following characteristics:

1. They are often species-typical, i.e., possessed by all members of the species. Of course, an adaptation may be sex-specific, species-typical. It may be present in only one sex (e.g., the human penis) or show different design between the sexes (e.g., sexual jealousy is cued by sex-specific environmental stimuli in men and women [see review in Buss 1994]). Similarly, adult and juvenile humans are differently designed in many regards; in this case, the adaptations are age-specific, species-typical. Evolutionary psychology emphasizes the universality of human nature, the collection of psychological adaptations that characterize the entire human species, some of which are age- and sex-specific.

Darwinian aesthetics leads to the hypothesis that the adaptations of aesthetic judgment are present in the brains of essentially all people; all people are aesthetes. Accordingly, the psychological inability to experience beauty is extremely rare in people. The absence of the psychological structure involved is as rare as the absence of a liver. Only occasionally would a human be born without beauty psychology. Indeed, the themes of beauty and ugliness seem to be universal among human cultures (Arnheim 1988; Brown 1991). This is seen, for example, in humans who everywhere structure their homes to increase positive aesthetic experience. In technologically advanced societies, offices, vacations, spare time, reading and TV watching are so structured as well. Moreover, there is much evidence that humans strive to structure their social relationships to enhance aesthetic experience: attractiveness is a desirable feature in one's mates and friends (for reviews, see Jackson 1992; Thornhill and Gangestad 1993). Many aesthetic judgments are made everyday by each human being.

2. Adaptations show little or no heritability (Thornhill 1990, 1997; Tooby and Cosmides 1990). Obviously, adaptations are *inherited* (i.e., they are passed between generations from parent to offspring), but they are not *heritable*. Heritability is a term that describes the extent to which the variation *among* individuals in a phenotypic trait (e.g., jaw size) is caused by genetic, as opposed to environmental, variation among individuals. The term heritability does not apply to the traits of any single individual, and the dichotomy of genetic versus environmental cannot be applied to the

features of an individual. The reason that adaptations show low heritability is that directional selection for a particular function fixes the genes involved in the function.

The view that Darwinism applies only to heritable traits (e.g., Lewontin et al. 1984) is erroneous, but not uncommon. In order for an adaptation to come into existence, there must, in the evolutionary past, have been genetic variation among individuals underlying the phenotypic variation in the feature. For example, the human psychological adaptation underlying habitat aesthetics demonstrates by its existence that there was heritable variation in feelings about habitats in human evolutionary history.

Research on the heritability of aesthetic judgments is in a very elementary stage. The heritability issue is central to some researchers. For example, Wilson (1993, p. 34) sees significant heritability as central in verifying notions that humans have adaptations to natural history, such as aesthetic habitat adaptation. Soulé (1992) feels the same way, and even suggests that to show heritability is the way to elucidate the genetic mechanism of human biophilic adaptation. Other researchers, working on aesthetics – for example, Heerwagen and Orians (1993) and Symons (1995) – emphasize correctly that human variation in aesthetic judgments will arise primarily not from genetic differences, but from differing conditions affecting species-typical psychological adaptation, i.e., from environmental, condition-dependent ancestral experiences. The experience may involve the past, as in a person's ontogeny or upbringing, or it may be solely due to cues of the moment. Heerwagen and Orians (1993) explain the adaptation-based, condition-dependent view of individual variation in aesthetic valuation of landscape as follows:

> *"Unlike many current approaches to landscape aesthetics, we do not assume that everyone will respond similarly to a certain environment... Variability in people's preferences and assessments is expected, but that variability is not random. Rather, it is a function of such biologically relevant factors as age, gender, familiarity, physical condition, and presence of others...[P]eople's preferences for spatial enclosure should move back and forth along a prospect [open spaces]-refuge continuum in accordance with their social, physical and emotional feelings of vulnerability. Thus, children and elderly people, as well as those who are physically ill or depressed, should prefer spaces offering refuge rather than open spaces where they can be readily seen by others"* (p. 165).

Heerwagen and Orians are saying that a species-typical adaptation for judging the value of prospect and refuge in landscapes will be sensitive to conditions, which include such subtle things as the perceiver's physical condition. This species-typical psychological adaptation – the one that we all have – is designed to evaluate the openness and refuge features of landscapes and to value these features (i.e., feel about them) based on circumstances that were consistently present and significantly influenced differential reproduction of individuals during human evolution.

Knowing the heritability of a trait in no sense implies knowledge of the genetic mechanism for the trait or of a direct genetic effect on the trait. Health is typically heritable because of the nature of the coevolutionary arms race between hosts and disease organisms (see Tooby and Cosmides 1990). Prospect-refuge valuation may show heritability because health is heritable and individuals adopt prospect-refuge behavior based on personal condition, and not because of heritability in the adaptation for judging prospect-refuge. The heritability of agoraphobia (fear of being in open or public places) is considerable (0.4, i.e., 40% of individual variation arises from genetic differences; Ulrich 1993), but such a finding does not detract from Heerwagen and Orian's hypothesis that valuation of openness and refuge in habitats arises from psychological design that is universal in people's brains.

For the most part, heritability is useful primarily in artificial selection of desirable features in domesticated plants and animals. Studies of heritability in evolutionary biology must be carefully framed and interpreted (see Thornhill 1990).

3. Adaptations are special-purpose, not general-purpose, in functional design. As Symons (1987) has emphasized, environmental problems bringing about selection are specific and not general problems. Thus, it is expected that phenotypic solutions to environmental problems – that is, adaptations – will be special purpose in design, because a general-purpose mechanism cannot solve a specific problem (see also Tooby and Cosmides 1989). Beyond this theoretical understanding of adaptation is the vast body of empirical data showing that adaptations in plants and animals are indeed special purpose in their functional design.

Thinking of adaptations as general purpose rather than special purpose is widespread. This general-adaptation concept stems, in part, from superficial thinking about functional level. If one concludes that the human heart's evolutionary function is to pump blood, it is not difficult to imagine that any general pump design will work. However, if the adaptive problem to be solved by selection is framed more specifically and appropriately as the need for a pump that will work in a human body (not a generalized primate or a baboon body), the general-adaptation concept is unreasonable. The same conclusion is reached by substituting habitat or mate for heart in the example. Choosing a human mate of high reproductive value or a human habitat that is safe and productive are highly specific adaptive problems, each requiring a specialized psychological adaptation.

It is often assumed in the social sciences that the human mind consists of a *tabula rasa* or only a few general-purpose learning adaptations (Symons 1987). However, given current knowledge of the functional specificity of the vast numbers of nonpsychological adaptations whose design is understood (e.g., the human heart is specially designed to pump blood in a human body), the vast numbers of psychological adaptations of humans and nonhuman animals whose design has been elucidated, as well as theoretical understanding of how selection works in molding adaptations (they are specific solutions to specific environmental problems), it is most likely that the human brain is composed only of highly specialized adaptations. The ontogeny of some of these adaptations is influenced by experiences most would refer to as learning, but these, like other psychological adaptations, are designed to process specific information, namely, the ancestral cues that guided learning toward adaptive ends in human evolutionary history (see Symons 1987 for a detailed critique of the idea that human psychology consists only of a few general-purpose adaptations).

The special-purpose design of human psychological adaptations forces rejection of the common view in aesthetics that there is one or a few principles of beauty that will work equally well for landscapes, human bodily form, ideas, etc. Such principles as symmetry, harmony, truth, unity, order, femininity or woman, etc., have been suggested (Kovach 1974). Choosing a habitat, choosing a mate, and choosing a belief system are very different information-processing problems. Each domain of aesthetic judgment is expected to coincide with special-purpose aesthetic adaptation.

Evolutionarily sophisticated scholars yawn at the debate among biologists in Darwin's time as to whether nonhuman animals have aesthetic preferences. However, the general position in philosophical aesthetics is unambiguous: only humans have aesthetic experiences (Kovach 1974). Darwin (1874) first demonstrated that animals in general – insects, birds, mammals, etc. – have tastes expressed as mate preferences. For example, female barn swallows have an aesthetic preference for male tail feathers that are symmetrical (Møller 1992). People have an aesthetic preference for symmetrical faces (see review in Thornhill and Gangestad 1999a). Female scorpion flies, rock lizards and humans perceive the scent of symmetrical males as more attractive than the scent of asymmetrical males (see Thornhill and Gangestad 1999b). In addition, mobile animals face environments that present variation in the habitat features impacting survival and reproduction. As Orians (1980) has pointed out, such species will have evolved faculties that judge habitats in terms of cues to survival and motivate them to settle in a suitable habitat. All mobile animals from amoebae to primates are environmental aestheticians. Humans and other animals that depend on foods that were limited in evolutionary history are culinary aestheticians.

This is not to say that humans, scorpion flies, flatworms, barn swallows or chimps make the same aesthetic judgments. Adaptationism says that they definitely do not. Each species has species-specific psychological machinery designed for aesthetic appreciation of species-specific features of the environment that impacted reproductive success in each species' evolutionary history. Indeed, the machinery will frequently be sex-specific in its design or function as well. In addition, it is not to say that nonhuman animals feel their aesthetic preferences, i.e., are consciously aware of their mental activity surrounding aesthetic judgments. It is not necessary to feel preferences in order for the judgments to guide motivation and behavioral action in adaptive directions.

4. An adaptation need not be currently adaptive, i.e., its current form or expression need not positively affect survival or offspring production. An adaptation was necessarily adaptive in the environments of its evolution to the state of being an adaptation. Current adaptiveness is not a legitimate theoretical expectation for any adaptation. The relationship between an adaptation and current reproduction depends on the similarity between the environment in which the adaptation is expressed currently and the environmental features that generated the selection that designed the adaptation. Often this correlation no longer exists for contemporary organisms (perhaps especially for humans, who now may all live in environments, social and nonsocial that are significantly different from the environments of evolutionary history). As a result, adaptations are not distinguished from nonadaptations by their effects on current reproduction. The sole criterion for distinguishing adaptation is evidence of functional design. Functional design is seen when a trait's organization solves an ecological problem (see Thornhill and Gangestad, this Vol. for further discussion on functional design of psychological adaptations).

Data on the current offspring-producing consequences of aesthetic judgment are totally irrelevant to the Darwinian approach to understanding aesthetic judgments that I discuss herein. With regard to aesthetics of ideas, one could ask questions such as, do individuals who adopt politically correct ideology have more children than those who do not? There are some interesting questions about the beauty of ideas, but they do not include how variation in beliefs covary currently with reproductive success. Instead, they pertain to the design of adaptations that assess alternative beliefs and motivate adoption of personal ideology. One could ask also, do more symmetrical Western men have more children than asymmetrical men? I doubt it, because of contraception use, etc. More symmetrical men are more sexually successful than asymmetrical men. They have more sex partners in a lifetime, and engage in more extra-pair copulations, achieve copulation earlier in relationships, and their mates have more orgasms of the type that retain sperm (Thornhill and Gangestad 1994; Thornhill et al. 1995; Gangestad and Thornhill 1997). These findings on male symmetry and sexual behavior imply that women find symmetrical men more aesthetically pleasing. Such men are rated more attractive (see review in Thornhill and Gangestad 1999a,b). Male symmetry may be positively or negatively correlated, or not correlated, with offspring production in any given human society. Regardless, there seems to be a bias in women's sexual aesthetic judgment that gives symmetrical males an edge in sexual competition. These biases appear to be the result of functional design for obtaining a developmentally healthy (symmetric) mate.

The Perpetual Problems of Traditional Aesthetics

Philosophical aesthetics yields a list of unresolved issues that aestheticians through the ages have attempted to treat (e.g., Kovach 1974). Darwinian aesthetics accounts for the central place of these issues in the minds of aestheticians and helps resolve them.

Strong Feelings

One perpetual concern has been why aesthetic judgments often involve very strong feelings, which range from sublime ecstasy when beauty is perceived to deep repugnance when ugliness is perceived. Visualize a human body covered with oozing sores, or better yet, view a picture of one in a book on skin diseases. Contrast the feeling that arises with the feeling felt when you view a healthy, professional model of the sex that you prefer for sexual intercourse. Similarly, contrast your feelings when you have listened to the cacophony of a jackhammer in operation with your feelings when listening to your favorite music. Or, view a rich, natural habitat with lush vegetation, a crystal-clear stream and abundant flowers, ripe fruit and game animals. Contrast that feeling with your feelings when you view a rat-infested city dump or a polluted pond with dead fish afloat. Viewing oozing sores or nonproductive habitats or hearing jackhammers and untestable, discovery-stifling ideas cause aesthetically negative mental experiences. Viewing a model's looks or productive habitats, or hearing a favorite tune or a scientifically meaningful idea lead to aesthetically pleasant mental experiences.

Aesthetic experience is affective, i.e., immediate, intuitive and creative of response and mood that may be profound (e.g., Kaplan 1992; Orians and Heerwagen 1992). The immediate affective response to the aesthetic value of a thing, when positive, can generate subsequent imagination and fantasy, e.g., when one views a beautiful landscape or hears a beautiful idea. The strength of aesthetic feelings identifies the importance of such feelings in human evolutionary history. By importance, I mean that the feelings covaried with and promoted adaptive judgments in human evolutionary history.

Feelings may have evolved in humans to promote immediate and focused interest in ancestral cues and the gathering of information for use in assessing alternatives. Most human mental activity is not felt. For instance, mental activity associated with digestion, respiration and blood circulation is continuous, but all this mental activity has been delegated by selection to unconscious and subconscious components of the human brain. The beauty experience is the physiological reward for having processed ancestral cues of promised evolutionary function, and promotes further acquisition of information about the habitat, potential mate, or idea that may then be cross-referenced with stored knowledge and decision rules to assess appropriate courses of action. Aversive feelings are the physiological punishment for having encountered ancestral cues of poor circumstances for survival and reproduction; they are designed to promote avoidance and the pursuit of better circumstances.

"Mystery" in landscapes, whether real or in landscape art, promotes gathering more information (Kaplan 1992). Mystery involves such items as a winding road that ends at the top of a hill, distant trees that appear to be bearing fruit, distant and vague savanna habitat features, distant people, distant flowers that may yield fruit or otherwise imply good habitat. Mystery is the promise of more useful information, and the mental inference of mystery draws us into the scenario for more information gathering. Mystery plays a role in domains of aesthetics other than habitat judgments (Kaplan 1992). For example, certain features of a stranger's physique or voice may promote closer inspection to gain further information about mate value or danger. Moreover, the mystery of a new idea can focus our thoughts and divert them from other channels.

Always with habitats, potential mates or other social allies, and with ideas, our information-processing about aesthetic value coincides with the existence of alternative courses of action. There are other habitats, potential mates, social allies, and ideas to evaluate in relation to evolutionary historical effects on personal survival and reproduction. The ancestral aesthetic cues guide us intuitively via our feelings about them.

Essence of Beauty

Another long-standing issue in aesthetics is the elusiveness of the essence of beauty (Kovach 1974). Said differently, why has it been so difficult to define beauty? The elusiveness of beauty led many aestheticians to the conclusion that beauty cannot be defined. Some of this sentiment arises from theological aesthetics: God alone knoweth the principle of beauty. However, aestheticians of many varieties have rejected the possibility of understanding beauty. After all, they themselves did not grasp it and all noted philosophers of recorded history have failed to comprehend the essence of beauty. In part, the elusiveness of beauty is paramount throughout aesthetics literature as a result of the strong feelings associated with tastes, which have implied to many aestheticians that personal introspection of aesthetic experience should uncover the essence of beauty (Kovach 1974).

Aesthetic value assessment is inaccessible to introspection. As Kaplan (1992) points out for landscape preferences, the preferences are made rapidly and easily, but study subjects' explanations of their choices have no relationship to the differences that consistently are associated with preferred and nonpreferred landscapes. The same is seen in people's judgments of physical attractiveness: they are made rapidly and effortlessly and without awareness of the features actually assessed or their relation to health (e.g., for facial attractiveness judgments, see Thornhill and Gangestad 1999a).

Fine art has been defined as "art produced or intended primarily for beauty alone rather than utility" (*American Heritage Dictionary*, 1976, Houghton Mifflin Co., Boston, MA). I believe that the utility in art is subliminal, but present to the extent it is judged as beautiful. The analysis of famous architectural structures, which provide evolutionary historical cues to refuge, to ease of exit from refuge, and of monitoring the outside environment (Hildebrand 1991) as well as the analysis of paintings (Appleton 1975; Heerwagen and Orians 1993) strongly support the view that art has unconsciously perceived cues to utility in human evolutionary history (see Heerwagen and Orians 1993 for a review).

Feelings are adaptations for guiding human motivation and behavior, and arise from ancestral cue processing by the mind. For the most part, humans are not designed to consciously comprehend the ancestral cues. Nor are they designed to perceive the evolutionary function of an aesthetic preference. A beautiful thing is one that has high personal, evolutionary historical reproductive value, but this value is totally out of reach of introspection. Only the scientific method can identify the cues involved in aesthetic judgment and the evolutionary function of the judgment.

Many aestheticians have struggled with the matter of where beauty resides – in the object, the beholder's mind, or the interrelationship of object with mind? Symons (1995, p. 80) put it succinctly: "Beauty is in the adaptations of the beholder." Beauty is the moving experience associated with information processing by aesthetic judgment adaptations when they perceive information of evolutionary historical promise of high reproductive success.

Beauty Principles: General or Individually Variable?

As mentioned earlier, aestheticians have frequently sought a single general principle, or a few general principles that will explain all aesthetic judgments. regardless of the thing perceived. Darwinian aesthetics leads to the conclusion that there will be many aesthetic rules, and they will be highly specialized in application. This follows from the modern understanding of adaptations, including psychological adaptations such as aesthetic value detectors, as special purpose in design.

While many traditional aestheticians searched for a general principle of beauty, many others have emphasized the absence of generality as implied by a lack of consensus among people of what constitutes aesthetic value in a given domain. The diversity of individual tastes is one of the central unresolved issues of aesthetics (e.g., Kovach 1974; Humphrey 1980). Darwinism may resolve the apparent contradiction between universal beauty and wide diversity of tastes between and within societies. Cross-cultural and individual variation in aesthetic value that arises from species-typical psychological adaptations for aesthetic valuation is expected to occur in all aesthetic domains. For example, although there is evidence of a universal preference for savanna habitat cues, there are individual habitat preferences as well, which may modify, but not eliminate, the universal preference (Kaplan 1992; Orians and Heerwagen 1992). Humans often feel positive about features of habitats of their upbringing, which is probably a manifestation of human adaptation for learning the rearing environment so as to increase success in that environment (see also Orians and Heerwagen 1992). Individual variation in valuation of open spaces and refuge was discussed above in Darwinian terms. As another example, people have adaptation for learning their cultural environment, which varies across and within societies and thereby generates variation in intellectual beauty (see next section). As a final example, individual variation in the importance of physical attractiveness in human mate selection appears to be due to condition-dependent expression of species-typical adaptation (Thornhill and Gangestad 1999a).

The Truth in Beauty

"Is beauty truth?" is a question that is salient in classical aesthetics analysis (Kovach 1974). Much of the discussion of beauty as truth in the mainstream aesthetics literature arises from the view that beauty is morally good and ugly is morally bad. To argue that beauty is goodness, or is our guide to goodness, commits the so-called naturalistic fallacy. The fallacious reasoning is the conclusion that what is actually true (determined by scientific procedure to be true), or is natural (occurs in nature or is the product of a natural process such as Darwinian selection), or is interpreted as the way of nature is therefore good or morally correct. It seems to be a fact that human physical beauty is a health certification. What is, is, and that's the end of it. No normative implications reside in this fact or arise automatically from it. The same can be said about any other scientific fact. Of course, for personal gain, humans often ascribe ideologically based values to scientific facts or to their view of what is natural (for reasons discussed below).

The naturalistic fallacy, applied to human physical attractiveness, has repeatedly led to oppressive and inhumane social policies, including the human breeding programs widely implemented under Nazism. The many eugenic programs of history have defined human beauty in ways that meet the political aspirations of the powerful. The programs view eliminating the ugly and attaining greater beauty as a moral obligation of civilized society. In these programs, the standard for attractiveness is set by interpretations based on mythology (e.g., the features of mythological Aryan heroes or the desires of a spirit) and by contorted interpretations of what is optimal under natural processes, including natural selection. Nose form, body shape, head size, various facial and body proportions, and skin and hair color have been used in discrimination and the establishment of human reproductive goals. In addition, asymmetries of face and body commonly are listed as undesirable in the eugenics programs of history, some of which give detailed accounts of body features that are to be eliminated, (see Hersey 1996 for an excellent historical account.)

People strongly favor attractive others (for review, Jackson 1992). The literature leads me to the conclusion that there is probably more prejudice in the context of physical attractiveness per se than on the basis of ethnic group and sex combined. Darwinism gives scientific insight into the beauty bias, but has nothing at all to say about the bias being right or wrong. A useful discussion of the naturalistic fallacy can be found in Symons (1979).

Beauty is truth in a number of ways. The signaling traits of animals, from the peacock's tail to the waist of women (Singh 1993), appear to be truthful indicators of phenotypic quality. In this formulation, beauty is truth, and so is ugliness. I suggest later that great art is a record of the design of human aesthetic adaptations, and thus artistic beauty may contain scientific truth about the nature of the human mind. In addition, a beautiful idea to a person is truthful in the sense that its beauty is in how the idea can promote the perceiver's status, ability to affiliate with a desired group or individual, or other social gain. A beautiful idea is one that identifies the most important path for pursuit of personal gain. The idea that a spirit created humans is scientifically unsupported, but it is the ideological truth of many people. Aesthetic judgments of ideas are based on the data of personal social experience that arises from holding a belief or from observation of the social success of others with the belief, and in this sense, beautiful ideas are truthful. The data of experience makes truthful to the theist the spiritual origin of humanity.

The aesthetic value of ideas varies cross-culturally, and often there is considerable variation in ideology among individuals in a culture. All people are immersed in attempts to have personal ideas viewed as beautiful, whether one is a teacher trying to convince students, a parent trying to persuade children, a scientist trying to persuade colleagues, or a politician trying to gain political support. Moreover, people are immersed in sifting through the diversity of ideas presented by others. This sifting, I suggest, is done by psychological information-processing machinery that is designed specifically for assessing the value to self (technically, to self's reproductive success in evolutionary historical environments) that alternative ideas can convey, in essence, an evolved intellectual beauty detector.

Ideology means (e.g., *The American Heritage Dictionary*), "The body of ideas reflecting the social needs and aspirations of an individual, group, class, or culture." Thus, a person's ideology is a significant component of its social strategy, i.e., how it presents itself, pursues social status and accesses social groups, friends and mates. I am arguing that intellectual aesthetic value represents a functionally based way of dealing with a cultural environment that is full of diverse ideas. In this perspective, cultural learning of values is not arbitrary. Learning mechanisms, in conjunction with feeling mechanisms and mechanisms of self-awareness that allow us to test how our ideas and behaviors are perceived by others, guide us through a maze of ideas toward intellectual beauty. Appropriate values will often differ between societies and within societies between social strata and individuals.

It has been my experience in teaching the Darwinian view of human nature to students that the greatest resistance to the view arises when it is applied to human values, i.e., to intellectual beauty. This is to be predicted: one's ideology is sacred, because it is truth in the sense of data-based conclusion as to how to proceed in one's social competition.

Traditional aestheticians have given humanity a voluminous literature containing a rich description of their interpretations. However, they have not done as much as they could have. In general, the literature of aesthetics views arbitrary cultural experience as the complete explanation of aesthetic preference. This is the mainstream theory of traditional aesthetics. Thus, aestheticians have contributed to the widespread image of humans as having brains that are equally likely to learn anything through cultural conditioning; this image erroneously assumes that the human mind consists of a *tabula rasa* or a few general-purpose learning mechanisms (Symons 1987). Fox (1989, p. 24) calls this scientifically inaccurate view of mind in our time "the whole secular social ideology." Certainly, it is the typical view of mind in the social environment in which I work and live, and it is fundamental to thinking in social sciences and humanities (see also Symons 1987; Brown 1991; Degler 1991; Tooby and Cosmides 1992).

The evolutionary psychological view on the issue of culture creating human aesthetic pleasures is that, causally, the matter is fundamentally the other way around, because there exists a complex and diverse set of psychological adaptations that comprise the human nature of pleasure-seeking. As I have emphasized, when culture plays a causal role, as is the case for individual and group differences in aesthetic preference, arbitrariness is not involved. Thus, there is no inconsistency between universal design for aesthetic valuation and cross-cultural and individual variation in the valuations.

Psychological Adaptations of Aesthetics

In this section, I attempt to dissect the aspect of the human brain that is responsible for the traditional topics of interest to aestheticians into a number of distinct aesthetic valuation adaptations using the principles of evolutionary psychology. One proceeds as follows: first, envision an information-processing problem to have been solved in human evolutionary history. For example, choice of a mate of high mate value, e.g., mature, developmentally sound, hormonally healthy and possessing disease resistance. Clearly, mate choice is a distinct domain in which special-purpose adaptation should exist (Thornhill and Gangestad 1993; Symons 1995). Then, one attempts to deduce the cues or stimuli that would have reliably provided the information needed to solve the adaptive problem in human evolutionary history. For example, it appears that facial symmetry is used in human mate choice because it cues high-quality development (Thornhill and Gangestad 1999a). In addition, facial and body features, facilitated by sex hormones (testosterone and estrogen) and important in mate choice, appear to cue hormonal health and overall phenotypic and genetic quality (see review in Thornhill and Gangestad 1999a; Thornhill and Grammer 1999). Obviously, mate choosers in human evolutionary history could not evaluate directly a potential mate's genetic quality or ability to resist disease or other environmental insults during development; they used reliable indicators of these traits. The deductions about hypothetical cues are the predictions that can be tested. When a prediction receives support, an ancestral cue is potentially identified. When a prediction is false, other potential cues are examined. Cross-cultural work is used to test universality (see Brown 1991).

Other researchers have addressed psychological adaptations of aesthetic judgment. Turner (1991) has attempted a broad treatment of the topic; also see Dissanayake (1988) and chapters in Rentschler et al. (1988b). Wilson (1993) has pointed out that biophilia arises from numerous evolved rules. Grammer (1993), Singh (1993), Thornhill and Gangestad (1993) and Thornhill and Grammer (1999) treat psychological adaptations of aesthetic valuation of human bodily form. The psychological adaptations associated with valuations of facial attractiveness have been treated by Johnston and Franklin (1993), Thornhill and Gangestad (1993, 1999a), Grammer and Thornhill (1994), Perret et al. (1994), Symons (1995) and Fink et al. (2001). Women's aesthetic preference for the scent of male symmetry (for developmental stability) has been investigated thoroughly (for a review of studies, see Thornhill and Gangestad 1999b). Research on environmental aesthetics has dealt in some detail with psychological adaptations associated with habitat assessments (reviewed in Kaplan 1992; Orians and Heerwagen 1992; Heerwagen and Orians 1993; Ulrich 1993) and with assessment of potential dangers (e.g., predators), weather and season (Orians and Heerwagen 1992; Heerwagen and Orians 1993; Ulrich 1993).

Before a taxonomy of aesthetic adaptations can be discussed, it is important to mention again that there is no objective way to delimit the aesthetic domain in Darwinian terms. Historical precedent in the field of aesthetics is my criterion for defining the types of experiences to which beauty is applied. I assume that when aestheticians show great interest in a domain of experience, it is because of the power of aesthetic feelings in the domain. All ten areas that follow have been considered in philosophical aesthetics throughout its history (e.g., Kovach 1974).

1. Adaptations for aesthetic valuation of landscape features. Beauty here is hypothesized to be the perception of ancestral cues to productive and safe habitats in the environments of human evolutionary history. Candidate ancestral cues that have received considerable empirical support are mountains, water sources, oasis, flowers, ripe fruits, savanna (open forests which give easy visual access), growth and leaf patterns of healthy savanna trees, closed forest canopy (shelter), and caves (with easy access to outside landscape). Considerable cross-cultural research has been done on some of these landscape cues (for reviews, see Kaplan 1992; Orians and Heerwagen 1992; Heerwagen and Orians 1993). Given the distinctiveness of many of the cues to suitable Pleistocene habitat, it is likely that there are numerous special-purpose adaptations involved.

Habitat scents have not been studied to my knowledge. I suggest that habitat scents become perfumes when the scents are ancestral cues to promised survival in human evolutionary history.

2. Adaptations for aesthetic valuation of nonhuman animals. Beauty here is hypothesized to be the perception of ancestral cues to available animal food and safety from predators as revealed in the behavior of nonhuman animals in one's environment. Animal food cues might be fish, large or small ungulates, certain rodents and birds (see Heerwagen and Orians 1993 for evidence for the cue of ungulates). Observing ungulates grazing (e.g., antelopes or cattle) without showing any alarm (not showing vigilance toward possible predators) can be very aesthetically pleasing, perhaps because it is a cue to a predator-safe habitat (see Heerwagen and Orians 1993 for evidence). Another similar candidate cue may be the presence of active, unalarmed birds. I predict that the behavioral actions (excluding the vocalizations, see below) of startled or nervous ungulates or birds will generate interest, apprehension and even alarm in humans (see also Orians and Heerwagen 1992). Aesthetic value in this domain could be studied further by having subjects self-report their feelings toward movies and still pictures. This could be coupled with the recording of physiological responses of subjects to stimuli predicted to be attractive or unattractive (see Ulrich 1993 for methods).

3. Adaptations for aesthetic valuation of the acoustical behavior of nonhuman animals. Productive habitats have a more diverse bird and acoustic

insect fauna than impoverished habitats. I predict that the sounds of these animals will often give rise to positive aesthetic experiences. Death and alarm calls of mammals and birds, buzzing noises (like those produced by bees) and growls are expected to result in cautious attention, apprehension and anxiety. This domain could be studied by having subjects judge different natural sounds and/or by measuring subjects' physiological responses.

4. Adaptations for aesthetic judgments arising from daily or seasonal environmental cues that signal a need to change behavior. Orians and Heerwagen (1992) have predicted that reliable environmental cues to the need to change behavior are responsible for certain strong emotional reactions that people show toward both the environment and landscape paintings. Weather events such as clouds may alert one to the need to seek shelter or indicate that there is time to move about before the weather changes. Long shadows in the afternoon indicate approaching darkness, and thus the need to change behavior. Sunsets reliably indicate approaching darkness as well. Light features after the clearing of storms, such as rainbows, have aesthetic appeal. They, like sunrise light, indicate opportunities to explore the habitat. The painter Maxfield Parrish appreciated the power of light effects indicative of diurnal light rhythms on human emotions (e.g., see his paintings *Sunrise, Daybreak*, and *Puss and Boots*). Portrayals indicative of seasonal shifts are, like weather and light effects, important ingredients in landscape art. The ancestral seasonal cues are predicted to be those environmental cues that reliably were correlated with seasonal resource changes in human evolutionary history (see Orians and Heerwagen 1992). Cues in this domain could be further studied using methods suggested for the previously discussed domains.

Categories 1–4 above constitute the topics discussed under biophilia (Kellert and Wilson 1993). There are, however, other domains of aesthetic valuation that arise from ancestral hominids' evolutionary historical interactions with life in the form of conspecifics, i.e., the domains of human biophilia. These are outlined below.

5. Adaptations for aesthetic valuation of human bodily form. Beauty here is the perception of cues to high reproductive potential in the body features of others. The others may be relatives, potential mates, or potential partners in nonmateship forms of social alliances. There is considerable evidence that physical attractiveness in adult humans is positively correlated with conditions that would have covaried with the bearer's reproductive potential during human evolutionary history (in women, youth; in men and women, sound development [bilateral symmetry], disease resistance, hormonal health, and overall health; for reviews, see Singh 1993; Thornhill and Gangestad 1993, 1999a, b; Symons 1995; Gangestad and Thornhill 1997; Fink et al. 2001). These reviews and numerous other works show that physical attraction to adults correlates positively with sexual interest and interest in romantic involvement (e.g., Jackson 1992).

Furthermore, there is some evidence than physical attractiveness positively affects the nepotism children receive (see Thornhill and Gangestad 1993), and considerable evidence that physical attractiveness affects the number of friends one has and one's social success in general (e.g., popularity, job opportunities, salary; reviewed in Jackson 1992). Social success, to an important extent, arises out of an evaluation by others that one is a useful partner in social reciprocity (see Thornhill and Gangestad 1993).

Studies of physiological reactions to habitats or to phobic cues has yielded interesting results (see Ulrich 1993). Similarly, recent research has examined physiological responses of people when viewing attractive versus unattractive others, and revealed strong emotional reactions to attractive others (see Thornhill and Gangestad 1999a).

Moreover, research that examines self-reported feelings of well-being might be illuminating in this domain as it has been in the domain of habitat preference. In comparison to when nonpreferred habitat features are viewed, when preferred habitats are viewed, subjects report more positive emotional states (Ulrich 1993). Even nature posters cue the mechanisms to the point that viewers feel better when the posters are visible (Heerwagen and Orians 1993; Ulrich 1993). I suggest that one reason for people's preference for attractive human forms in pictures in their environment is that the pictures affect a state of well-being arising from the perception that they are amidst healthy conspecifics.

6. Adaptations for aesthetic valuation of status cues. Almost constantly humans seek to possess, control or be associated with status markers. Markers of social rank are of many types: accents, music, ideology, education, friends, mates, pets, automobiles, homes, cellular telephones, recreation, clothes, etc. I suggest that, in general, status markers confer status because they reliably indicate social rank by their rarity, expense, or the difficulty by which they are achieved or obtained. Ascription of aesthetic value occurs with each type of status marker. Beauty here is the perception of cues to increased status.

People seem to have species-typical adaptation for identifying and adopting status markers (Alexander 1979; Flinn and Alexander 1982; Barkow 1989). This special-purpose machinery will interact with other special-purpose adaptation to yield preference. As an example, attractive people are judged to be of higher status than unattractive people, and association with attractive others is perceived as status enhancing (for review, see Jackson 1992). Our status-pursuit machinery apparently motivates us to pursue attractive others based on information processing by the psychological adaptation for valuation of physical attractiveness as phenotypic quality. Ideological preferences seem to work similarly. Humans pay more attention to and more readily adopt the views of people of social rank, compared to the views of lower ranking individuals (e.g., Wright 1994). The psychological mechanism of intellectual beauty evaluation assesses ideas for their effects on personal interests, including personal status. This adaptation uses inferential information provided by the status-pursuit machinery.

7. Adaptations for aesthetic valuation of social scenarios. In technological societies, social scenarios commonly are provided by literature, theater, movies, TV, music and everyday life. Although much human interest in this domain involves status pursuit, I suggest that there is aesthetic judgment independent of pursuit of rank involved. Beauty here is the perception of information that gives solutions to social problems. Human interest in social scenarios, both real and acted, seems to function in allowing people to practice thinking about relevant social scenarios (Brown 1988; Alexander 1989; Barkow 1992). Romance novels and soap operas may provide women with information pertaining to sexual problems faced by our adult female evolutionary ancestors, such as the benefits and costs of sexual infidelity and how to secure and keep male investment and commitment and win female sexual competition in general. Scenarios involving male–male competition or group competition (sports and warfare) may provide men with information that was central to success in male sexual competition in human evolutionary history.

8. Adaptations for aesthetic valuations based on skill. Brown (1991, p. 116) points out that appreciation of tool-making and language skill can evoke positive aesthetic experiences. Athletic skills, including the postural and coordinational skills of dance, and musical skills also provide pleasure to people. I suggest that skill is aesthetically pleasing when it would have positively covaried in human evolutionary history with phenotypic quality, especially neural and nonneural physical development. Because assessing skill in language, athletics, music and tool-making requires the successful processing of very distinct information, it is likely that the human brain contains numerous, functionally specialized, psychological adaptations for skill detection. Miller (2001) has developed in detail the argument that art is a skill that signals phenotypic quality.

9. Adaptations for aesthetic judgments of food. Humans are predicted to make emotional judgments of foods based on their nutritional value as perceived visually and olfactorily (also see Rozin 1988; Turner 1991). Beauty here is the perception of cues of high nutritional value in human evolutionary history (e.g., foods containing sugar, salt, fat and spices that kill food-borne pathogens; for discussion of the function of spice use by people, see Billing and Sherman 1998).

10. Adaptations for aesthetic judgments of ideas. Earlier I discussed hypothetical design features of the intellectual beauty detector.

Undoubtedly, this list of categories of human aesthetic assessment adaptations that may underlie the typical topics of interest to aestheticians is incomplete and will need refinement as knowledge of the relevant beauty detectors increases. For example, the subject matter of experimental aesthetics, the aesthetics of shapes and patterns is not covered in the above taxonomy. Humphrey (1980) has attempted to deal with experimental aesthetics in Darwinian terms. He argued that human aesthetic pleasure happens upon viewing shapes and patterns when solutions are seen to classifying objects in our environment, because ". . . an activity as vital as classification was bound to evolve to be a source of pleasure. . . ." (p. 64). There is no reason to expect all vital activities to be tied to aesthetic feelings. Digestion is vital, but not consciously experienced. Moreover, classification is what perception itself entails – that is, perception is the active psychological process by which we seek order. Clarity in perception, i.e., classification does not automatically lead to aesthetic pleasure.

I suggest that the truth that Humphrey touches upon is that the inability to classify generates the adaptive inference of mystery when certain cues of historical reproductive significance appear to be present. I have discussed above the relevance of mystery in human aesthetic feelings. Experimental aestheticians have shown the importance of features such as novelty, complexity, surprise, ambiguity, asymmetry and incongruity in shapes and patterns to viewers' interest, arousal and curiosity (Berlyne 1971; Solso 1994). I am suggesting that it will not be these features per se that generate mystery; instead mystery will arise when these features are used to confuse the perception of ancestral cues. Rubin's cup is a good example of how patterns can give the inference of mystery: the viewer is motivated to discern the lateral profiles of two human faces facing and in close proximity to one another from the background of a prominent cup that delimits the fronts of the faces. The mystery arises from the need to discern the presence of conspecifics that may be in an intimate interaction, negative or positive. Possibly another factor promoting mystery in Rubin's cup is that the faces are intersexes. The eyebrow region of the faces is projected outward as in adult males, whereas the chins and lower faces are short as in adult females and children (see Symons 1995 for a discussion of sex differences in facial traits). I also suggest that the use by artists of various filtering techniques and artificial distortions in visual art will have effects when they create mystery in the human mind. These ideas are largely consistent with the interpretation of such alterations of shape and pattern by Rentschler et al. (1988a). Mystery generated by incongruity may be the reason for the prominence of Modigliani's art in which there is exaggeration of the asymmetry in faces of people who were otherwise above average in attractiveness (e.g., *Chaim Soutine Seated at a Table*). Modigliani also painted a large lower jaw on an otherwise beautiful female body form (e.g., *Seated Nude*). A small lower jaw is attractive in the adult female face (Johnston and Franklin 1993; Perret et al. 1994; Symons 1995). Dali's *The Enigma of William Tell* is the epitome of incongruity in phenotypic quality: highly asymmetrical buttocks depicted on a man of otherwise above-average body attractiveness. These ideas could be examined empirically by manipulating features in visual art. Self-report evaluations as well as physiological responses would be appropriate dependent variables.

Turner (1991) lists a number of candidate aesthetic adaptations that I have not discussed, but feel can be accounted for by adaptations in the above ten categories, with the possible exception of Turner's music adaptation. For example, human aesthetic valuations of colors probably arise from special-purpose adaptations for assessing colors in relation to Pleistocene adaptive problems such as the ripeness of fruit and suitability of vegetables (see Zollinger 1988). Music appreciation is a human universal (Brown 1991), but it is unclear whether there are ancestral cues contained in music that are independent of the status, ideology and skill considerations I've discussed.

Darwinian aesthetics has much to offer the scientific study of aesthetic value. Two groups of scholars are most likely to advance scientific understanding of aesthetics. First, evolutionary psychologists may figure importantly, because they are most interested in strong emotions, which are the most easily studied in terms of large empirical effects. In addition, evolutionary psychologists assume the special-purpose design of psychological machinery, an assumption necessary for identifying the appropriate experiments and observations for analyzing the design of each beauty detector. Second, I believe that the arts and the humanities, in general, will play a crucial role in the scientific study of aesthetics. The criterion that distinguishes significant art and humanistic rhetorical scholarship from the vulgar and ordinary in these fields is the profundity of emotional effect in the former. Thus, great art, in the broad sense of art, is hypothesized to be one important record of the design of our aesthetic adaptations. Darwinian aesthetics is the method to determine the cues in great art that make it great – that is, determine the actual information that human aesthetic mechanisms process during aesthetic valuation of art.

Summary

Aesthetics does not delimit objectively a part of nature. The adaptations that underlie aesthetic experiences, as well as all other adaptations, interact with the environment and prefer certain states to others. Although there is no way to define objectively the aesthetic domain, there is utility in treating the various topics of traditional aesthetics (a branch of philosophy) in a Darwinian framework. This framework is explained and its usefulness in clarifying and illuminating various puzzles of traditional aesthetics is discussed. A taxonomy of ten categories of human aesthetic adaptation is presented.

Acknowledgements. This paper is dedicated to the new sources of beauty experiences in my life: Sophie Bess Thornhill, Reed Randolph Thornhill, and Ella Jean Thornhill. Anne Rice's assistance with manuscript preparation is greatly appreciated. I thank an anonymous reviewer for helpful comments on the manuscript.

References

Alexander RD (1979) Darwinism and human affairs. University of Washington Press, Seattle, WA

Alexander RD (1989) The evolution of the human psyche. In: Mellars P, Stringer C (eds) The human revolution. University of Edinburgh Press, Edinburgh, pp 456–512

Appleton J (1975) The experience of landscape. Wiley, New York

Arnheim R (1988) Universals in the arts. J Soc Biol Struct 11:60–65

Barkow JH (1989) Darwin, sex and status: biological approaches to mind and culture. University of Toronto Press, Toronto

Barkow JH (1992) Beneath new culture is old psychology: gossip and social stratification. In: Barkow JH, Cosmides L, Tooby J (eds) The adapted mind: evolutionary psychology and the generation of culture. Oxford University Press, Oxford, pp 627–638

Berlyne DE (1971) Aesthetics and psychobiology. Allerton Century Crofts, New York

Billing J, Sherman PW (1998) Antimicrobial functions of spices: why some like it hot. Q Rev Biol 73:3–49

Brown D (1988) Hierarchy, history and human nature: the social origins of historical consciousness. University of Arizona Press, Tucson, AZ

Brown D (1991) Human universals. McGraw-Hill, New York

Buss DM (1994) The evolution of desire: strategies of human mating. Basic Books, New York

Cosmides L, Tooby J (1987) From evolution to behavior: evolutionary psychology as the missing link. In: Dupre J (ed) The latest on the best: essays on evolution and optimality. MIT Press, Cambridge, MA, pp 277–306

Cronin D (1991) The ant and the peacock: altruism and sexual selection from Darwin to today. Cambridge University Press, New York

Darwin C (1874) The descent of man and selection in relation to sex. Rand, McNally, Chicago (republished by Gale Research, Book Tower, Detroit, 1974)

Degler CN (1991) In search of human nature: the decline and revival of Darwinism in American social thought. Oxford University Press, Oxford

Dissanayake E (1988) What is art for? University of Washington Press, Seattle, WA

Fink B, Grammer K, Thornhill R (2001) Human (*Homo sapiens*) facial attractiveness in relation to skin texture and color. J Comp Psychol 115:92–99

Flinn MV, Alexander RD (1982) Culture theory: the developing synthesis from biology. Human Ecol 10:383–400

Fox R (1989) The search for society: quest for a biosocial science and morality. Rutgers University Press, New Brunswick, NJ

Gangestad SW, Thornhill R (1997) Human sexual selection and developmental stability. In: Simpson JA, Kenrick DT (eds) Evolutionary social psychology. Lawrence Erlbaum, Mahway, NJ, pp 169–195

Grammer K (1993) Signale der Liebe: die biologischen Gesetze der Partnerschaft. Hoffmann and Campe, Hamburg

Grammer K, Thornhill R (1994) Human (*Homo sapiens*) facial attractiveness and sexual selection: the role of symmetry and averageness. J Comp Psychol 108:233–243

Greenough H (1958) Form and function: remarks on art, design and architecture. University of California Press, Berkeley, CA

Heerwagen JH, Orians GH (1993) Humans, habitats and aesthetics. In: Kellert SR, Wilson EO (eds) The biophilia hypothesis. Island Press, Washington, DC, pp 138–172

Hersey GL (1996) The evolution of allure: art and sexual selection from Aphrodite to the incredible hulk. MIT Press, Cambridge, MA

Hildebrand G (1991) The Wright space. Washington University Press, Seattle, WA

Humphrey NK (1980) Natural aesthetics. In: Mikellides B (ed) Architecture for people. Studis Vista, London, pp 59–73

Jackson LA (1992) Physical appearance and gender: sociobiological and sociocultural perspective. State University of New York Press, Albany, NY

Johnston VS, Franklin WM (1993) Is beauty in the eye of the beholder? Ethol Sociobiol 4:183–199

Kaplan S (1992) Environmental preference in a knowledge-seeking, knowledge-using organism. In: Barkow J, Cosmides L, Tooby J (eds) The adapted mind: evolutionary psychology and the generation of culture. Oxford University Press, Oxford, pp 581–600

Kellert SR, Wilson EO (eds) (1993) The biophilia hypothesis. Island Press, Washington, DC

Kovach FJ (1974) Philosophy of beauty. University of Oklahoma Press, Norman, OK

Lewontin RC, Rose S, Kamin LJ (1984) Not in our genes: biology, ideology and human nature. Pantheon Books, New York

Miller GF (2001) Aesthetic fitness: how sexual selection shaped artistic virtuosity as a fitness indicator and aesthetic preferences as mate choice criteria. Bull Psychol Arts 2:20–25

Møller AP (1992) Female swallow preference for symmetrical male sexual ornaments. Nature 357:238–240

Orians GH (1980) Habitat selection: general theory and applications to human behavior. In: Lockard JS (ed) The evolution of human social behavior. Elsevier, Chicago, pp 49–66

Orians GH, Heerwagen JH (1992) Evolved responses to landscapes. In: Barkow J, Cosmides L, Tooby J (eds) The adapted mind: evolutionary psychology and the generation of culture. Oxford University Press, Oxford, pp 555–580

Perret DI, May KA, Yoskikawa S (1994) Facial shape and judgments of facial attractiveness. Nature 368:239–242

Rentschler I, Caelli T, Maffei L (1988a) Focusing in on art. In: Rentschler I, Herzberger B, Epstein D (eds) Beauty and the brain: biological aspects of aesthetics. Birkhäuser, Berlin, pp 181–218

Rentschler I, Herzberger B, Epstein D (eds) (1988b) Beauty and the brain: biological aspects of aesthetics. Birkhäuser, Berlin

Ridley M (1993) The red queen: sex and the evolution of human nature. Macmillan, New York

Rozin E (1988) Aesthetics and cuisine – mind over matter. In: Rentschler I, Herzberger B, Epstein D (eds) Beauty and the brain: biological aspects of aesthetics. Birkhäuser, Berlin, pp 315–326

Singh D (1993) Adaptive significance of female physical attractiveness: role of waist-to-hip ratio. J Personality Social Psychol 65:293–307

Solso RL (1994) Cognition and the visual arts. MIT Press, Cambridge, MA

Soulé ME (1992) Biophilia: unanswered questions. In: Kellert SR, Wilson EO (eds) The biophilia hypothesis. Island Press, Washington, DC, pp 441–455

Symons D (1979) The evolution of human sexuality. Oxford University Press, Oxford
Symons D (1987) If we're all Darwinians, what's the fuss about? In: Crawford C, Smith M, Krebs D (eds) Sociobiology and psychology: ideas, issues and applications. Lawrence Erlbaum Associates, Hillsdale, NJ, pp 121–146
Symons D (1992) On the use and misuse of Darwinism in the study of human behavior. In: Barkow J, Cosmides L, Tooby J (eds) The adapted mind: evolutionary psychology and the generation of culture. Oxford University Press, Oxford, pp 137–162
Symons D (1995) Beauty is in the adaptations of the beholder: the evolutionary psychology of human female sexual attractiveness. In: Abramson PR, Pinkerton SD (eds) Sexual nature/sexual culture. University of Chicago Press, Chicago, pp 80–118
Thornhill R (1990) The study of adaptation. In: Bekoff M, Jamieson D (eds) Interpretation and explanation in the study of animal behavior, vol 2. Westview Press, Boulder, CO, pp 31–61
Thornhill R (1997) The concept of an evolved adaptation. In: Bock GR, Cardew G (eds) Characterizing human psychological adaptations. Wiley, West Sussex, pp 4–22
Thornhill R, Gangestad SW (1993) Human facial beauty: averageness, symmetry and parasite resistance. Human Nat 4:237–269
Thornhill R, Gangestad SW (1994) Human fluctuating asymmetry and sexual behavior. Psychol Sci 5:297–302
Thornhill R, Gangestad SW (1999a) Facial attractiveness. Trends Cognitive Sci 3:452–460
Thornhill R, Gangestad SW (1999b) The scent of symmetry: a human sex pheromone that signals fitness? Evol Human Behav 20:175–201
Thornhill R, Grammer K (1999) The body and face of woman: one ornament that signals quality? Evol Human Behav 20:105–120
Thornhill R, Gangestad SW, Comer R (1995) Human female orgasm and mate fluctuating asymmetry. Anim Behav 50:1601–1615
Tooby J, Cosmides L (1989) Evolutionary psychology and the generation of culture. Part I: Theoretical considerations. Ethol Sociobiol 10:29–50
Tooby J, Cosmides L (1990) On the universality of human nature and the uniqueness of the individual: the role of genetics and adaptation. J Personality 58:17–67
Tooby J, Cosmides L (1992) The psychological foundations of culture. In: Barkow JH, Cosmides L, Tooby J (eds) The adapted mind: evolutionary psychology and the generation of culture. Oxford University Press, Oxford, pp 19–136
Turner F (1991) Beauty: the value of values. University of Virginia Press, Charlottesville, VA
Ulrich RS (1993) Biophilia, biophobia and natural landscapes. In: Kellert SR, Wilson EO (eds) The biophilia hypothesis. Island Press, Washington, DC, pp 73–137
Williams GC (1992) Natural selection: domains, levels and challenges. Oxford University Press, Oxford
Wilson EO (1984) Biophilia. Harvard University Press, Cambridge, MA
Wilson EO (1993) Biophilia and the conservative ethic. In: Kellert SR, Wilson EO (eds) The biophilia hypothesis. Island Press, Washington, DC, pp 31–41
Wright R (1994) The moral animal. Why we are the way we are: the new science of evolutionary psychology. Pantheon Books, New York
Zahavi A (1975) Mate selection – a selection for a handicap. J Theor Biol 53:205–214
Zollinger H (1988) Biological aspects of color naming. In: Rentschler I, Herzberger B, Epstein D (eds) Beauty and the brain: biological aspects of aesthetics. Birkhäuser, Berlin, pp 149–164

II From Is to Beauty – From Perception to Cognition

The Beauties and the Beautiful – Some Considerations from the Perspective of Neuronal Aesthetics

OLAF BREIDBACH

Patterns and Beauties

What is evolutionary aesthetics about, but the evolution of aesthetics? If such an evolution exists, there must be an evolutionary procedure which sorts out which aesthetic would be advantageous in evolutionary terms and which would not. It would, therefore, be reasonable to assume that such an optimizing procedure would result in something that could be described as "better", that is, in evolutionary terms, more successful, and in terms of aesthetics, more pleasant or "beautiful". In this study, the central task will be to describe such an optimizing procedure. Such an approach would allow us to make sense of terms such as "beauty" and "beautiful". If it were possible to define the neuronal mechanism by which a certain pattern of beauty is selected, questions such as the definition of beautiful might be answered. Eventually, an evolution of schemes of beautiful could be outlined, providing us with the opportunity to describe the essentials of human cultural development in biological terms.

In the late nineteenth century such an idea was already widespread (Romanes 1882; Cajal 1906). One of the most elaborated attempts to describe the physiological background for our perception of beauty was worked out by the Austrian physiologist Sigmund Exner (Exner 1894). Exner was only part of a community which was, by the end of the nineteenth century, discussing the impact evolutionary theory might have on our understanding of cognition, aesthetics and emotion. Common to these approaches to cognition was the attempt to get rid of spiritual qualities by describing such qualities as the outcome of physiological mechanisms. Around 1900, scientists from the psychological school in Würzburg believed that they had proven that the phenomenological base for a self did not exist. For them, the situation had become quite clear: every human cognitive activity is nothing but brain activity. Aesthetics could thus be regarded as simply the output of a nervous system which had evolved to behave in this way. This view gained such currency that even the art historian Alois Riegl asked the physiologist Sigmund Exner to define beauty.

Here, I will argue against the idea that biological reductionism is helpful in addressing the question of aesthetics. In a certain sense, Baumgartner had already anticipated such a view. If aesthetics is just 'seeing', then biology would indeed be the right science to look to for a solution to the problem. When aesthetics, however, is recognition, then the situation is very different. In this case, aesthetics is not just the process of sensory imprinting, but also the process *interpreting* such sensory imprints. Such interpretation may even entangle that faculty formerly called fantasia. Aesthetics, therefore, are not easily analyzed in neurobiological terms. Nevertheless, Exner developed an idea of the mechanism by which cultural adaptations are manifested physiologically. He argued that sensory stimuli and memory traces were associated in a parallel processing device formed by the neuronal network of the human brain (Breidbach 1996). Exner, however, did not work out the details of his theory. In particular, he did not develop an explicit evolutionary perspective, which he could have done at his time, following the idea of an evolutionary development of mind as had been put forward by George Romanes (1883, 1888), the former friend of Charles Darwin.

In my view, one has to establish a more diversified phenomenology of aesthetics before dealing with the question of beauty. Furthermore, the question is, whether "beauty" is the object aesthetics must deal with. One can show that in the arts it has been the most interesting, not the most beautiful, which has captured the most interest. No one could reasonably describe the results of the body art of the early 1970s as actually 'beautiful'. In fact, the artists themselves were explicitly fighting against the simple idea of identifying art with beauty.

Similar problems can be addressed in relation to other periods of our culture. Thus, beauty is not necessarily the central term the history of arts and aesthetics has to deal with. The recent idea of an image anthropology (Bild-Anthropologie; Belting 2001) developed out of attempts to describe how, in our various cultures, we handle sensory objects. Similarly, neuronal aesthetics (Breidbach 2000) from the first attempts in the mid 1990s onwards has tried to describe the mechanisms of perception in both its cultural and the physiological dimensions, and to develop new ways of understanding the evaluation of perception and aesthesis (Böhme 2001). The present text will describe such an alternative approach. It will discuss evolutionary aesthetics from the perspective of neuronal aesthetics, and will develop some perspectives for a theory encompassing not just the beautiful, but which also attempts to deal with the sublime.

History

In 1894, when Exner published the monograph which sought to demonstrate the physiological basis of psychological phenomena, he was basing his ideas on a view of the human brain as a complex neuronal network.

Certain sensory inputs are somehow associated with pleasure, thus the elucidation of the events which caused this response would allow one to describe the idea of 'beauty' in neurophysiological terms. Exner believed that a person's idea of beauty arises from the association of a specific sensory input with particular memories which transport a multitude of pleasant experiences.[1] Such associations were described as complex and culturally formed, resulting from whole sets of experiences.

Consequently, following Exner, aesthetics cannot be explained by categorizing the objects of perception, but in describing the way such objects are perceived. Thus, a biological approach results in an analysis of the observer, not an analysis of that which is perceived. According to Exner – who followed the argument of James Mill – an analysis of the internal capacities of an observer would be simple. One should just denote the mechanisms by which the signal transfer is achieved within the nervous system, and describe how the signal input is overlaid onto the activation patterns of the sensory system to understand how more complicated behavior originates.

Certain experiences would automatically elicit a train of associations which are perceived as pleasant. Accordingly, the objects which elicit this response could be regarded as beautiful. Following the ideas of Exner, one can readily deduce that the categories of beautiful are, therefore, the product of our education, and were encoded in our brains. Everything we are familiar with fits quite well into such pre-established schemes of information transfer. However, it would be too simple just to look onto the neuronal mechanisms of signal transfer. Here, evolutionary aesthetics contradicts Exner. It is worth looking at the external, nonneuronal predispositions of the way we experience beauty. For evolutionary aesthetics, it is the signal itself, not the way it is processed, which defines the quality of beauty. Thus, a male actually signals his beauty to a female. It would be insufficient for the female to simply imagine that the male is beautiful. In such a case, the female's engagement with the male would not guarantee anything for the next generation. Thus, evolutionary theory requires that

---

[1] In 1882, the Viennese physiologist Exner had attempted a 'physiological analysis' of the figures of Michelangelo. Today, we may be a little disturbed that his interest was just in analyzing the putative biomechanics of the construction of angels in Michelangelo's paintings. Describing these, he pointed to a disjunction between our cultural response to the paintings and the physiological viewpoint. He explained this in a somewhat far-fetched way, by pointing to a putative mechanism of 'enculturation' which he thought to be manifested in the associative organization of our visual perceptions: as we are used to describing Michelangelo's works as being beautiful, we would not actually see the biomechanical difficulties in his constructions. He explained this through his idea of an associative neuronal network that our brain uses to evaluate our sensory experiences. This argument is far from consistent, but it gives a good idea of how a former physiologically based conception of aesthetics might have been constructed.

there must be an objective base for beauty. Beauty is not an idea, but a characteristic of the observed. These characteristics act as signals. These signals thus represent something that has a biological reality. Their existence is thought to be a product of selection by the environment in which they are found. Only those attributes that guarantee a success will be selected in the course of evolution because of their adaptive value in a certain environment. In this way, both features and modes of communication were thought to have been selected. As we are social beings, our environment is, to a large extent, our social system and the culture we live in. This culture predisposes our experiences. Within this context, we are conditioned to follow rules of behavior that optimize our enculturation. Accordingly, we adopt certain culturally specific judgements. Cultural styles and cultural tastes – ranging from music to meals – are the result of this. If we experience something as pleasant, we do so by activating that cultural background. Beautiful is classified according to the categories of culture one is accustomed to: things already known will be more easily coupled into a train of association. Thus, the preference for a certain style within a certain period of cultural development can be explained. The mechanism by which such cultural preferences were selected can, therefore, be explained in terms of an evolutionary aesthetics.

The mechanism of enculturation, according to evolutionary aesthetics, is not an abstract one. It is based on the biological mechanisms of the species *Homo sapiens*, which we are dealing with here. The idea of basing cultural evolution on the evolution of the nervous system has some plausibility. It allows us to describe the physiological bases of the cultural ecosystem and define it as the outcome of a natural history, thereby assigning a fitness function to aspects of cultural life. The nervous system, according to this schema, simply reflects the natural situation. It is designed to function successfully in its particular environment and its reactions can be described as 'adaptations' to the situation in which it is found. The structure of the nervous system can be described as optimal for the production of such behavior. However, this implies that the nervous system is merely an implemented mechanism within a person. The subject himself, accordingly, has to be omitted. Its objectivity would be nothing else but a mirror for the outer world. Its internal organisation, thus, must become optimized not by the internal needs of the subject, but by the external features of the surroundings and their evolutionary effect. The brain had to be understood as encoding such an evolution. Wilson (1999) has argued for such a concept of self and culture, recently describing the whole complex of cultural interactions as the result of the evolutionary optimization of behavioral features of *Homo sapiens*. Such a neurophysiological approach is based on the idea of beautiful not in the brain, but in the culture in which the brain is formed (Breidbach 1997b).

In contrast, modern neuroscientists such as Zeki (1993, 1999) have outlined how the brain forms certain preferences in visual orientation which match the experiences one can have of artwork. Nevertheless, it is hard to test whether such preferences reflect objective categories or whether they simply illustrate the extent to which our sensory system is able to sort out certain combinations of stimuli. If we could identify a certain preference in our visual orientation, then we could predict that something would be perceived in a specific way. We already know that our brain can decode complex visual stimuli in a very specific and economic way. What we don't know, however, is the extent to which the filtering functions which encode certain types of visual input already responded to such stimuli (independently of our consciousness). The sum of various experiences which have been decoded in our brain does not automatically describe a category of beauty (cf. Fredrigo 2000).

Pattern Evolutions

Around 1900, ideas about the evolution of our concept of beautiful were put forward by the biologist Ernst Haeckel, who thought that beauty was simply what was natural (Krauße 1995; Kockerbeck 1997). In his art forms of nature, which had a significant impact on culture around 1900, especially on naturalism in literature and on the decorative style in art nouveau and the 'Jugendstil' (Krauße 2000), he developed these ideas further. Haeckel tried to show that beautiful is simply a recognition of the symmetries by which nature evolved its various forms of life. This argument looks somewhat contaminated by typological thinking, as outlined earlier by Goethe. Indeed, Haeckel aimed at a synthesis of the typological idea of a metamorphosing nature, put forward by Goethe, with the analytical concept of evolution, after Darwin. Haeckel argued that evolution was simply a process which gave rise to more and more complex biological constructions. This development could be analyzed by describing the symmetries in the different organic structures found (Breidbach 1998). Thus, beautiful is just the more complex form of nature. Its complexity could be described by studying those symmetries from which the different forms are constructed. Beautiful is simply a description of the degree of symmetrical complexity in a morphological structure.

This idea of symmetries forming the basis of nature, and thus the nature of beauty, has been followed until recently (Weyl 1952; Rössler and Hoffman 1987; Hahn and Weibel 1996). It was even lent support by certain views on physics in which subatomic organizations were thought to represent the basic characteristics of symmetrical order (Finkelstein and Rubinstein 1968). From these considerations, however, two totally different ideas of symmetry were developed. The former Haeckelian concept, in fact is a very simple perspective, and a denouncing of mirror symmetries as the essence of beauty (See Thompson 1917, for early criticism). Haeckel's concept of a gradual extension of symmetries in the course of evolution, an idea which he weaves into his phylogenetic trees (Haeckel 1866), has been shown to be nothing but a typological interpretation of Darwinian evolution, and as such is scientifically rejected (Breidbach 2002). The evolution of forms cannot be reduced to a series of crystallizations, thus beauty cannot be understood by such an approach.

The evolution of aesthetics cannot be described in such simple terms. Attempts to fit the formation of aesthetic preferences to predetermined patterns are reductive, and ignore the basis of evolutionary development, which is an accidental one.

Cultural diversification cannot be reduced to certain basic patterns. The evaluation of sensory experiences cannot be merely a matter of their interpretation by a system that is encoded to anticipate all possible sensory inputs. As the nervous system is highly dynamic and the outer world hugely complex, the limited capacities of any nervous system are apparent. Furthermore, evolution does not 'order' characteristics, apart from ranking highly those that are effective at the moment of evaluation (selection). Order is consequence of, not a precondition for, the evolution of a natural system. Attempts to elucidate a preconceived system of aesthetics will fail in evolutionary terms. If we could identify visual preferences, we cannot reduce them to the physiological reactions described. Their analysis must be more complex and examine both neurophysiological conditions and cultural situations.

Neuronal Art History

In his neuronal art history, Karl Clausberg (1999) outlined examples of the visual preferences underlying our perception of art. He also attempted to outline how far our perception of art was overlaid by neurophysiological preferences. The series of different examples he used to analyze the relationship between cultural pre-formation and neurophysiological orientation demonstrate that it is complex. One cannot simply dissect beautiful on the basis of a standardized scheme from our cultural experiences.

Memories are secured by certain physiological mechanisms. However, memories cannot be reduced to pure physiological reactions. While physiology guarantees maintenance of the memory, memories have a meaning that is independent from their mode of storage – whether in print, electronic form or physiological reaction. In organisms that mechanism is guaranteed by the brain. Without the brain, there is neither perception nor memory. Accordingly, one has to go into neurophysiology to understand perception and memory formation. While these physiological mechanisms determine how we interpret experiences, they also determine what is likely to be experienced. Such mechanisms are objects that can be used and played with. On the basis of his various case studies, Clausberg argued that artists are not only working on the basis of such mechanisms, but actually playing with them. Every canvas that represents something does this. Magritte's pipe is not actually a symbol, it is the pipe he is presenting, although it is also just a colored piece in outline ('piece' implies a three-dimensional form rather than painted representation). It is not only the canvas which can be played with, as Cannaletto has done, but our way of handling such representations, which can be played with.

If the pointillist school demonstrates to us the way our retina perceives an object, it also presents us with a new visual experience. In his paintings, Seurat demonstrated not only a theory of color perception. In presenting a picture which showed what our visual receptors actually present to the brain, he dissected a part of our visual experience; that representation provides a new and thrilling experience. It confronts us with our own mode of perception, presenting the latter as an object to look at. Thus, Seurat gave us both a statement about how to dissect our perceptions and an insight into the way we look at the world.

Neuronal Aesthetics

Neuronal aesthetics is interested in such dissections. It looks on both sides, the reactions of the nervous systems and the cultural phenomenology of cognition. Thereby, neuronal aesthetics is aware that our modes of perception are not something to be taken for granted. The neurobiological perspective told us a principal message: the world is not as it is, the world is as our brain has designed it for us (Breidbach 2000).

For neurophilosophers like Dennett (1984) that is no problem. Dennett (1985) knows that our brain has been formed by an evolutionary process. For him this meant that the brain evolved as an adaptation to our specific ecological niche: the environment is represented in the brain only in so far as it is necessary to optimize our behavior. What however is optimizing in such a scenario? What we optimize is not explicitly our way of looking at the world. It is not a question of whether what we perceive is true or not, it is only the question of whether that particular mode of perception is helpful to us. A certain benefit arising from a particular way of looking on the world may strengthen the correlation of a sensory input mode to a brain area that controls certain motor patterns. Whether this is of evolutionary advantage or not depends on whether such a change in the functional network of the brain will enhance the survival of an individual or its offspring. It is irrelevant whether this correlation actually outlines a world as that world really is. It is only necessary that the right action will be performed at the right time.

Such functional optimization is not necessarily based on an objective representation of the world as it actually is. Our sensory system already filters out much of the physical structure of the outside world. Such mechanisms of selection do not occur only in the higher cognitive functions, but are effective from the very first steps of sensory integration. The different physical properties recorded by a sense organ first make such a preselection, the encoding capacities of the underlying sensory neurons then make the next selection possibly already effected by signals from the central brain which interact with the stimulus responses of the peripheral nervous systems. Experiences do not, therefore, simply mirror the outer world. Experiences are defined by the capacity of the neuronal hardware and its interactive properties. This can be demonstrated by very basic experiences. For example, it is possible for us to view actions in a film as a continuous performance. We do not see the sequence of the pictures that actually constitute a film, as time resolution for us is limited by our visual perception. This is used by Hollywood and others to give us an impression of reality which, in fact, does not exist.

Internalizations

Searle (1990), in his portrait of a human computer acting in the famous Chinese room, has described the level at which an evolutionary advantageous correlation might have become established in our brain. An operator was required to combine certain input signals to certain output signals using a set of rules. The operator had no way of knowing the meaning of the incoming signals, but just had to obey the rules. Correctness was not correlated with understanding. In Searle's example, a human operates according to certain rules that were given to him. If the person adopts those rules, he will react in a way judged as proper by an external observer. However, as Searle described, the person himself knows only about the rules and that he has to obey them; he does not know what purpose these rules serve. Now we have to ask whether this scenario is adequate to describe brain operations, or whether it just describes the limitations of certain philosophical concepts about brain operations.

Neuronal aesthetics tries to sort out the more basal brain functions by looking at the actual physiological processes involved. It tries to define the principal operations of an associative network, looking into the basic mechanisms of brain functions. Then it tries to reconstruct how far such basic modes may be effective in producing higher behavioral features. Thirdly, it combines ideas about the mechanism of basic brain functions with concepts derived from the humanities, trying to understand how far the one matches the other and to identify the inconsistencies in such different ways of understanding cognitive behavior. Fourthly, it works on the different vocabularies used, trying to sort out where different research traditions overlap and whether such an overlap is significant or accidental.

Accordingly, it asks, what we really do know about our experiences when combining the approaches of different sciences (Breidbach 1999b). In the description of brain reactions and in the interpretation of the implications of such physiological reactions, we have to be extremely careful as the brain does not present an outer world, stored away somewhere in its memory board. Instead the brain, firstly and secondly represents its internal activity patterns.

Nagel's (1974) famous paper on what it is like to be a bat pointed to the question of what behavior actually represents (Tye 1997), and how far a brain is able to view the principal activation patterns implemented within it. Martin Heisenberg (Heisenberg and Wolf 1984) asked whether the brain is just a reflex machinery or whether, even in animals of lower phyla, it decoded internal reflex patterns. He asked whether a small animal such as the fruit fly *Drosophila* might already have some primitive intentionality. He thought he could find one: if the insect he was studying was not simply reflecting actual sensory input, but integrating it over a longer period, it becomes to a degree independent of external sensory input. It is not then acting as a simple reflex machinery. Even a simple nervous system can be shown to respond to internal dispositions. How far such internal dispositions match the mechanism of higher cognitive functions, and contribute to intentionality is another question.

Nevertheless, we now know from various experiments on more diverse brains that indeed such an action, by which the brain evaluates the momentary activation mode, by referring to former activity modes it has worked on, is very important to understand how a set of stimuli is integrated into the brain (Vaadia et al. 1995). In fact, the actual stimulation due to sensory experiences is overlaid onto an internal activation pattern (Breidbach 2000). The brain is not physiologically inactive until a stimulus is perceived and the stimulus itself cannot simply be stored away. It is acting on the whole complex of neuronal connections within the brain. Any new experience will extend these activation modes, so that gradually, a reference pool of new experiences is created.

Sensory input is integrated into the internal reaction mode of a functional brain. At the level of a certain neuron there is no physiological difference between the reaction elicited by an internal activation and the reaction caused by a sensory stimulation. Thus, internal activation and externally induced activation are combined to form the new reactions of the neuron. To understand our way of perceiving and recognizing patterns, it is essential to understand how far new inputs fit into preexisting reaction patterns within the system. The overlay of the internally and the externally induced activation reveals how far the new stimulus fits into the former pattern of brain internal activation, or whether it does not fit. The new stimulus is 'compared' with the internal activation modes of the system. This 'comparison' is essential to the very first acts of neuronal representation in sensory perception. There is no preconception reflecting the outside world stored away within the brain. Any activation is a function of both internal and external activation modes. Thus the brain, at the outset, is an associative device.

Associations

How can one understand this? James Mill, in describing the phenomena of the human mind, outlined the principles of such an associative brain in the first half of the nineteenth century (Mill 1869). Like Exner, Mill saw signal identification and association as the result of endogenous brain activation modes. Signal induction elicits activation in certain neuronal pathways; these are superimposed on internal oscillations, and, thereby stimulate more complex reactions in the brain. The result is a complex coactivation of signal pathways established by former impressions. This concept dominated neuroscience until 1900, when the neuroanatomical base of a functional morphology of the brain became dubious due to failings in the neuroanatomical techniques used at that time (Breidbach 1997a).The ideas were reestablished in the 1950s with the idea of neuronal networks (Hebb 1949) and dominate current theoretical considerations of the principles of brain activation patterns (Gardner 1985).

The main idea in relation to understanding perception is that something is not just represented as a kind of photography, in which an external world has impressed itself on the brain. Rather, sensory stimulation is treated according to the internal functional organization of a brain (Breidbach 2000). Thus, a visual signal is not just stored away, it is not even described as a single event. It is perceived according to the internal mode of the system. It is described in reference to internal reactions. Moreover, it is characterized in relation to representations of the other things already perceived. Thus, the internal representation of an impression is evaluated in regard to the internal organization of the system.

Internal Representations

What is actually handled in the brain are internal representations. These are defined in reference to the internal state of the system. This state is the result of former impressions and the associations elicited by them. They are propagated within a genetically determined hardware that, nevertheless, varies in its functional connections according to the experiences of the system. An externally mediated impression is not the energetically dominant activation mode of the system. In fact, its actual effect is small in relation to the energy of the internal coactivation patterns of the neuronal network.

Any input into the system is superimposed on existing internal activation patterns. The new input will be either integrated in the activation modes already present in the system, or will skip the internal activity states (the attractors) that characterize the functional potential of a certain system (Engel et al. 1997; Singer 2000). Thus, the effect of the input is calculated with regard to the internal dynamics of the system. Insofar as it is possible to speak of an internal evaluation of the impression received by the system, this 'evaluation' can be described as a principal mechanism at the level of neuronal activation patterns. Such evaluations are sufficient to categorize the various impressions which act on the system and which in turn influence the internal order characteristics of the system itself. The details of this concept of internal representation have been outlined more extensively recently (Breidbach 1999b, 2000).

The principal mode of sensory activation, therefore, is not an imprinting of the outer world into the brain. Contrary to Michael Ende's legend (1979), there is no archive of invariant impression-representations in the brain. The world is perceived according to the internal state of the system: representations do not work from the outside in, but from the inside out. What is seen is categorized by the internal activity functions defined by the innate coupling characteristics of the neuronal elements and their experience. These internal characteristics are not directly connected with the actual input that is evaluated by that internal activity mode.

In a computer simulation, it can be shown that numerically and qualitatively simple artificial neuronal networks, with 16 interconnected elements, operate in the way described, if they are provided with local coupling characteristics (Holthausen and Breidbach 1997). Furthermore, it was shown that such small systems allow different input signals to be sorted according to the actual activity modes of the system. As these activity modes are not set at random, but can be characterized by a limited set of attractors, there is a kind of functional fingerprint for a certain system. A new input will either fit into this structure and activate something already known, or it will not fit into these structures. In the latter case, it may be insufficient to elicit a new type of response, for example, if the system is confronted with something new. This new item is defined by the fact that it is not part of any 'known' activity mode. Thus, it is defined negatively in relation to the set of internal representations which have already been defined. Following this line of argument, a whole new series of inputs can be implemented and defined in relation to each other, and the functional characteristics of the system in general.

If more and more elements are added to the network of internal representations, the definitions of the functional states of the system will gradually increase in elaboration. Thus, the network of relations by which the single internal representations are defined extends its specifications. The more specified a set of negative attributions is, the better the internal representations are characterized. Model simulations show that increasing the inputs into the system does not result in an overflow of the functional characteristics of the system, but in a more and more refined landscape of attributes. The system extends its set of functional states and allows for a greater degree of precision.

What is gained by such relative attribution? Such qualifications work without an external observer. Accordingly, the putative experiences of such a system as the human brain are not perceived via an expert system, but established as the internal world of an associative system. This meant that someone starting to dissect a new structure or to uncover objects never seen before, had to know beforehand what to expect. Such a system which established its internal order characteristics without an external observer, thus, is a truly self-organizing one.

Any input into the system is defined by the way it fits into the system. Its classification is dependent on internal system parameters. Accordingly, the categories by which a system defines its world view are internal categories. Thus the concept of internal representation, the former concept of an objective information as defined by Shannon and Weaver (1949), is replaced by a concept of subjective information (Holthausen and Breidbach 1997, 1999).

The classical definition compares a system's intrinsic response with an objective idea gained from an expert or an expert system. In such an approach the classification devices are derived from a world that is essentially already known. A concept of representation based on this classical idea of information will describe signal propagation within the brain simply as a transfer of something from the outside into the brain. Internal representation works the other way round, starting with those things the brain actually knows its internal activation patterns. Thus, the concept of internal representation does not use an external observer who can correct the system response, as the system is able to do this itself using internally defined categories.

This may seem reasonable as neurobiological theory works only on the basis of what we know of human physiology and not on the idea of a supreme power. Internal activation patterns are the only coins neuroscience can deal with (Breidbach 1999a).

Encyclopedia of Cognition

Until recently a different, encyclopedian approach to cognition was in use (Breidbach 2001a). Our idea that cognition operates like an expert system is in this tradition. The encyclopedian concept of cognition is an old one. It dates back to the sixteenth century and in some essentials might be even older (Breidbach 2001b). In other words, it is derived from a nonneurobiological perspective. It was established before the concept of brain tissue organization was available. Rather, it reflected neurobiological ideas based on philosophical and theological concepts. According to this idea, there is an absolute knowledge that is represented in cognition. If the brain is the organ of cognition, then that absolute knowledge must be integrated into the brain. The idea of an expert system is based on such an approach. This idea of knowledge representation is based on the idea that within our brain there is a complex archive of sensory experiences. Cognitive sciences, and artificial intelligence in particular, developed the concept that cognition can be modeled by a system which stores every possible experience. Such a system could associate any new experiences with the pool of material it had in its archive and, thus, correlate any new experience to those things already known. Identification and orientation in its surroundings would not be a problem for such a system. This idea could not be transformed into an operating system, however. A closer historical analysis demonstrates that such schemes are based on an outdated concept of cognition. They go back to the encyclopedian tradition of the sixteenth century (Breidbach 2001b). We cannot assume the existence of any neuronally implemented expert system. We have to look at an operating associative brain. The idea of a representation of the outer world representation, thus, has to be encoded into the language of such an associative brain (Breidbach 2001b,c). According to the encyclopedian concept, such an association implies a hierarchy of order characteristics. This concept would describe a topological order of internal representations. That topological order was then interpreted as reflecting the objective organization of our knowledge of the world. Classical AI approaches tried to sort out the details of this storage program (Gardner 1985). To do this, a control function has to be provided either by an external observer, or by a secured expert system. Neither of these exists, however, for or within an associative brain. Neither an external observer, nor an expert system would be the best source of such representations because we have to know beforehand what we looking at. We could not say whether a certain observation matched part of an expert system. Here, it is useful to extend the idea of a control function by replacing the aforementioned external observer by a mechanism that would guarantee a fit of the external to the internal representations. That control function might have been established by evolution, which resulted historically in a co-adaptation of the external world and the internal order characteristics of the human brain.

Evolutionary reasoning then might just work on an incorrect concept following the idea of an encyclopedian organization of knowledge acquisition centers in the brain (Breidbach 2001b). Nevertheless, the principal idea of an evolutionary optimization of the functional organization of the brain remained unaffected by this argument. If indeed evolution designed the brain to best fit the environment, we should ask how this optimization is to be defined? One should outline the neurobiological equivalent of an external observer, thus avoiding the need to adopt metaphysical explanations in order to try and understand the functional optimization of the brain.

There is, however, considerable evidence that certain brain centers adapted to perform precisely defined functions (Zeki 1988, 1993). The most striking examples are the face-detecting centers found (so far) only in higher primates (Grüsser et al. 1990). Such centers are directly coupled to a certain input channel. These are brain areas where a certain subtype of information is filtered off from a multitude of visual inputs.

Already the sensory organs are selective in their response to physical signals. For example, we only perceive light within certain wave lengths. Our visual orientation has a preference for certain movements or shapes. The brain's ability to filter numerous and complex signals was probably important for the survival of an individual. It is of no value to a rabbit to define the specific species of an approaching bird of prey. It would be better for it to run and question later whether it had been a hawk or an eagle. The brain's various filter functions are able to elicit certain types of behavior. One would expect, therefore, that the various signal filtering functions also preselect much of the brain's sensory input. The internal signal room within the brain which represents our surroundings is thus a scattered one. Only parts of it actually proceed to the highest brain centers; another part is used directly in particular brain regions to optimize reaction modes to certain sensory stimuli. Such fragmentation of functions displays a topological organization of functional morphology of the brain, in which filter functions that are coupled to fast motor reaction patterns are implemented downstream, near the region a certain input is delivered into the system. If a neurophysiologist remained close to that input area in his studies, he would experience a more or less defined topological fragmentation of functional items in signal propagation. That functional topology would be much more complicated, however, if he moved away from that input channel. In the history of neurophysiology, for practical reasons, research was at first concentrated in those areas close to the input channel (Breidbach 1997a). Here, with the methodologies to hand, one could detect and characterize single elements in the hierarchy of signal processing units. Away from that region, these methods were, until very recently, insufficient to give reliable results (Palm and Aertsen 1986; Abeles 1991).

Objectivity

If the assumptions outlined above regarding the overlay of input activation on a preexisting internal activation mode are right, one would expect that a physiological reaction to a certain stimulus would not be constant throughout the history of an individual brain (Freeman 1988). In fact, that is the case. Nevertheless, such associative regions were not the main focus of neurophysiological studies until recently. Consequently, only those areas which fitted into the aforementioned conceptual scheme were studied. That scheme seemed to be correct, and so it was, but only in regard to the input channels found within the entrance region of sensory pathways into the brain. However, even there, neuroanatomy tells us the importance of central feedbacks for a proper evaluation of sensory inputs (Braitenberg and Schüz 1991).

Unfortunately, a God-like evolution that – in a Leibnizian sense – offers the best of all possible brains as our cognitive organ is quite unlikely. Moreover, even if evolution had taken such a course, it would be quite difficult for us to tease out just how far evolution has taken us in a direction that really tells us something about the outside world. Evolution, remember, is just a history by which a certain functional type is secured. Furthermore, this process of adaptation is an opportunistic one. It is not the best or most 'truthful' solution, but only a relatively better solution which is secured. Thus, optimization does not necessary lead to a truer and more reliable concept of the outer world. What is established is just a relative optimum. While up to now a certain solution is better, it cannot necessarily be regarded as aiming for the truth (Bowler 1996; Ghiselin 1997).

The mechanism of evolution is a relative one, sorting out those systems which are a little better than other ones. In this respect, it cannot be regarded as an objective measure of the internal flexibility of a system. One has to remember that the reaction capabilities of an organ such as the brain are tremendously complex, and thus may be difficult to correlate directly with an optimizing function (Breidbach and Kutsch 1995).

An organization which is relatively good, i.e., better adapted to its surroundings, might be favored and so gradually becomes established. Thus, some structures can only be guaranteed to be better than those of their contemporary competitors. In theoretical terms this means that within such a process, the structure is only locally optimized. Thus, its specific state cannot be proved to be the absolute optimum, but just a relative optimum. The correlation between an inside activation pattern and the respective signal processing is no guarantee of a direct link between internal order functions and the structures of the outside world. Seen from within, and that is the only perspective possible for a brain, one cannot determine the accuracy of an internal representation in regard to the categories of an outside world. This could be done only in a case where the outside is already known, and thus, can be compared with the representations found within the brain. However, in this case, there would be an external observer which was not as opportunistic as Darwinian evolution.

Objectivity could not be guaranteed by an evolutionary mechanism that was opportunistic, nor can we assume a goal-orientated external observer. The concept of the co-adaptation of brain function and the objective world will arise therefore, only in a teleological misinterpretation of evolutionary processes.

Gestalt or Pattern Recognition

For the concept of internal representation, it has been shown that the relativistic interpretation of an input into the brain, qualified only in regard to certain of its attributes, is sufficient to describe a diversified world on which the behavioral program of the human being can be adapted. The internal representation is sufficient to describe categories by which an internal world is created and according to which certain behaviors can be properly directed. Thus, internal representations work only on relative evaluations: an entity is defined by its relative position to other entities characterized within the system. In this way, a complex network of interrelations develops in which relative definitions are evaluated by system internal characteristics. These relations form a kind of defining network of relative local positions. The order parameters defined are thus subjective ones. Nevertheless, within an expanding, diverse and integrated network of interconnections, the contours of something like a world might be described (see Breidbach 1999b for details).

Such a subjective entity of relations may be transferred to a second one without losing its identity. A network of relations transferred to a second network will not lose its relative determinism. The relative relationship between entities a and b in the first network, will not lose its integrity by a transfer into a second network, in which b becomes related to ε, δ, or ϕ. Irrespective of these new relations, the relative distance between a and b is conserved, even when the absolute dimensions of the respective distance functions are radically changed. (see Breidbach 2001c for details).

Thus, there is something like a relative fingerprint of data sets which allows something to be categorized; as relative distance functions can be transferred from one system to another one (making a translation between two systems possible). Patterns can be defined by relative definitions. This has been shown in practice, for example, in a system for clinical EEG diagnosis which utilizes the principles outlined (Breidbach et al. 1998; Schramm et al. 2000). This relative contour diagnosis demonstrated that a structure could be characterized using relative distance functions. For a relative definition of one structure in regard to other structures, a matrix is defined into which the various distance functions of the data series that constitute the EEG pattern are plotted.

Two structures were characterized as similar when the overall distance functions were closer to each other than to any other combination of characters within the system. Thus, a pattern is defined internally by calculating the relative distances to other system characteristics. In this way, a measure of internal similarities is also gained.

Similar structures can be identified irrespective of alterations in their absolute dimensions. If a structure shrinks or is expanded, the relative distance between the various items characterizing it is conserved. Thus, not only a certain pattern, but types of pattern can be identified by this approach. More differences and similarities in pattern types can be identified. Internal representation thus allows the characterization of types of pattern, and families of such pattern types. The mechanisms described as characteristic of a parallel distribution system, with local coupling characteristics that work on internal representations, can be interpreted as a gestalt recognition system.

So far, neuronal aesthetics have aimed to describe some essentials of our way of looking at the world (Breidbach 2000). It outlines principal characteristics of perception and aims at a better understanding of our ideas of the morphologies that are essential for our understanding of perception. Special interest needs to be directed to the correlation between language and perception. It is assumed that this correlation is a tight one, as we learn to look at the world by following rules adopted in our early childhood, when we were trained in the respective language of our society. In this way, we adopt the particular categories implicit in the language system of our society. Neuronal aesthetics, nevertheless, is far from demonstrating the actual mechanisms in the ontogenesis of our cognitive system. Yet it can already show that the perception of images and the perception of a text are not so different as has often been thought. We are, therefore, aiming at a neuronal semantics that is not only applicable to the analysis of syntactic peculiarities of a text, but also allows the structural characteristics of images to be compared (Breidbach 2000).

As can be seen in this short sketch of neuronal aesthetics, within the framework of our understanding of the basis of human perception, there is no concept of beauty. We just encounter the problem of categorizing our perceptions, but we do not evaluate the resulting categories.

Aesthetics and the Beautiful

If we deal with aesthetics in evolutionary terms, the situation is different. From the outset, within an evolutionary framework, there must be an evaluation of categories in our perception. Evolutionary theory has need of an evaluation system. This has consequences. First, in such an approach, one has to ask whether there is, or has been, a decision about what is better or worse in terms of aesthetic considerations in the evolution of a species. If the aesthetical ground is neutral in evolutionary terms, it cannot be dealt with in evolutionary aesthetics. This would mean that in evolutionary aesthetics something can be described as 'good' or 'bad' in evolutionary terms or it would not be integrated into such a theory.

An evolutionary aesthetics, taking itself seriously, has to apply reason to such evaluations. Thus, evolutionary aesthetics aims to elucidate the 'good' principles established within aesthetics in the course of evolution.

Such a positively evaluated scheme could thereafter be defined as a measure of beautiful. Beautiful in that sense means something that is in evolutionary terms advantageous.

As a consequence of this there is, from the outset, a considerable difference between an evolutionary and a neuronal aesthetics. In the latter, one is only interested in understanding the mechanisms by which we classify our sensory perceptions. Thus, evolutionary aesthetics appears very much in front: it is no longer dealing with the question of how far our way of looking at objects can be accurately described at the level of brain activity. Although evolutionary aesthetics is working on how certain brain activation patterns were established, it is not looking at the question what these activity patterns really reflect. The meaning is not part of this theory. Thus far, beauty cannot be defined in regard to cultural impact, but can be described only in regard to certain behavioral patterns.

One of the essential positions of neuronal aesthetics is to understand what is really meant when we refer to something as being objective. Here, certainty is gained in regard to the network of internal representations manifested in our brain. Objectivity, thus, is a subjective construct. It would not make sense to define such a subjective mode of evaluation to provide a measure of beautiful. To detect something, and to be sure of how it can be categorized in view of our recent experiences, does not qualify an item as beautiful. Such a qualification would require qualification within the different categories in which we organize our internal representations. If one of such an item has meaning only in relation to other items in the world view of a subject, an aesthetic distinction within objects of this inner world becomes difficult. Nevertheless, as has been shown (Breidbach 2001a), a subject working within such qualifications does not degenerate to a solipsistic existence, but is able to communicate. In fact, many of its categories may have been gained by adopting other networks of interrelationship, as is done in language acquisition, for example.

Thus, neuronal aesthetics argues that it is not the outer world we represent in our way of looking at things; rather, it is our own understanding and our own categories through which we establish our world view. This sentence can be proved both in regard to neurophysiological measurements, and to the principles of the function of an associative network with local coupling characteristics. Thus, such a neuronal system does not work on fixed categories, but is quite flexible as it is continually integrating new input.

What does this mean for our attempt to define beauty? Is beauty a category of its own? Can we sort out those characteristics that make something beautiful and describe principles for the evaluation of beauty? Neuronal aesthetics had shown the extent to which the categories we try to adopt are impacts of a particular culture. The strong correlation between our language and our way of looking at things already provides an indication of the importance of cultural influences. Language, in fact, is a kind of externalized network of associations which extend the categorizing functions of the human brain (for an explicit discussion, see Breidbach 2000).

A sentence structures our picture of the world we react to, the relationships within it and the way we classify it (Glasersfeld 1997). Language provides the mechanism by which an individual can optimally identify and represent something new. Using language, the complex physical signal spectrum that is a flower can be precisely described by the two terms which describe the objects species and color. Using these terms, correlations can be drawn to other flowers noticed before, and thus activate a whole complex of knowledge within a reasonable time frame. Secondly, the picture of such a flower may elicit a complex of memories, even characterizing a certain atmosphere at a certain place. Nevertheless, it is obviously much easier to compare certain memories on the basis of language classification than to search our personal memory for different stored scenarios.

Furthermore, the correlation between language and the brain is not one way. As language allows us to draw certain comparisons, the acquisition of language channels our capacity to make comparisons. Our linguistic skills thus predetermine our way of looking at objects (see Friederici 1998) and language channels our view. Already then, our perceptions are guided by our culture. Even neurophysiologically this has meaning (Zeki 1999); sensory input is not only filtered by the innate wiring pattern of our neuronal tissue, but also passes through a cultural filter mediated by language, which predisposes us in particular ways and weighs certain aspects of our experience. Understanding how we perceive cannot, therefore, be reduced to a neurophysiological description, but has to integrate cultural and neuronal studies.

Has that a bearing on our analysis of beauty? Is beauty just a cultural filter by which we sort out a complex group of objects we denote with specific behavioral attitudes? Simply by treating such objects as precious and setting them apart from others in our surroundings which we notice simply for survival (such as the colors of poisonous snakes, flying arrows or high speed cars) are we making them beautiful? According to such reasoning, beauty is a cultural category and can only be analyzed as such.

If so, beauty does have a meaning as it is an entity within a cultural system. Such a meaning, however, cannot be identified just by a syntactic analysis of object representation in our internal representations. Within a culture, such an internal view is externalized. This means, it is exposed to various interpretations. In essence, it will perform a similar function in different individuals with a common 'enculturation'. Accordingly, such a category will direct various subjects in a common way, at least in so far as it predisposes one to a particular way of perceiving and handling certain objects. First consequence is, it cannot be dealt with in physiological terms that describe just one reflection of such an entity, its internal representation. It is just a response to the external world, but not this world itself. It allows a set of categories to be described which would allow an evaluation of the details of that world. Evolutionary theory generally claims to identify such categories using the concept of selection, adaptation and fitness. Thus, evolutionary aesthetics must describe a correlation between the certain structure of an organism that is selected and its resulting adaptations. Thus, there must be an adaptive value of beauty or – more explicitly – evolutionary aesthetics must provide a fitness function for beauty.

Evolution of Beauty or Evolution of Beauties

What would constitute an evolutionary impact of beauty? It might establish preferences in our perception of patterns. What evolutionary aesthetics is doing is sorting out behavioral patterns that can be directly correlated with a benefit. These patterns are found in sexual behavior. Thus, sensory preferences have to be interpreted in terms of whether they are correlated with sexual behavior. Thus, the categorization of patterns would be due to just one certain type of behavior (i.e., sexual behavior). This aspect of our behavior is easily connected with selective advantage: the most attractive is selected as the most promising partner for the propagation of ones own genes (Wilson 1975; Dawkins 1976). Actually, evolutionary aesthetics is arguing this line (see Voland, this Vol.).

Evolutionary aesthetics reduces itself to the study of preferences in partner selection. This assumes that theory aims to qualify what the benefit of beautiful is. There must be a mechanism by which we can recognize the best partner to choose. One must, therefore, look for those orientation preferences that might have become stabilized during our evolution. The outcome of such an evolution would be a scheme to select the most promising partner.

Evolutionary aesthetics is narrowing its perspective on beauty to the description of fitness functions. Where does such a reduction lead? As there is no direct test of breeding capacities in mating behavior, selection will concentrate on secondary attributes correlated with those capacities. The result is pretty clear: it is a conception of the 'pretty' woman and the 'handsome' man, most promising to share my genes with. Following this idea, evolutionary aesthetics can tell us a lot about such beauties. The question is, however, will it present a consistent concept of beauty which is not reduced to that being presented as a 'beauty' in the sense of *Vogue* or *Cosmopolitan*.

Here, I might have a specific personal deficit, but I do know that I am not able to look on the 'Mona Lisa' as just another type of Claudia Schiffer. On the contrary, I think it important to separate the representation of something we find attractive from the idea of beauty. A picture – or a composition, a poem or even a mathematical formula – might be something more than just a form of documentation or a substitute for reality.

To Be Fit Is to Be Beautiful?

Evolutionary biology, including aspects of population biology and sociobiology, has concentrated for some years now on the question of how to describe mating success (Maynard-Smith 1989, pp. 12–13). The idea was to look at behavioral strategies and to sort out the fitness value of certain behavioral activities. Mating and breeding are such behaviors that can be readily evaluated in terms of success. Furthermore, this type of behavior is central for the maintenance of a species. Accordingly, it was a pretty good choice for evolutionary ethology to concentrate on this type of behavior. The outcome is a simple model. The partner to be selected will adopt a strategy of looking the best without consuming too much energy. The selecting partner will, however, be most pleased by the best show: that is, they will be impressed by the courtship behavior requiring the greater amount of energy. Certain structures such as the tail of a male peacock can be interpreted in this way. Evolutionary aesthetics follows this idea, simply extending the paradigm by defining those structures that are effective in enhancing mating success as beautiful.

If the female has a choice, she will be attracted by the most magnificent. Now the question for both partners is how costly will mating success be. The female, being the one who becomes pregnant, normally takes the greatest risk and will put more energy into the next generation. If the male is just spinning around without any strong partnership, his preferences will be less developed and he will put sperm into any female he can get hold of. The situation changes, however, when mating success becomes costly for him too. Then he also has to be selective in choosing the most suitable partner to put his energy into. Hence, even the male might develop preferences.

This is the scenario that at least part of evolutionary aesthetics has taken on board; beauty simply signals whether a choice is good. In humans these preferences may be culturally manipulated, as the fitness landscape to which one must adapt has changed drastically over the historical development of human society.

One is tempted, therefore, to measure the correlation between attractiveness and genetic fitness. Those beauties that are the most desirable humans, thus, might establish a partnership of greater benefit to them – allowing for the spread of their genes under most optimal conditions. No effort in that respect might be too great; neither the palace built for a queen nor the costly clothes worn to attract the right male. Thus far it seems feasible to describe beauty as the outcome of a special strategy for optimizing mating success. The best performance is that which achieves maximum success with a minimum of risk to their own constitution. Sports – even on an international scale – would be such a playground for attracting the right partner. Sex symbols should develop out of such sports, especially those in which a bright brain is not put at risk.

Opera Stage or Sporting Arena?

Why would such a way of selecting the most promising partner result in an aesthetic outcome? The best thing for an ancient society would have been a direct test of strength and skill, and by implication, a direct test of the genetic makeup of a certain individual. Why should I focus, therefore, on such things as poetry or art? Especially as those people so engaged might be unfit to succeed in social competition for honor, gold and luxury. It would much more promising to sort out the beautiful guys directly. Accordingly, it would be a pretty good idea to look at the presumptions underlying such sports, especially the Olympic Games, which were quite early manifested in society to sort out the best man, even if the female had no direct access to the playground.

Later, the knights had comparable tests of skill to determine who was the best. The male beauties could have, therefore, been selected directly. Thus, it can be asked whether evolutionary aesthetics might be convincingly applied to sports, at least in the broadest sense; even an opera – at least in Verona – has a sportive dimension, though it is certainly not reducible to a match without missing those elements we seek in attending such a performance.

Theories of the Best

As evolutionary aesthetics tries to sort out their concept in an evolutionary theoretical approach, it uses the vocabulary of an established biological theory. Thus, it has to think about the rules by which something has been established within evolution. Attributes which have a greater impact on human behavior might, therefore, have an adaptive function. Otherwise, they are not comfortably dealt with by evolutionary theory. Consequently, evolutionary aesthetics aimed at a fitness function, describing how something might have come to be regarded as beautiful.

What could such a behavioral function be? Normal behavior might result in establishing something we like, and neuroimmunology tells us that only stress-free surroundings are optimal for the establishment of our immune system. Thus, a warm atmosphere might be advantageous in behavioral terms, for example, by providing a comfortable and congenial environment for social interaction. What is the driving force for such interactions? Sociobiology tells us that the primary purpose of life is to establish itself, which will only be effective over a sequence of generations. The appropriate behavior to guarantee this is mating behavior.

So far the evolutionary view is consistent. However, the difficulty is in interpreting various activities we connect with beauty, with the optimization of courtship strategies. Voland (this Vol.) has tried this approach. One principal advantage of the failure of theories which focus on the way we perceive and orient ourselves in our surroundings, is that an evolutionary approach actually may define what a beauty is in terms of the sexually most attractive partner. Such attractiveness comes from a morphology that reflects, for example, the breeding capacities of the female, and the male's strength to support both the female and her offspring. In more advanced cultures, such direct means may be replaced by insignia of power and social success that are not directly connected with an individuals morphology. Thus, a king might be much more attractive because of his social prestige and the guarantees of breeding success that this implies, over that of an ordinary solder, with little reference to the physical attributes of either individual.

This means that the gold medal winner of the Olympic game may not in fact be the best candidate for mating. This argument is in line with modern research on male preferences by women, which have pointed to more delicate strategies for selection of the optimal partner (Eberhard 1996; this volume). Nevertheless, there remains the problem of determining how far something that does not contribute directly to mating success, such as poetry or art, is really attractive to a female.

Just pointing to the honor such poets, composers or painters may command in society is not sufficient, in which case Rosemarie Pilcher would be much more attractive than Else Lasker Schüler. Furthermore, what about the honor that is guaranteed long after the best time to beget and secure one's own offspring. The history of the arts shows that the social conditions of such men were not good until the nineteenth century. One need just think of Goethe, one of the very few normal citizens productively involved in the arts, who enjoyed a favorable social position at his time, to see the difficulty with this argument.

Or would it be just the 'condottiere', the renaissance military man, who could afford to hire a Leonardo, who would be selected by the most fitting woman? Beautiful would thus be only a subordinate indication of luxury and nothing else. Beautiful itself would be without meaning; only useful in demonstrating that one could afford to spend money. The arts would be – then – something comparable to an Indian potlatch festival, in which costly things were thrown around to show how much power one had.

This view is inconsistent. Such an argument allows one to define sexual attractiveness and also to define what a beauty is. However, it does not allow for a definition of beautiful. Furthermore, in regard to our behavioral orientation, it is too reductionistic to define every behavioral pattern in terms of optimal partner selection. Just imagine someone hurrying though the Louvre to sort out the optimal scheme for his potential partner. He would be better advised to visit the restaurant of a dancing hall.

Beauty Naturalized

Beautiful, and thus beauty, cannot be elucidated through an evolutionary perspective that relies on optimizing strategies and a conceptual framework gained from evolutionary theory. A concept of beauty cannot be established in such a way. As has been described, the Haeckelian way of identifying beautiful just with those things nature offers, likewise elicits difficulties. In particular, mating success did not guarantee beauty. That is shown by parasites. It might be interesting to sort out the way nature has evolved such creatures, but such reasoning is clearly out of focus of the aforementioned Haeckelian perspective. Accordingly, beauty cannot simply be regarded as something that guarantees mating success.

My proposal, derived from neuronal aesthetics, is to describe beauty actually in cultural terms. These are not reducible to the neuronal settings in which we install our ways of looking at nature, as culture itself is not genetically determined. We have to learn our culture. That is we learn our language and adopt our culture's ways of looking at things. Nevertheless, there is a neuronal disposition that influences our way of looking at the world, to a large extent, but which cannot simply be reduced to a neuronal pattern. Thus, neuronal aesthetics is restrained in its ability to discuss the arts or even beauty. Here, it meets with an image anthropology put forward by art historians, who likewise have tried to sort out our way of thinking about pictures that is not directly aimed at arts and beauty (Stafford 1999; Belting and Kamper 2000; Belting 2001).

Some Scholastic Considerations

Nevertheless, there might be an aspect of beauty one could discuss even in terms of neuronal aesthetics. This is the question of whether there is a connection between beauty and truth. The idea followed up here is essentially a scholastic one. It has been termed *ens et verum et bonum convertuntur* (the being, the truth and the beauty somehow coincide) This was not a problem in a world view in which the ultimate position was seen to determined by God. Being the Absolute was nothing less than being the measure for everything.

However, internal representation has denied the absolute perspective of an external observer to evaluate brain functions. In the concept of internal representations, the idea of subjective information was developed which accepted only relative evaluations of that which was represented in the mind. Thus far, the external world is only indirectly accessible. It is known in the formation of internal representations that are interwoven to such a degree that they form an interrelated system of references, a scaffold of descriptions. This is explicable in the syntax of a language that is, in fact, able not only to describe things actually thought to have been perceived, but also things which are only imagined. If such an imagination is a good one – and as methodologically sound as it is in science – it may result in a proposal that could actually be verified.

Nevertheless, such a concept of a world view is a very restricted one. Therein, truth will not be defined by correlations of representations of the objective world, characterized by a preformed expert system or judged by an authority which is not internally represented. Truth here is defined by the consistency of arguments within the mind. There is, however, one aspect in that consistency which may allow an extended view of the word in such relative terms.

Such a coherence of internal representation might be a momentary one, describing even the atmospheres that we recollect in our memory. That momentary consistency, in which we condense the impressions of various senses, might be a very intense one. It characterizes only a moment in our stream of consciousness, however. Yet, there is another truth for such an internal representation, which may even result in a formula for describing a whole series of particular impressions.

Science is dealing with such formulae. As modern physics tells us, such formulae must not be descriptive ones. They may – as they do in the Schrödinger equation, for example – outline a consistency within the theory. This means that a whole set of arguments is combined within another set of arguments. The theory enfolds therein within a theory. Thus, our arguments work out an internally defined consistency that does not directly refer to an impression. It extends our views which are based on temporary impressions. It is tested in relation to the set of any other putative representations available. The equation, or other formulae, (even such ones given in speech, tunes, or arts), extends the set of correlations of internal representations into new dimensions. Its use reveals that the network of relations itself can be extended. A continuity of putative descriptions becomes feasible in which the network of discrete interactions seems to be condensed to form a perspective that surpasses the former surface of intercorrelative determinations. Thus, it seems to be that a truth is found, (secured by a network of internal correspondences).

Such a truth is not just an orientation put forward by an external authority. This truth reveals something within the subject that seems to point beyond the series of distance functions in which something is evaluated internally. What is touched on in this short sketch is the notion of Einstein's that a formula is beautiful. What he meant was not the simple symmetry underlying the mathematics he used. He actually thought of a new dimension of clarity that surpassed that of his former notions. It did so by developing a new dimension in which the series of former observations could be interpreted. The interesting thing for us is that such notions do not prevail endlessly. Even the formula of Einstein is a historical one. As new discussions in theoretical physics show us, the foundations of quantum physics are under discussion. Thus, the new dimension Einstein discovered has led to a new series of observations that, eventually, will have to be set in a different framework.

The idea of something surpassing a temporary set of arguments is not seen as the manifestation of some absolute truth accidentally put into our brain. Even that fresh focus will soon be an old perspective. It is a cultural entity, and thus, it is bound into a certain historical perspective. Nevertheless, it allows us to work out new order systems by which former observations will be set into a new explanatory framework. Such a new framework will not deny the former series of observations; it will simply put them into a new perspective.

For someone trained in the theory of science history, the idea of two different levels of truth sorted out by an individual mind closely correlates with Kuhn's idea of a matrix of scientific arguments, each surpassed in a scientific revolution which finds a new focus, and by which a new matrix is outlined (Kuhn 1970; cf. Bach 2001, pp. 6–42). As Kuhn had worked out, such a matrix could be not an absolute one, but is simply one step in an historical process defining our culture of science. Here, it is argued, something similar can be found in a much simpler form in the individual mind, as it takes part in those judgements and diversifications offered by culture which allows someone to categorize the world.

One has to work out in much more detail what the mechanisms of such a subjective revolution might be. However, this can be done only in the vocabulary of internal representations. Such an internal revolution is part of our evaluation of internal references. Subjectivity – as described in a previous monograph (Breidbach 2001a) – is not a hindrance, but allows access to the truth.

Aesthetic Transcendentalism?

The coherence within internal representations could be defined as a reflection of an outer world. In that coherence, the internal reference system is superimposed on itself. The subject actually catches a glimpse that surpasses its own categorizing functions. In that phase of reorientation, the internal world is seen as an entity. Furthermore, it is seen to be defined irrespective of the different matrices of arguments applied. Thus, the internal world is manifested to be a world of its own that surpasses not only the reference system by which it was ordered before, but even the memories and the momentary impressions of a certain subject. The subject surpasses itself and its temporary conditions. Thus, through the consequences of a subjective mind, some idea of objectivity becomes feasible. That objectivity is not externally supplied, but is based within the subject. As has been outlined before, the consequence of an externally supplied objectivity would be the belief in an authority. The consequence of internal representation is the objectivization of the subjective viewpoint. That might be beauty. Yet that beautiful is far from the beauties.

Such beauty might be without a direct adaptive value, nevertheless; it is essential in a self-definition of the subject. Only when such a subject secures itself, is it really able to act. In other words; it can perform instead of just react. Beautiful, which is only the rococo of a partner finding selection program, will be in quite a different situation.

Neuronal aesthetics thinks the world to be a projection, describable in terms of internal representations. Thus, there is no objective evaluation for something that is beauty. However, there is truth in the internal view in which the mere subjective is installed as an objective reality, in that the subject is secured. Thus, the subject does not claim an absolute perspective. It is just on its own, but it is secure in being so. Therein, it surpasses its internal categorizations. In that action it effaces something sublime. Such a sublime would not be something absolute, as it is bound to a certain cultural world view and, thus, to a certain historical context. It is, nevertheless, transcendental as it shows the culture the mind is embedded in and the objectivity the subject is founded in.

Summary

If aesthetics were just 'seeing', biology might offer a solution of the problem of defining beauty and establishing a new type of aesthetics. When aesthetics, however, is recognition, then, the situation is different. In such a case, aesthetics is not just the process of sensory imprinting, but also the process of interpreting such sensory imprints. These interpretations may even entangle that faculty formerly called fantasia. Aesthetics, thus, cannot be explained by or reduced to a biological analysis.

In my view, one has to establish a more diversified phenomenology of aesthetics before dealing with the question of beauty. Furthermore, the question is whether beauty is, in fact, the object aesthetics must deal with. One can show that in the arts it is not the most pleasant, but the most interesting which deserved most interest. Beauty is not necessarily the central term in the history of art and history of aesthetics. The recent idea of an image anthropology (Bild-Anthropologie) arises from trying to describe how in our cultures, we handle sensory objects. Similarly, neuronal aesthetics from their first attempts in the mid 1990s onwards, tried to describe the mechanisms of perception in both the cultural and the physiological dimensions, and to access the way we evaluate perception and aesthesis. The present text will describe such an alternative approach. It will discuss evolutionary aesthetics from the perspective of such neuronal aesthetics and will develop some perspectives for a theory entangling not just the beautiful, but even, hopefully, the sublime.

References

Abeles M (1991) Corticonics. MIT-Press, Cambridge
Bach T (2001) Biologie und Philosophie bei C.F. Kielmeyer und F.W.J. Schelling. Schellingiana vol 12, Frommann-Holzboog, Stuttgart
Belting H (2001) Bild-Anthropologie. Fink, München
Belting H, Kamper D (eds) (2000) Der zweite Blick. Bildgeschichte und Bildreflexion. Fink, München
Böhme G (2001) Ästhetik. Vorlesungen über Ästhetik als allgemeine Wahrnehmungslehre. Fink, München
Bowler PJ (1996) Life's splendid drama. University of Chicago Press, Chicago
Braitenberg V, Schüz A (1991) Anatomy of the cortex. Springer, Berlin Heidelberg New York
Breidbach O (1996) Vernetzungen und Verortungen – Bemerkungen zur Geschichte des Konzepts neuronaler Repräsentation. In: Ziemke A, Breidbach O (eds) Repräsentationismus – was sonst? Vieweg, Braunschweig, pp 35–62
Breidbach O (1997a) Die Materialisierung des Ichs. Zur Geschichte der Hirnforschung im 19. und 20. Jahrhundert. Suhrkamp, Frankfurt
Breidbach O (1997b) Bemerkungen zu Exners Physiologie des Fliegens und Schwebens. In: Breidbach O (ed) Natur der Ästhetik – Ästhetik der Natur. Springer, Berlin Heidelberg New York, pp 221–224
Breidbach O (1998) Kurze Anleitung zum Bildgebrauch. In: Haeckel E (Reprint) Kunstformen der Natur. Prestel, München, pp 19–29
Breidbach O (1999a) Neuronale Netze, Bewußtseinstheorie und vergleichende Physiologie. Zu Sigmund Exners Konzept einer physiologischen Erklärung der psychologischen Erscheinungen. In: Exner S (Reprint) Entwurf zu einer physiologischen Erklärung der psychischen Erscheinungen. Deutsch, Frankfurt am Main, pp I-XXXVIII
Breidbach O (1999b) Die Innenwelt der Außenwelt – Weltkonstitution im Hirngewebe? Zur Konturierung einer Neuronalen Ästhetik. In: Breidbach O, Clausberg K (eds) Video ergo sum. Hans Bredow Institut, Hamburg, pp 34–60
Breidbach O (1999c) Internal representations – a prelude for neurosemantics. J Mind Behav 20 (4):403–420
Breidbach O (2000) Das Anschauliche oder über die Anschauung von Welt. Springer, Wien, New York
Breidbach O (2001a) Deutungen. Velbrück, Weilerswist
Breidbach O (2001b) Hirn und Bewußtsein – Überlegungen zu einer Geschichte der Neurowissenschaften. In: Pauen M, Roth G (eds) Neurowissenschaften und Philosophie. Fink, München, pp 11–58
Breidbach O (2001c) The origin and development of the neurosciences. In: Machamer P, Grush R, McLaughlin P (eds) Theory and method in the neurosciences. University Of Pittsburgh Press, Pittsburgh, pp 7–29
Breidbach O (2002) The former synthesis – some remarks on the typological background of Haeckel's ideas about evolution. Theor Biosci 121:280–296
Breidbach O, Kutsch W (eds) (1995) The nervous system of invertebrates: an evolutionary and comparative approach. Birkhäuser, Basel
Breidbach O, Holthausen K, Scheidt B, Frenzel J (1998) Analysis of the EEG data room in sudden infant death risk patients. Theor Biosci 117:377–392

Cajal SR (1906) Studien über die Hirnrinde des Menschen. 5. Heft: Vergleichende Strukturbeschreibung und Histogenese. Anatomisch-physiologische Betrachtung über das Gehirn. Struktur der Nervenzellen des Gehirns. Barth, Leipzig

Clausberg K (1999) Neuronale Kunstgeschichte. Selbstdarstellung als Gestaltungsprinzip. Springer, Berlin Heidelberg New York

Dawkins R (1976) The selfish gene. Oxford Univ. Press, Oxford

Dennett DC (1984) Elbow room: the varieties of free will worth wanting. MIT Press, Cambridge, MA

Dennett DC (1985) Brainstorms. Philosophical essays on mind and psychology. MIT Press, Brighton

Eberhard WG (1996) Female control: sexual selection by cryptic female choice. Princeton University Press, Princeton

Ende M (1979) Die Unendliche Geschichte. Thienemanns, Stuttgart

Engel AK, Roelfsema PR, Fries P, Brecht M, Singer W (1997) Binding and response selection in the temporal domain – a new paradigm for neurobiological research. Theor Biosci 116:241–266

Exner S (1882) Die Physiologie des Fliegens und Schwebens in den bildenden Künsten. Braumüller, Wien

Exner S (1894) Entwurf zu einer physiologischen Erklärung der psychischen Erscheinungen. Engelmann, Leipzig

Fedrigo G (2000) Válery et le cerveau dans les cahiers. L'Harmattan, Paris

Finkelstein D, Rubinstein J (1968) Connection between spin, statistics and kinks. J Math Phys 9:1762–1779

Freeman WJ (1988) Nonlinear neural dynamics in olfaction as a model for cognition. In: Basar E (ed) Springer series in brain dynamics 1. Springer, Berlin Heidelberg New York, pp19–29

Friederici A (ed) (1998) Language comprehension. A biological perspective. Springer, Berlin Heidelberg New York

Gardner H (1985) The mind's new science. Basic Books, New York

Ghiselin MT (1997) Metaphysics and the origin of species. State University of New York Press, New York

Glasersfeld E v (1997) Radikaler Konstruktivismus. Suhrkamp, Frankfurt

Grüsser O-J, Naumann A, Seek M (1990) Neurophysiological and neuropsychological studies on the perception and recognition of faces and facial expression. In: Elsner N, Roth G (eds) Brain – perception – cognition. Proceedings of the 18th Göttingen Neurobiology Conference. Thieme, Stuttgart, pp 83–94

Haeckel E (1866) Generelle Morphologie der Organismen. vol 2. Reimer, Berlin

Hahn W, Weibel P (eds) (1996) Evolutionäre Symmetrietheorie. Selbstorganisation und dynamische Systeme. Hirzel, Stuttgart

Hebb DO (1949) The organization of behavior. A neurophysiological theory. Wiley, New York

Heisenberg M, Wolf U (1984) Vision in *Drosophila*. Springer, Berlin Heidelberg New York

Holthausen K, Breidbach O (1997) Self-organized feature maps and information theory. Network Comput Neural Syst 8:215–227

Holthausen K, Breidbach O (1999) Analytical description of the evolution of neural networks: learning rules and complexity. Biol Cybernet 81:165–175

Holthausen K, Breidbach O, Scheidt B, Frenzel J (1999) Clinical relevance of age-dependent signatures in the detection of neonates at high risk for apnea. Neurosci Lett 268:123–126

Kockerbeck C (1997) Die Schönheit des Lebendigen. Ästhetische Naturwahrnehmung im 19. Jahrhundert. Böhlau, Wien

Krauße E (1995) Haeckel: Promorphologie und "evolutionistische" ästhetische Theorie – Konzept und Wirkung. In: Engels E-M (ed) Die Rezeption der Evolutionstheorien im 19. Jahrhundert. Suhrkamp, Frankfurt, pp 347–394

Krauße E (2000) Zum Einfluß Ernst Haeckels auf Architekten des Art Nouveau. In: Breidbach O, Lippert W (eds) Die Natur der Dinge. Springer, Berlin Heidelberg New York, pp 85–96

Kuhn TS (1970) The structure of scientific revolution. Chicago Univ. Press, Chicago

Maynard Smith J (1989) Evolutionary genetics. Oxford University Press, Oxford

Mill J (1869) Analysis of the phenomena of the human mind. Longmans, Green, Reader and Dye, London

Nagel T (1974) What is it like to be a bat? Philos Rev 83:435–450

Palm G, Aertsen A (ed) (1986) Brain theory. Springer, Berlin Heidelberg New York

Rössler OE, Hoffman M (1987) Quasiperiodisation in classical hyperchaos. J Comput Chem 8:510–515

Romanes GJ (1882) Animal intelligence. Kegan, London

Romanes GJ (1883) Mental evolution in animals. Kegan, London

Romanes GJ (1888) Mental evolution in man. Kegan, London

Rusch G, Schmidt SJ, Breidbach O (eds) (1996) Interne Repräsentationen. Neue Konzepte der Hirnforschung. Suhrkamp, Frankfurt

Schramm D, Scheidt B, Hübler A, Frenzel J, Holthausen K, Breidbach O (2000) Spectral analysis of electroencephalogram during sleep-related apneas in pre-term and term-born infants in the first weeks of life. Clin Neurophysiol 111:1788–1781

Searle JR (1990) Ist der menschliche Geist ein Computerprogramm? Spektrum Wiss 3/1990:40–47

Shannon CE, Weaver W (1949) The mathematical theory of communication. Univ. of Illinois Press, Urbana, IL

Singer W (2000) Ein neurobiologischer Erklärungsversuch zur Evolution von Bewußtsein und Selbstbewußtsein. In: Newen A, Vogeley K (eds) Selbst und Gehirn. Mentis, Paderborn, pp 333–352

Stafford BM (1999) Visual analogy. Consciousness as the art of connecting. MIT Press, Cambridge, MA

Thompson D'Arcy W (1917) On growth and form. Cambridge University Press, Cambridge

Tye M (1997) The problem of simple minds: Is there anything it is like to be a honey bee? Philos Stud 88:289–317

Vaadia E, Haalman I, Abeles M et al. (1995) Dynamics of neuronal interaction in the monkey cortex in relation to behavioural events. Nature 373:515–518

Weyl H (1952) Symmetry. Princeton University Press, Princeton

Wilson EO (1975) Sociobiology: the new synthesis. Belknap, Cambridge, MA

Wilson EO (1999) Consilience. The unity of knowledge. MIT Press, Cambridge, MA

Zeki S (1988) The functional logic of cortical connections: Nature 335:311–317

Zeki S (1993) A vision of the brain. Blackwell, London

Zeki S (1999) Inner vision. An exploration of art and brain. Oxford University Press, Oxford

The Role of Evolved Perceptual Biases in Art and Design

RICHARD G. COSS

Introduction

The seminal ideas about the relationship of aesthetic appreciation and evolutionary theory emerged initially with the Darwinian construct of sexual selection that emphasized the importance of mate choice and physical attractiveness (Darwin 1885). Anthropomorphic linkage of processes of human intelligence and those of other species (Romanes 1886), coupled with emphasis on the role of natural selection in adjusting human intelligence (Spencer 1888), set the stage for describing behavior in terms of evolutionary history. The idea that innate knowledge accumulated over successive generations and influenced current behavior pervaded nineteenth-century tomes that characterized behavioral relics as atavistic (Nietzsche 1909), especially the fearful behavior of children (Hall 1897). Jung (1916, 1972) continued the development of these ideas with his construct of the archetype as a species-typical pattern of thought that might account for cross-cultural similarities in mythology and graphical symbolism.

Katesa Schlosser (1952) was among the first art historians to link the expressive components of drawings and sculpture of artists from preliterate societies with the ethological construct of "sign stimuli," or visual releasers of innate behavior. Schlosser noticed similarities in conspicuous animal signals and the way artists exaggerate human body features, such as the eyes, projecting breasts, hips, and legs. Many artists tended to simplify these specific features into their essential, recognizable form. The portrayal of vigorous activity like running, for example, was characterized by elongation of the limbs in dynamic postures.

More recent applications of evolutionary theory to aesthetics have emphasized the adaptive value of art as social communication in which artistic devices are used to increase the attractiveness of the body through ornamentation (Dissanayake 1974) and rituals are used to enhance group cohesiveness (Dissanayake 1979, 1992). Similar functional views from an ethological perspective have been expressed by Eibl-Eibesfeldt (1989). Darwinian or evolutionary psychologists have adopted a more specific epistemology that emphasizes the costs and benefits of behavioral adaptations in promoting inclusive fitness as heuristic guides for discovering innate psychological mechanisms (Cosmides and Tooby 1987; Tooby and De Vore 1987). A topic of major interest to evolutionary psychologists is human mate choice and physical attractiveness.

This chapter reflects my initial theoretical conception of studying the role of ecologically important visual shapes and textures that might explain cross-cultural similarities of certain aesthetic elements in art and design. My empirical approach (Coss 1965, 1968) was triggered by an epiphany of the heuristic value of envisioning various historical circumstances in which human ancestors had to deal with specific dangerous situations. Failure to behave appropriately in these situations would have constituted the sources of natural selection for pattern recognition. Choice of shapes and textures for long-term experimental study emerged through this process of visual imagery, as did the value of conducting comparative studies of the behavioral development of nonhuman species. It became evident from these comparative studies that innate visual pattern recognition was not simply a reactive mechanistic process as initially conceived by comparative ethologists (e.g., Tinbergen 1951) and adopted as an explanation for some aesthetic effects (Aikens 1998a). Alternatively, innate visual pattern recognition appeared to embody contextual information in which many facets of an organism's relationship with its environment were incorporated immediately into its rapid decision making processes (e.g., Coss and Owings 1985; Coss 1993; Coss and Ramakrishnan 2000). In essence, information about the nature or behavioral propensities of the perceived object is unveiled almost immediately as part of the process of pattern recognition. As will be discussed further with regard to predator recognition by inexperienced perceivers, one texture on a partially concealed object might connote the potentiality of a sudden threatening burst of chasing and climbing capability while another texture invites inspection, but only within a certain radius because of the expectation of sudden dangerous lunging. This innate information about the behavioral properties of perceived objects is not part of the visual stream initiated by retinal input, but emerges as the equivalent of a memory-like process of what each object affords (Coss 1991b). Because this latter aspect is subject to modification by experience and is amendable to cultural interpretations, various shapes and textures can have a different meaning relevant to their application in art and design.

I have organized this chapter into five sections, the second of which explores the possible role of imagery and concomitant emotional sensations in mediating aesthetic expression. The aim of this section is to extend the construct of evolved perceptual biases to incorporate ideas of how intense experience and cultural information might be manifested in art and design. To provide this link, the relationship of innate perception and evolutionarily prepared learning will be discussed as it relates to issues of neural plasticity. In two subsequent sections, I will review the sources of natural selection on specific pattern-recognition abilities and describe experimental studies of habitat perception and species recognition as they apply to human preferences, physiological arousal and aesthetic applications in art and design. Although relevant to the overall theme of this chapter, I will not review studies on the aesthetics of face perception (e.g., Coss and Schowengerdt 1998) which are topics emphasized by other chapter contributors.

Perceptual Processes Influencing Aesthetic Expression

Mental Imagery and Preattentive Processes

While studies have linked early experiences and personality attributes to innovation and the production of artwork (reviewed by Simonton 1999), little is known about how mental images of objects that emerge endogenously or are retained after initial visual fixation are translated into physical design (see Willats 1987; Johnson 1993). Both preattentive or unconscious manipulation of imagery as well as deliberative manipulation likely contribute to the realization of thought into the physical reality of artwork. Aesthetic preferences are thought to play an important role in artistic expression and repeated exposure to specific styles coupled with artistic training provide a partial explanation for the discrepancies in preferences by some artists for art forms with simplicity and symmetry contrasted by complexity and asymmetry by other artists (McWhinnie 1968; Coss 1977). Differences in cognitive styles between artists and nonartists might also explain the motivation for artistic expression (Bilotta and Lindauer 1980) and choice of aesthetic features for display in painting and sculpture.

Any perceptual biases that might occur in aesthetic expression would not be independent from the general perceptual processes used to guide behavior in natural settings that are filled with objects and surface textures with different degrees of salience to perceivers. Recent evidence using functional magnetic resonance imaging (fMRI) to reveal the loci of synaptic activity during perceptual tasks (Logothetis et al. 2001) suggests that the contours and surface textures of objects are segmented into coherent groups by preattentive perceptual processes without conscious attention given to them in the visual array (O'Craven et al. 1999). Such preattentive processes of image manipulation might be subject to the meaningfulness of specific configurations that have acquired significance during ontogeny or have had historical significance during human evolutionary history. Because of their vividness as mental images and ability to pop out from the background environment of natural settings, some ecologically relevant shapes and textures might be represented in artwork or designs at frequencies well above the likelihood of encountering them in natural settings.

The Role of Natural Selection on Visual Pattern Recognition and Physiological Arousal

Natural selection would operate unquestionably on failure to perform visual tasks that impact fitness. These are: (1) the specificity of initial visual-pattern detection, (2) the recognition of visual-pattern significance in the appropriate organism-environment context, (3) the emotional significance of the visual pattern that emerges in this context, and finally, (4) the successful expression of appropriate action in dealing with this visual pattern (see Coss and Owings 1985). For critical contexts, such as predator detection and avoidance, failure of one or more of these processes would provide the primary source of natural selection. Despite the final outcome of failure to perform appropriately, natural selection would operate most effectively on the perceptual processes of pattern detection and recognition that require great precision in neural integration. Natural selection would operate much less directly on the looser coupling of the emotional significance of a pattern and choice of ways of dealing with it (Coss 1999).

Precision in visual pattern recognition can be viewed in the context of two temporal phases: (1) relatively immediate pattern recognition and (2) pattern recognition that emerges progressively in the stream of dynamic cognitive processing in which more information is processed and inferences derived. In many ways, this initial phase of pattern recognition has theoretical properties consistent with James Gibson's (1979) ecological conception of "direct perception", the automatic, noncognitively mediated perception of what objects afford for guiding action. The preattentive segmentation of texture from background, the alignment of edges producing subjective contours, and grouping of contours by proximity and similarity constitute the inherent aspects of form perception with general utility (Grosof et al. 1993). Because these fundamental processes are essential for guiding behavior, perceptual assessment of these so-called scene primitives (Willats 1992) has likely been under a consistent regime of natural selection since early vertebrate evolution. As such, a strong pleiotropic entrenchment of genes would be built up over time within the mammalian lineage that would likely insure the reliable embryonic development of these early visual processes as well as their evolutionary conservation across species (see Tyler et al. 1998). A similar form of phylogenetic conservation has been reported for diverse species of insects (Buschbeck and Strausfeld 1997). In particular, strong pleiotropic interactions have been shown for *Drosophila* mosaics, in which it is estimated that two thirds of all vital genes play an essential role in the development of this insect's visual system (Thaker and Kankel 1992).

Pattern-recognition by primates involves an expansive perceptual organization that can be seen specifically in macaque inferotemporal and prefrontal cortex which exhibit more dispersive intercolumnar connectivity than that found for early vision in primary visual cortex (Levitt et al. 1993;

Fujita and Fujita 1996). The level of interneural connectivity in inferotemporal cortex is also much greater than in primary visual cortex (Elston and Rosa 1998), a property consistent with the less specialized role of inferotemporal cortex in extracting contour information from scene primitives and its broader capacity for identifying objects. A high level of neural circuitry complexity in this brain region would afford the direct perception of object meaningfulness based on the organization of evolutionarily old neural circuitry and that remodeled by experience. Within the primate lineage, some facets of this process, such as the detection of texture regularity, the completion of partially occluded form, and size-invariant shape recognition (Kovacs et al. 1995a) illustrate the value of having large visual centers. Neurophysiological studies of macaques indicates that the initial phase of form recognition has preattentive properties (Kovacs et al. 1995b); albeit, the longer cognitive phase clearly integrates more information based on prior experience (e.g., Miller et al. 1996). Similar temporal effects are evident in measurements of event-related potentials (ERPs) during face perception by humans. For example, face recognition involves initial assessment of specific facial features, including direction of gaze, followed by longer ERP latencies associated with the assessment of facial orientation, facial familiarity and motion in facial expression (Eimer 2000; Puce et al. 2000; Taylor et al. 2001).

Öhman (1986, 1994) believes that the automaticity seen in the early phase of perception presumably is a primitive precursor to recognition operating outside conscious cognition, whereas intentionally controlled cognition would be conscious and amenable to Pavlovian conditioning. In this view, the primitive system has automatic processing without interference from higher levels of cognition where intentions are started. Theoretical arguments on these issues focus on the distinction between automatic and controlled perceptual processing, time scales of emotions, and the locus of behavioral control. The debate centers on: (1) whether recognition of an object's significance is congruent temporally with changes in sympathetic nervous system arousal and precedes conscious evaluation of its full contextual properties (e.g., Zajonc 1980), or (2) whether recognition of an object's significance and concomitant arousal occurs only after its full contextual evaluation (Lazarus 1981). Study of provocative words using rapid presentations followed by masking images of meaningless letters suggests that consciousness is not entirely necessary for differential evaluation of the emotional or nonemotional meaning of words (van den Hout et al. 2000). Rapid presentations of images of snakes and spiders followed by neutral masks also engender changes in sympathetic nervous system arousal, suggesting that recognition of important threats can emerge unconsciously or preattentively (Öhman and Soares 1994; Robinson 1998). While the adaptive value of rapid pattern recognition and the triggering of behavioral automaticity, such as freezing and startling, would seem obvious, the second phase of pattern recognition which incorporates a greater amount of previous experience would also be subject to natural selection because it would foster cautious investigative behavior or flight (Coss 1993; Coss and Ramakrishnan 2000). Such capability might promote survival in a natural setting in which a person freezes or jumps back as she or he was about to step on a venomous snake. Indeed, an informal survey of snake-experienced field researchers in this situation suggests that nearly all were aware of the snake's presence moments before they exhibited involuntary freezing or leaping away. More deliberative reasoning of the dangerousness of their situation quickly followed, illustrating the adaptive value of this second, higher-order phase of contextual assessment.

Although the perceived dangerousness of an object is coupled with changes in physiological arousal and the possible expression of behavioral automatism in emergency situations, the major neocortical pathways leading from perception to action in mammalian brain follow a sequence indicating that pattern recognition would precede changes in subcortically mediated sympathetic nervous system arousal (see An et al. 1998). Elevated physiological arousal, however, might prompt broader redirection of visual attention to initiate tactical changes in behavior. A sudden surge in physiological arousal would also be accompanied by a stress-related epinephrine response which is thought to facilitate memory formation by enhancing circulating glucose levels (Gold and McCarty 1995). Ecologically provocative experiences are known to produce vivid memories of arousing events that persist for years (Brown and Kulik 1977; Pillemer 1984).

Innate Perceptual Processes

To be subject to natural selection, pattern-recognition capabilities would require a relatively long history of exposure in which both the pattern was perceived in a relatively invariant fashion and the resultant behavior affected fitness. Although patterns of biotic movement do provide spatiotemporal-recognition cues, static views of biotic or abiotic objects would engender more consistent contour and surface texture processing in which the relationships of features could acquire meaning. It seems reasonable to assert that the absence of motion would permit the evolution of more precise feature-assessment capabilities involving the ventral (mostly parvocellular) visual cortical stream than moving features that require saccadic glances to maintain visual fixation. Because recognition processes are most efficient when objects are fixated rather than attended to obliquely (e.g., Eimer 2000), the invariant properties of objects can become important recognition cues, even permitting recognition to occur under partly occluded conditions.

One example of an ecologically important static image is a fully exposed or partially occluded face in which two facing eyes are exposed. In the need to maintain their visual fixation on potential prey, mammalian ambush predators risk detection from concealed vantage points when they expose their eyes to the view of prey. Because the essential information characterizing a face is present in the static schema of two facing eyes, this pattern has acquired significance in species as divergent as humans and house mice (*Mus musculus domesticus*), irrespective of whether this schema is embedded in an oval or circular facial surround (Coss 1965, 1970; Topál and Csányi 1994). There is emerging evidence that recognition of two facing eyes by macaques is an adaptive neural specialization (e.g., Bruce et al. 1981; Perrett et al. 1982; Desimone 1991; Tanaka 1996) relevant for modulating social interactions and possibly useful for the detection of predators.

Evidence that pattern recognition is innate can be accepted with reasonable confidence in visually precocious species by examining their ability to recognize ecologically relevant shapes the first day they are born or hatch (for a critique of this view, see Nelson 2001). Similar confidence can be obtained from developmentally altricial species the first day they show evidence of using visually guided behavior to avoid obstacles. For example, newborn human infants appear to be sensitive to face-like schemata, and even patterns with two facing eyes (e.g., Jirari 1971; Goren et al. 1975; Johnson et al. 1991; Easterbrook et al. 1999), a topic relevant to artistic expression that will be dealt with further below. Innate recognition of snakes is evident in some visually altricial rodents. On the first day they can see, California ground squirrel pups (*Spermophilus beecheyi*) are frightened by the appearance of a Pacific gopher snake (*Pituophis melanoleucus catenifer*) or even a textured strip superficially resembling the staggered spots on gopher snake scales (Coss 1991a). Although less capable visually, wood rat pups (*Neotoma albigula*) also show visual recognition of a gopher snake on the first day they appear to see (Richardson 1942).

A third method of examining innate pattern recognition involves the application of "Kasper Hauser" protocols of developmental deprivation unique to each species (e.g., Curio 1975). Developmental deprivation in laboratory settings can be used successfully to study predator recognition at different periods of development to determine, as in nature, whether the visual system awaits initial predator exposure and successfully initiates antipredator behavior. Natural deprivation of these awaiting states can be studied by examining populations living in habitats where historical predators are rare or absent and comparing the antipredator behavior of these populations with those that experience these predators (cf. Curio 1993; Coss and Biardi 1997; Blumstein et al. 2000). Inferences of innateness in experimental studies of pattern recognition are more tenuous for species in which behavioral expression is developmentally delayed, but pattern exposure during development is likely. This problem is particularly relevant to studies of the innateness of social perception because even short-term social deprivation can disrupt normal processes of "experience-expectant" trimming of unused synaptic connections during imprinting in birds (e.g., Wallhäusser and Scheich 1987). Long-term social deprivation is more severe and can cause the arresting of "experience-dependent" neuron growth in both birds and mammals (cf. Black and Greenough 1986; Rollenhagen and Bischof 1994) and behavioral abnormalities in primates (reviewed by Mitchell 1970). For humans, studies of the effects of facial expression would need to be conducted early in neonatal development to circumvent strong experiential influences (e.g., Meltzoff and Moore 1977). One indirect method of quantifying any evolutionary constraints on human behavior is by comparing pairs of identical and fraternal twins to determine the effects of shared environment and to estimate the heritability of particular traits. Studies of twins are especially valuable in identifying the unique experiences causing differences in identical twins (Kendler et al. 1992).

Learning as Adjustment of Instinct

As intimated above for the whole organism, evidence that innate pattern-recognition systems function in different age classes the first time they receive appropriate input indicates a form of developmental stability that is probably reflected at lower levels of neural organization (e.g., Coss 1991b). One source of stability appears to be the dendritic spine, the locus of synaptic activity which is assumed to be the smallest unit of neuronal integration and the primary site of synaptic plasticity in the brain (Fischer et al. 1998; Matus 2000; Smart and Halpain 2000). Numerous studies pinpoint the dendritic spine as one component of neural circuitry that changes dynamically with experimental stimulation or following single and repeated behavioral experiences (reviewed by Coss and Perkel 1985; van Rossum and Hanisch 1999; Yuste and Bonhoeffer 2001). Studies of behavior coupled with computational simulations of synaptic activity suggest that shorter dendritic spines do not exhibit the same degree of long-term morphological change with repeated synaptic activation found in longer spines (Coss 1985). Morphometric measurements of spines of varying length indicate that short spines retain their shape more persistently than long spines following expressions of innate behavior by honeybees and fish, whereas longer spines show both rapid and persistent changes in morphology with learning in these taxa and in birds (Brandon and Coss 1982; Coss and Perkel 1985; Patel and Stewart 1988).

The differential adjustability of dendritic spines could provide the major substrate for information preservation, and computational simulations of synaptic input suggests that localized clusters of spines can adjust their properties and interact in ways that have analog pattern-matching abilities (Yang and Blackwell 2000). Variation in spine length can be affected by genetic disorders, diet, crowding, and developmental deprivation which would affect the timing of arrival of presynaptic terminals (see Papa et al. 1995). Since genetic expression is greatest during embryogenesis, genetic regulation of cell signaling would insure a much greater precision in the timing of synaptogenesis and overall spine length on growing proximal dendritic branches than later in development when more distal dendritic branches are formed stochastically by experience (cf. Black and Greenough 1986; Dailey and Smith 1996). In support of this argument, experimental stimulation of dendrites can cause the localized formation of relatively long filopodia or protospines (Maletic-Savatic et al. 1999). Mammalian spines on neocortical neurons are typically longer on distal higher-order dendritic branches growing later in development than on proximal first- and second-order dendritic branches growing earlier in development (Peters and Kaiserman-Abramof 1970). Moreover, evidence that snakes are recognized visually by ground squirrel and wood rat pups shortly after eye opening implicates the proximal dendritic branches in the visual system as one source of developmental stability in snake recognition because distal dendritic branches would not be likely to be present at this early age (see Coss 1991b). Other attributes of neural circuit stability would presumably be the shorter dendritic spines on proximal dendrites which would exhibit only small changes in their biophysical properties with repeated activation; albeit, changes in the biophysical properties of longer spines with experience might begin to approximate the stable properties of shorter spines in ways that promote the emergence of learned behavior with reliable, instinctive-like properties (Coss 1985).

Learning Preparedness

Evidence that localized neural activity promotes the growth of nearby synaptic connections (Rusakov et al. 1995; Engert and Bonhoeffer 1999; Toni et al. 1999) sets the stage for addressing issues of evolutionarily "prepared" learning in which innate propensities in visual perception might achieve greater refinement with repeated experience (Nelson 2001). However, taken further, these perceptual processes might require experience to link them to specific emotions and action patterns. Social species like humans would more likely be exposed to danger in the presence of others than alone and thus benefit from observing how more experienced individuals deal with specific situations. Learning in this context might be very rapid and persistent, and the social circumstances sufficiently uniform that natural selection could operate on the evolution of neural circuitry that facilitates associative learning of the appropriate relationships between objects and situations and specific consequences. Therefore, applications of specific graphical patterns in art and design might be maintained culturally or adopted by other cultures as iconic symbols because they engender perseverant imagery and emotional sensations due to social modeling.

Seligman (1970, p. 408) defined learning preparedness as the amount of input or frequency of exposure necessary to obtain a reliable acquisition of learned behavior. In this continuum, prepared learning would require few trials with reinforcement, whereas "unprepared" learning would require numerous trials. In this construct, the tendency for rapid learning of some relationships and difficulty with others have been viewed as the by-product of differential selection (Davey 1995). Repeated failure to make specific learned associations is labeled "contrapreparedness" and reflects a lack of neurological coupling or "belongingness" between perceptual inputs and reinforcements that were unlikely to have occurred together during phylogeny. Seligman (1971) later extended the preparedness construct to explain differences in the frequency of specific fears or phobias in humans.

The term "simple phobia" refers to a large, clinically rare category that is characterized by persistent, seemingly irrational fears leading to the avoidance of specific objects and situations. Simple animal phobias are unique from other phobias because they typically emerge by about 5 years of age rather than in adulthood (Marks and Gelder 1966). Specific situational phobias show onsets with much wider age ranges, with some starting before 5 years of age. Because of their persistence, simple phobias, like the fear of small animals, can influence the topics selected or avoided in creative expression or restrict exposure to natural settings that might inspire the production of artwork. Fortunately, fear of some animals, such as dogs and snakes, tends to diminish in both sexes by the age of six (Jersild and Holmes 1935) and a general fear of animals drops sharply between 9 and 11 years of age, especially in boys, remaining steady at this low level throughout adulthood.

Studies of fear in identical and fraternal twins can produce heritability estimates by comparing pair-wise intraclass correlations for each twin category. The highest heritability (0.47) in a study by Skre and colleagues (2000) was for the common fear of small animals. Heritability estimates for these phobias can be as high as 0.66. Inanimate object fears exhibit much lower intraclass correlation coefficients, indicating that preparedness as an explanation for the onset of phobias has relatively restricted assumptions. Complementary research on 8- to 9-year-old twins indicated that both heritable and environmental factors, such as trauma, vicarious associative learning, and social modeling produce differences in the level of general fearfulness and might influence the appearance of specific phobias (Lichtenstein and Annas 2000). Multivariate genetic analysis of phobia subtypes in female twins, such as agoraphobia, social phobia, animal phobia, and situational phobia, indicates that simple animal phobias are related to situational phobias and that both phobia subtypes are shaped by specific environmental experiences (Kendler et al. 1992). It must be noted, however that animal phobias do exhibit the highest loading on the common genetic factor (Kendler et al. 1992). Low heritability of proneness for animal phobias does suggest, however, the influence of natural selection and fitness maximization. On the whole, the individual experience of twins, rather than familial factors which are partly genetic and include shared environments, is a major contributor to the risk of phobia onset (Kendler et al. 1999). Because different indices of fearfulness toward small animals are usually positively correlated, heritability estimates tend to reflect common attributes of fearful emotions or the perceived dangerousness of small animate creatures that move suddenly, rather than any unique properties of their configurations or behavior.

Sources of Selection on Recognition of Habitat Affordances

This section provides a general review of early hominid evolution and the historical sources of natural selection for perception of habitat attributes affording survival. Relationships with historical predators will be emphasized since they governed the use of habitat features as refuge sites and processes of vigilance and rapid decision making necessary for avoiding dangerous plants. Research on preschool children will be discussed to pinpoint the developmental origins of some perceptual biases possibly influencing the aesthetic applications of adults. To bolster the theoretical arguments for such aesthetic applications, attention will be given to experimental studies examining differences in attitudes and physiological arousal as indices of object significance.

Hominid Exposure to Predatory Risks

Throughout the early to Middle Miocene of Africa, trees in moist forests provided hominoid apes with consistent refuge from predators, sources of water and edible fruits for nutrition, and shade to prevent dehydration. Around 7 Ma (million years) ago in the eastern rift valley, Kenya, water became a seasonally scarce resource due to the emergence of strong Asian monsoonal conditions (Quade et al. 1989). This shift toward seasonal precipitation initiated a rapid decline in wet forests and the emergence of deciduous woodlands (Jacobs and Winkler 1992; Jacobs and Deino 1996) in a time frame roughly coincidental with the emergence of bipedal locomotion in the earliest hominids (Senut et al. 2001). Under these water-stressed conditions, early bipeds would have been forced to abandon their forest sites to search for water in nearby patches of closed forests, traversing open wooded grasslands concealing ambush and cursorial predators. Evidence that the chimpanzee-sized hominid (*Orrorin tugenensis*) fell victim to predation in the rift valley 6 Ma ago has been inferred from a grouping of bones with tooth marks (Pickford and Senut 2001).

A diversity of mammalian carnivores appear in the faunal records of eastern and southern Africa between 7 and 5 Ma ago that could have taken hominids in their night nests and stalked them in the open (Fig. 1). These

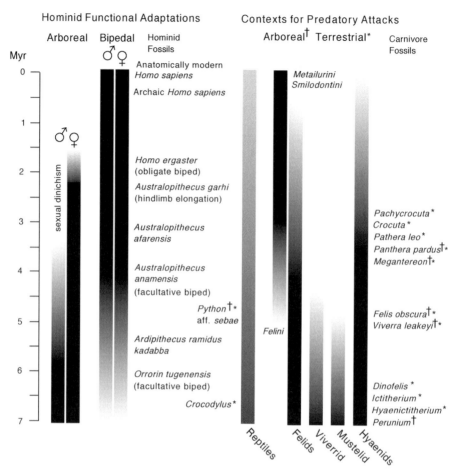

Fig. 1. Time frame of hominid evolution in Africa spanning the Late Miocene to the present is shown with the corresponding appearances of likely hominid predators, excluding large eagles. Leopards (*Panthera pardus*) have constituted the predominant threat since the Middle Pliocene, whereas crocodiles and pythons have been consistent threats throughout this entire time frame. Sexual dinichism in which females retained a more arboreal lifestyle than heavier males emerged slowly with improved bipedalism, terminating relatively abruptly with hindlimb elongation in *Homo ergaster*. Trees retained their importance for shade and emergency refuge from predators following this evolutionary advancement

records indicate the presence of medium-sized predators of which the pantherine-like mustelid, *Perunium*, short-sabered cat, *Felis obscura*, and possibly the giant civet-like viverrid, *Viverra leakeyi* (Hunt 1996), conceivably attacked early hominids asleep in night nests. Another Late Miocene hominid, *Ardipithecus ramidus kadabba*, lived in moist, well-forested areas of Ethiopia (Haile-Selassie 2001; WoldeGabriel et al. 2001) and would have been a candidate for such nighttime predatory attacks. The robustly built short-sabered cat, *Dinofelis* (Hendey 1974; Leaky et al. 1996) could have been an occasional threat in open grasslands if surprised by blundering encounters as are lions today. *Dinofelis* is closely related to *Metailurus*, a known primate killer from the late Miocene of Greece (Zapfe 1981). Additional threats on the ground might have come from packs of cursorial hyenas, such as *Ictitherium*, and pythons (*Python* aff. *sebae*). Nile crocodiles (*Crocodylus niloticus*, Leakey et al. 1996; Senut et al. 2001) could have constituted majors threats to hominids drinking from ponds, streams and lakes.

Throughout the Early to Middle Pliocene, predators such as *Dinofelis*, *Viverra leakeyi* and possibly *Felis obscura* (see Howell and Petter 1975; Barry 1987), continued to share gallery forests and dry, open woodlands with the hominid, *Ardipithecus r. ramidus*, from Ethiopia (White et al. 1994) and hominid, *Australopithecus anamensis*, from Kenya (Leakey et al. 1995). Extensive contact with predators can be inferred from many of the remains of *A. anamensis* that show carnivore damage (Leakey et al. 1998). Additional predatory threats are evident about 3.5 Ma ago with the appearance of the giant hyena (*Pachycrocuta*), cursorial hyena (*Chasmaporthetes*) and leopard-sized saber-toothed cat, *Megantereon*. Modern predators also appear at this time, represented by the bone-crushing spotted hyena (*Crocuta crocuta*) and the lion (*Panthera leo*) and leopard (*Panthera pardus*) that suffocate their prey (Barry 1987; Turner 1990; Turner and Wood 1993; Werdelin and Lewis 2000).

Although powerfully built for tackling smaller ungulates, *Megantereon* appears to have been sufficiently agile that it could hunt in trees like leopards do (Turner 1997), possibly attacking hominids seeking refuge in trees or asleep in their night nests. Evidence that primates were a major part of its diet comes from the fossil tooth enamel of *Megantereon cultridens* and those of leopards, spotted hyenas, and *Dinofelis* from the Plio-Pleistocene boundary, approximately 1.7–1.8 Ma ago (Lee-Thorp et al. 2000). Carnivore diets can be inferred from their tooth enamel, which reflects the $^{13}C/^{14}C$ ratio of the dietary carbon in prey tissues. With the exception of *Dinofelis*, which appears to have targeted large ungulates that grazed in arid-adapted grasslands rather than large primates as speculated by Brain (1981), the $^{13}C/^{14}C$ ratios of the spotted hyena, leopard, and *Megantereon* were remarkably similar to those of baboons (*Papio* spp.), *Parathropus robustus*, and *Homo ergaster* that favored rootstocks, fruits and nuts. Less ambiguous evidence that hominids were victims of leopards comes from tooth punctures in one skull characterizing a leopard's gripping bite (Brain 1970, 1981). If knowledge of the current circumstances that promote human depredation by the large cats applies to historical circumstances, such as the ease that old cats experience in rasping flesh off human bones (Corbett 1948; Noronha 1992), man-eating leopards and lions were likely to have pillaged hominid communities relentlessly (see Patterson 1986; Thomas 1990).

Within the australopithecine lineage, *A. anamesis* and its presumed descendant, *A. afarensis*, exhibit a comparable degree of gorilla-like body size dimorphism (Leakey et al. 1998); albeit, the body weights of both species approximate those of bonobos. The female *A. afarensis* is estimated to have weighed between 29–32 kg, approximately 62% of male weight (McHenry 1992a,b). The retention of long powerful arms with a well-developed grasping capability and short hindlimbs with a residual grasping foot point to morphological adaptations for slow vertical tree climbing (Stern and Susman 1983). As in male chimpanzees that are reluctant to forage on thin branches (Hunt 1992), heavier *A. afarensis* males might have been endangered much more than females when they attempted to reach fruit on unstable terminal branches. This form of endangerment might have provided a major source of selection for the evolution of sexual dinichism (i.e., sex difference in habitat use), with males adopting a less arboreal pattern of foraging and nest building than females (Susman et al. 1984, p. 149). Based on later fossil evidence of sexual dimorphism, this inferred pattern of sexual dinichism spanned at least a 2-Ma-time frame (Fig. 1) and likely includes the 2.5-Ma-old transitional species, *A. garhi*, which exhibits elongated hindlimbs coupled with the retention of ape-like arms (Asfaw et al. 1999). Sex difference in habitat use probably diminished relatively abruptly with the advent of marked expansion of hindlimb length, producing the tall, linear body form of *H. ergaster* and a much improved biomechanical efficiency for running and walking long distances (Brown et al. 1985; Wood 1992); this important terrestrial adaptation eventually led to a dietary shift from mostly plant to animal resources (Shipman and Walker 1989). Analyses of cut marks on bones and discarded stone tools indicate that hominid ranges between 2.5 and 1.8 Ma ago were still relatively restricted to areas with water, shade, and rocky outcroppings (de Heinzelin et al. 1999; Potts et al. 1999). Despite the development of lithic culture and the ability to mob predators, nighttime in the savanna woodlands would have been treacherous.

Perception of Trees and Landscape Attributes

Studies of tree preferences are important for shaping arboriculture both in urban and wilderness settings and for optimizing the aesthetic appeal of landscaped environments. Aesthetic evaluation of tree attributes has been useful as a method for selecting appropriate street trees for residential and commercial areas (Sommer et al. 1992, 1993). The study of tree configurations, especially preferences for crown structure, is relevant to energy conservation in urban and recreational settings (McPherson and Rowntree 1993) and for characterizing the scenic quality of landscaped parks and gardens (e.g., Schroeder 1986). To introduce the idea of evolved perceptual biases to landscape architecture, Orians (1980, 1986, 1998) posited the *savanna hypothesis* to predict what features would make one landscape more attractive than another. Orians (1986) described a hierarchical structure for habitat choice in which the lowest level involves responses to specific attributes that attract attention and are mildly arousing, such as trees that characterize important habitat qualities that include the presence of water. Cross-national research on tree preferences conducted in Argentina, Australia, and Seattle, Washington, showed that individuals in all three areas rated photographs of Umbrella Thorns (*Acacia tortilis*) from Kenya as the most attractive trees aesthetically. These acacias exhibited broad, moderately layered canopies and trunks that bifurcated near the ground. Among the set, trees with skimpy or very dense canopies and tall trunks were rated as least attractive (Orians and Heerwagen 1992; Orians 1998). It is ironic indeed that Umbrella Thorns with wide layered canopies are rated as aesthetically attractive because these acacias can be hazardous due to their long thorns. As will be discussed further with other examples, there appears to be a conceptual link between the assessment of danger and perceived physical beauty.

In another collection of preference studies, Sommer and Summit (1995) and Summit and Sommer (1999) used schematic drawings, providing complementary evidence that bole height and crown configuration were indeed important attributes used in the aesthetic assessment of trees. Further study of a greater diversity of cultures in biomes in which adults as children had observed a large variety of trees yielded complementary evidence of tree-shape preferences consistent with Orian's savanna hypothesis (Sommer and Summit 1996; Sommer 1997; Summit and Sommer 1999). Low preferences were given to tall trees or trees with narrow columnar and conical canopies as characterized by palm trees and conifers even though exposure to these trees during childhood mitigated some of this dislike (Sommer and Summit 1996; Sommer 1997).

Aesthetic preference for trees with wide crowns appears to be an adult phenomenon. Coss and Moore (1994) examined the preferences of four tree silhouettes by 3- to 5-year-old children in Israel, Japan, and the United States and with a larger sample of 3- to 4-year-old children exclusively in the United States. In both studies, the prettiest tree selected by the largest percentage of children was an Australian pine (*Pinus nigra*), a common ornamental tree with a tall, conical canopy. However, in agreement with previous studies, the prettiest tree selected by the largest percentage of adults was an Umbrella Thorn from the African savanna with a wide, thick canopy; few adults selected the Australian pine as the prettiest tree (Fig. 2A). In contrast, the largest percentage of children selected the Umbrella Thorn with thick canopy as the best tree for shade (Fig. 2B), the best tree to climb to hide (Fig. 2C), and the best tree to climb to feel safe from a lion (Fig. 2D). Binary Item Factor Analysis conducted on each tree (Coss and Moore 2002) revealed an association between choice of the Umbrella Thorn with thick canopy as the prettiest tree and choice of this tree to feel safe from the lion, but not its choice for shade. It is conceivable

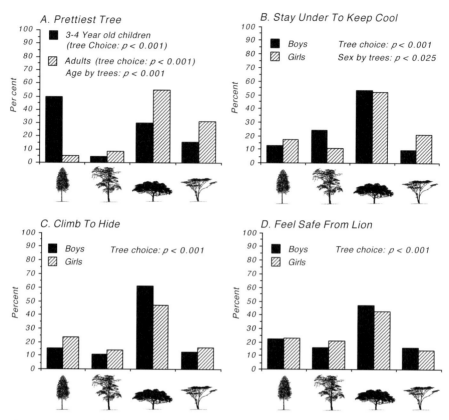

Fig. 2. Frequency of choice of four tree silhouettes by 3- to 4-year-old children and adults from the United States. From *left* to *right*: Australian Pine (*Pinus nigra*), Fever Tree (*Acacia xanthophloea*), unbrowsed Umbrella Thorn (*Acacia tortilis*) and browsed Umbrella Thorn. A Multinomial log-linear analyses revealed that the interaction of children and adults differed significantly in the choice of prettiest tree ($n = 110$ children and 93 adults). The unbrowsed Umbrella Thorn was selected by the largest percentage of children ($n = 49$ males and 63 females) as the best tree to stay under to keep cool on a hot day (**B**), as the best tree to climb to hide (**C**), and the best tree to climb to feel safe from a lion that has escaped from a nearby zoo (**D**)

that this linkage between tree aesthetics and refuge might be the childhood precursor to the adult aesthetic preference for trees with wide crowns. Moreover, this study of tree perception by preschool children who are too young to be experienced climbers (Coss and Goldthwaite 1995) suggests a precocious understanding of tree affordances useful historically and currently to avoid predators and to prevent dehydration. Furthermore, to consider this issue of climbing hazards, it is important to consider from a cost-benefit perspective that most trees in dry savanna settings are thorny and relatively brittle, which would have mitigated their utility for hominids as refuge sites unless the risk of injury was outweighed by predatory risk. Such an assessment can lead to emergency tree climbing, resulting in severely shredded flesh (see Patterson 1986).

The question remains as to whether adult aesthetic preferences for trees with wide crowns influence the design of landscaped environments. Orians (1980, 1986, 1998) has suggested that in nontropical biomes, some landscape designers intuitively mimic the appearance of savanna-like settings. For instance, Japanese gardens emphasize the display of trees and shrubs, many of which are pruned in ways that exploit the malleability of limb growth. These gardening practices yield trees and shrubs with layered canopies and branches radiating from the ground that superficially resemble savanna acacias. Similarly, American landscape architects in the middle nineteenth century, such as Frederick Law Olmsted, adopted the landscaping practices of English manors, producing numerous parks with turf meadows, ponds, and savanna-like clusters of trees. The primary aims of these design practices were to block perception of the cityscape and foster tranquilness in individuals who promenaded in these broad open spaces surrounded by shade trees (Hiss 1991).

The idea that ecologically natural properties of the environment might have restorative qualities is an old one in Western thought and is manifested by the desire of city dwellers to visit wilderness areas and send children to camp (Tuan 1978). Research by Ulrich (1984) using hospital recovery data for gallbladder patients has shown that patients with window views of trees and bushes recovered more quickly than patients who viewed a brick wall. It is conceivable that recovery from surgery engenders bodily stress analogous to that of a predatory attack; thus faster recovery might reflect a greater sense of well-being and security in patients viewing a veil of tree canopy foliage from upper floors. Although there is no experimental evidence to my knowledge that pictures have similar restorative properties, there is some evidence that nature photographs have the potential of lowering physiological arousal that affects a person's well-being. In several studies using pupillary dilation as an index of sympathetic nervous system arousal, Hess (1975) reports that, compared with provocative pictures, landscape pictures engendered very little arousal and were therefore used as experimental controls. It is evident that high degrees of aesthetic preference are given to images of spacious landscape scenes with greater apparent depths of field (Ulrich 1977, 1983; Clearwater and Coss 1991), especially distant mountain ridges with low hills and sparse trees in the foreground (Buhyoff et al. 1982; Hull and Buhyoff 1983; Patsfall et al. 1984).

Water Perception and the Choice of Surface Finish for Artworks and Consumer Products

With the retraction of moist closed forests in east Africa during the Late Miocene, the migratory behavior of early hominids was under strong selection for evaluating the safety of traversing open areas via sightable evidence of water and arboreal refuge sites and routes to circumvent occluded landscape features to avoid blundering into predators (Woodcock 1982; Appleton 1984, 1996). Dispersal of ancestral hominids was likely restricted to warmer and wetter climatic conditions with the ready availability of fresh water, a view supported by evidence that hominid fossils and artifacts are frequently found near the margins of ancient lakes (Brown et al. 1985; de Heinzelin et al. 1999). Abundant water in rivers and streams during the wet season might have shaped migratory routes and small lakes and rain pools would have provided additional resources during the dry season, permitting the passage through more arid regions. Failure to prevent dehydration by the inability to find drinking water on a daily basis (Newman 1970) would have provided a constant source of selection during the past 7 Ma, the outcome of which is the current cognizance about the perceptual information signifying the presence of water in diverse microhabitats (Coss and Moore 1990).

Water provides both static and dynamic optical properties that could attract the attention of perceivers. Still water can act as a natural mirror, reflecting the hue and macrostructure of the overhead environment (Gibson 1979). Turbulent, foamy water is conspicuous in natural settings because of its luminous white appearance. The dynamic, glinting properties of the smooth rippling surface of a pond or stream reflected by the sun or moonlight is even more salient in terms of attracting attention (Koenderink 2000). This sparkling property of water could constitute a historically consistent recognition cue, easily detected at a distance in full view and when partially occluded by tall grasses and shrubs. Glittery dew seen in the early morning sunshine would provide another indicator for obtaining water by licking and sucking moist leaves (Coss and Moore 1990).

Several studies indicate that scenes depicting water or moist vegetation are preferred over those that are more arid (see Zube 1974; Balling and Falk 1982; Lyons 1983). Similarly, Ulrich (1981) found in comparing projected slides of nature scenes versus city scenes of similar complexity that nature scenes depicting water engendered the strongest positive ratings as well as the greatest amount of "wakefully relaxed" cortical activity measured via electroencephalographic (EEG) recordings of alpha wave amplitude. As with the study of pupillary dilation by Hess (1975), landscape scenes of water in the foreground yielded the lowest physiological arousal compared with close-up views of landscape features (see Coss et al. 1989; Clearwater and Coss 1991). Several cross-sectional studies of children's preferences for landscape pictures report congruent evidence that scenes with water are preferred to those depicting harsh, dry habitats (Zube et al. 1983; Bernáldez et al. 1987). For example, children 8–11 years of age find savanna pictures taken during the wet season more appealing than those taken during the dry season even though no bodies of water are apparent (Lyons 1983). Desert scenes in particular are clearly the least attractive among a series showing the African savanna and relatively dense coniferous and deciduous North American forests (Balling and Falk 1982; Lyons 1983).

Although children as young as 6 years of age can indicate that the presence of water enhances the scenic value of landscape pictures (Zube et al. 1983), infants and toddlers behave in ways that imply an unlearned ability to distinguish wet and dry surfaces. A number of 7-month-old infants to 24-month-old toddlers have been observed to selectively mouth and lick metal and plastic toys that are not drinking vessels (Coss and Moore 1990). Follow-up experimental study of whether the glossy surface finish of toys promoted mouthing was conducted in ten day-care facilities in northern California. Two blue translucent plastic plates were placed among toys in these facilities to facilitate handling. One plate exhibited a glossy surface finish that reflected highlights in the manner of a wet surface; the other plate was sanded evenly to provide an optically dull, but smooth surface finish that reflected more diffused light as measured photometrically. Of the 143 children observed, 46 children handled both plates a minimum of five times, permitting a repeated measures statistical analysis of mouthing frequency as a function of age class. Only the youngest age class of 7–12 months differed significantly in the frequency that they

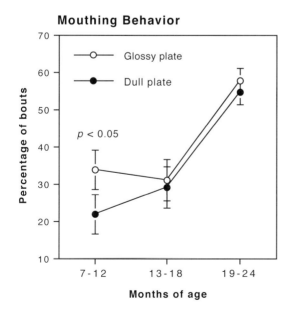

Fig. 3. Frequency of infants and toddlers mouthing and licking translucent blue plates with glossy and dull surface finishes. The plates differ significantly for only the youngest age class. Mean and standard error values are shown

mouthed and licked the two plates, with the glossy plate receiving the largest percentage of mouthing (Fig. 3). Although most children held the plates to mouth them, it was not infrequent to observe toddlers on their hands and knees mouthing and sucking the center of the glossy plate as if drinking from a rain pool. This precocious activity in children of nursing age illustrates the early onset of an adaptive behavior historically useful in much older individuals.

Because glossy and sparkling surfaces afford proximal and distal optical information that might attract perceivers, it was important to determine in adults if glossy and sparkling surface finishes connote wetness (Coss and Moore 1990). Four whitish surface finishes (mat, glossy, sandy, and sparkling) were mounted on 10×12-cm panels that were held and tilted in different directions by 138 subjects who then rated the panels using a semantic differential questionnaire that assessed wet and dry connotations as well as other physical and cultural attributes. Analyses of data showed that the glossy surface finish appeared wetter than the sparkling surface finish and that both these surface finishes were perceived as much wetter and markedly less dry than the mat and sandy surface finishes (Fig. 4A). However, evidence that the sparkling surface finish was not as convincingly wet as the glossy surface finish suggests that, as a distinct cue, glitter affords less reliable optical information about the presence of water, possibly due to its appearance as shimmering crystalline quartz embedded in dry granitic minerals.

The differential semantic pair (sensuous)–(chaste) was used to examine the cultural attributes and experiences with gloss and sparkle on fabricated surface finishes. For this sensuousness dimension, a strong interaction of sex and surface finish was apparent, with women rating the sparkling panel as significantly more sensuous than men did (Fig. 4B). The ubiquity of glossy and sparkling surface finishes on consumer products and decorative artifacts, such as fabrics and jewelry (see Rivers 1995, 1999), might account for this sex difference in the United States mediated predominantly by advertising and fashion. On the whole, both the study of infant and toddler mouthing and adult connotations of surface finishes (Coss and Moore 1990) support the general conjecture that both glossy and sparkling surfaces appear wet rather than dry. Furthermore, such information could be vital in dry habitats for differentiating a glistening body of rippling water from a mirage or detecting well-illuminated rippling water behind heavy vegetation.

The implications that water perception influences aesthetic expression are profound from a history of art perspective because some of the oldest artifacts of anatomically modern humans show evidence of polishing to achieve a glistening luster. For example, bone artifacts older than 70,000 years in Blombos Cave, South Africa, were shaped intentionally by grinding and polishing for transformation into awls and symmetrical points (Henshilwood et al. 2001a). In the Upper Paleolithic site of Lourdes,

The Role of Evolved Perceptual Biases in Art and Design 89

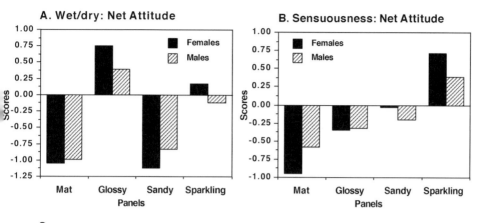

Fig. 4. Mean values of semantic differential questionnaire examining the connotations of surface finishes mounted on panels. A For the dimension of wetness and dryness, the combined mean score for the glossy and sparkling panels was significantly higher on the wetness component ($p<0.0001$) than the combined mean scores for the mat and sandy panels. B The sparkling panel was rated as significantly more sensuous by women than by men ($p<0.05$). C The Experience Music Project building, Seattle (Frank O. Gehry, Architect, 1997) clad in glistening metal from which reflected light pulsates in a rippling fashion as the viewer moves about

France, about 14,000 years ago, an ivory horse shows evidence of hand polishing as does the well-known "Venus" figure of carved mammoth ivory from the French site of Lespugue (Dennell 1986). The grinding and polishing of stone tools that were resharpened rather than discarded became a ubiquitous practice with the advent of agriculture in the early Neolithic. Thus, the polishing of surfaces in the quest for luster seems to have shaped the choice of materials for decorative objects, and might be implicated in some of the earliest use of malleable copper for jewelry in the early Neolithic (Mellaart 1967) followed later by the choice of gold and silver. In more contemporary times, the positive preferences for lustrous, gleaming surface finishes are clearly reflected by the strong cultural attributes and historical importance given to polished and faceted surfaces, especially those of rare minerals. As revealed by the strong preference for water in landscape scenes (e.g., Zube et al. 1983; Herzog 1985), perhaps the ancient evocative cues for water engender a rewarding emotional state reflected elsewhere in the creative desire to manufacture gleaming, lustrous surface finishes for display, adornment, and advertisement. Clearly, the ubiquitous presence of the glistening surface finishes of modern sculpture, product design and architecture (Fig. 4C), and its simulation in advertising and computer graphics, illustrates this powerful attraction to water as a vital resource.

Sharp and Piercing Forms

In humans, the mouth is recognized as a distinct component of the face as evinced by the "Margaret Thatcher illusion" in which an upright mouth in an upside-down face retains its expression and is still recognizable as belonging to a specific person (Thompson 1980; Parks et al. 1985). The mouth has provided a phylogenetically old and important schema that has undergone marked transformation in shape with dental changes affected by diet and social structure. In primates, canines are used mostly as weapons during agonistic encounters, but they may serve a nontrophic display function like grimacing or yawning which can defer violent confrontations (Hooff 1967; Packer 1979). Canine length and sharpness play an important role in acquiring and maintaining dominance rank in primates where fighting ability determines rank (Hamilton and Bulger 1990). Primates are not alone in responding to sharp points. According to Guthrie and Petocz (1970), the horns and antlers of ungulates are sufficiently important that many species have evolved automimetic facial patterns that exaggerate the size of these weapons.

Canines are salient features of fossil hominoid skulls during the early Miocene, diminishing markedly in size in Middle Pliocene hominids (cf. Leakey et al. 1995, 2001). Large canines have not been a perceptual component of the hominid mouth for the past 2 Ma (Asfaw et al. 1999), yet these conspicuous dental features are very provocative to human observers

(Coss 1965, 1968). As an example of this potency, Coss and Towers (1990) measured the pupillary dilation elicited by 16 sets of 20 photographic slides of different topics, with each set observed by 20 different subjects. The most arousing slide in the entire series of 320 slides was a painting by H.R. Giger depicting an incongruous close-up view of gruesome baboon-like canines emerging from a cluster of serpentine forms (see painting in Masello 1986, p. 81).

Recognition of the danger of piercing forms is not restricted to canines or the avoidance of the thrashing horns of wounded or mobbing ungulates. The African savanna is replete with plants with thorns that mitigate plant predation or seeds that stick to fur for transport. Inadvertent blundering into spiny thicket due to inattention or with the automaticity of fleeing danger would have caused severe incapacitation and possible entrapment that would have acted as a consistent source of selection since Miocene times. Most *Commiphora* species and some acacia trees (e.g., *Acacia nilotica*) have long narrow spines that can puncture the feet, eyes, and abdominal wall deeply. Recognition of this danger might have prompted some of the earliest application of hand-worked defensive technology. Kortlandt (1980) suggests that early hominids might have carried about the thorny branches of acacias and thorn bush as weapons to ward off predators and conspecific attacks. He argues further (Kortlandt 1980, p. 89) with regard to spiny vegetation that: "Some species could be used as a beating weapon, others as a stabbing weapon, like Neptune's trident, and still others could be flung towards the enemy in order to catch him and stick to him."

As part of a series of studies on the effects of vegetation, I examined the subtle behavioral responses of pedestrians to shrubs with pointed leaves, recreating the historical circumstance in which attentiveness to spiny and serrated plants might have provided safe passage. In the most recent study, walkers and joggers were video-taped as they passed by two plants juxtaposed on each side of a 142-cm-wide path in the arboretum on the University of California, Davis, campus (Fig. 5A). A Spanish Dagger (*Yucca gloriosa*) from the southwest United States was selected as the provocative plant because its long, dagger-shaped leaves radiate from one central stem. The Spanish Dagger was compared with a rounded-leafed Crêpe Myrtle (*Lagerstroemia indica*) from East Asia of identical height (110 cm) and similar canopy width (93–110 cm). Neither plant intruded into the path and plant locations were reversed half way through the study. Analysis of the distance of the right foot from right edge of the path revealed that, despite its dramatic appearance, both walkers and joggers were physically attracted to the Spanish Dagger (Fig. 5B). This result was unexpected because two previous studies of pedestrian behavior on narrower sidewalks showed a significantly larger frequency of deflection from plants with rapier-like leaves, such as an African Tiger Lily (*Agapanthus africanus*) and Blue Draceana (*Draceana indivisa*). The difference in context

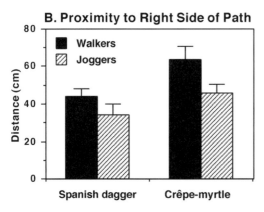

Fig. 5. Attraction of spiny looking plant on jogging path. **A** View of Crépe Myrtle on left side of pathway and Spanish Dagger on the right as positioned during the second half of the study. **B** In comparison with the Crépe Myrtle, the Spanish Dagger engendered a significantly closer approach for walkers ($n = 59$, $p<0.0005$) and joggers ($n = 47$, $p<0.05$), as measured from the edge of the right foot to the right edge of the path. Mean and standard error values are shown

between narrower and wider paths might indicate that plants are inspected in subtle ways from optimal distances that are still safe, with the Spanish Dagger on the left side luring pedestrians away from the less important Crépe Myrtle of the right side.

A striking implication of the recognition of the piercing and slicing properties of sharp forms is the emergence of Oldowan lithic technology in the Late Pliocene (Kibunjia 1994; Semaw et al. 1997). The inadvertent fracturing of rocks while pounding tissue and bones scavenged from predator kills could have led to more systematic fracturing of rocks to obtain flakes and cores with sharp edges for cutting and scraping (Fig. 6). Recognition of the utility of these surfaces for extracting resources might have been prompted by both innate recognition of the cutting potential of sharp edges and painful experiences with sharp rocks and ubiquitous thorns. To explore this idea of perceived functionality, the attributes of four black silhouettes with pointed and rounded contours were examined using an attitude survey with a semantic differential scale. Two teardrop silhouettes were identical except that their points faced upward or downward; the proportion of this silhouette was similar to flake-blades possibly

Fig. 6. Illustration of an early Oldowan chopper dated at around 2.5 Ma from strata at Gona, Ethiopia. (Semaw et al. 1997)

used as knives (Fig. 7) by anatomically modern humans at one of their oldest South African sites (Singer and Wymer 1982). Large attitudinal differences were evident for attributes that characterized tool-like qualities (Fig. 8). The downward-pointing teardrop was rated as having a significantly greater piercing, sharp, and dangerous appearance than its upward-pointing counterpart. From an aesthetic consideration, all the pointed shapes were rated as significantly more beautiful than a rounded oblong shape (Fig. 8D).

Experimental study of the arousing properties of pointed shapes further supports the contention that these forms are visually provocative. Coss (1965, 1968) presented a series of slides depicting a variety of sharp-looking and rounded shapes of equivalent complexity and measured pupillary dilation with different devices. The result of one study of ten

Fig. 7. Pointed chalcedony flake-blades from the Middle Stone Age at Klasies River Mouth Cave in South Africa. (Redrawn from Singer and Wymer 1982)

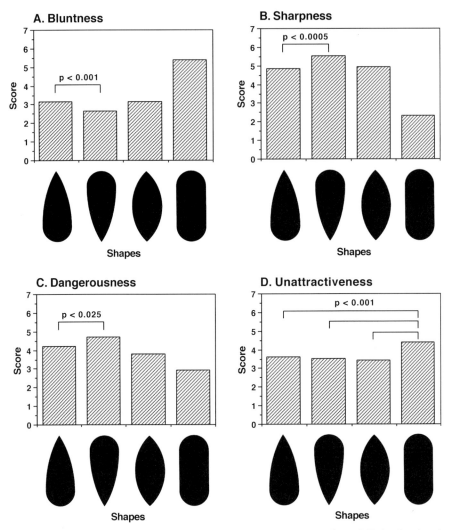

Fig. 8. Silhouettes with pointed and rounded contours examined as full-sized printed images (12.5 cm length × 5.5 cm max. width) on white background using a semantic differential scale (n = 53 men, 80 women). These are: **A** blunt = high vs. piercing = low, **B** sharp = high vs. dull = low, **C** dangerous = high vs. safe = low, and **D** ugly = high vs. beautiful = low). Note that the downward pointing teardrop exhibits a more stabbing tool-like appearance than the upward pointing teardrop

men using a two-factor repeated measures design is shown in Fig. 9, illustrating that a pair of zigzag patterns with pointed contours engendered a significantly greater amount of pupillary dilation with large effect size than a pair of undulating patterns with rounded contours. Together, these aforementioned studies suggest that, in addition to elevating physiological

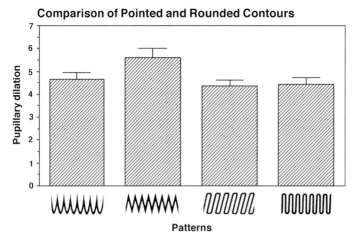

Fig. 9. Comparison of pupillary dilation scores after a 5-s-exposure to projected slides. Mean and standard error values are shown. The main effect comparing the sets of sharp and rounded zigzag patterns is significant. ($n = 10$ men: $p = 0.025$; Cohen's $d = 1.79$)

arousal that possibly enhances memory, pointed shapes can be physically and aesthetically attractive as well as engender caution which at the extreme is expressed as aichmophobia, the dread of sharp objects like scissors, knives, and needles (Berggren 1992). As such, it is not surprising that these shapes abound in art and design.

The emotional sensations felt during the visual fixation of sharp forms in circumstances other than one requiring rapid evasive action are probably abstruse, although with a subtle subjective dangerousness prompting their noticeability. One of the most salient and well-described series of twentieth-century modern sculptures is Constantin Brancusi's *Bird in Space* (Fig. 10), which consists of an upward thrusting lanceolate shape of polished bronze or marble symbolizing soaring flight (see Varia 1986). Although art historians emphasize this flight symbolism in commentaries, it is difficult to avoid the conjecture that the glistening, rapier-like appearance of these pieces inspires their prominence. In addition to their frequent occurrence in biomorphic sculpture, pointed forms are prominent in graphical design, notably in theatrical contexts in which Roman text can have exaggerated serifs that connote mystery and danger. An example of the use of provocative vegetation in this manner (Fig. 11) can be seen in the pre-Raphaelite flavor of book illustrations by Aubrey Beardsley in the late nineteenth century (Desmarais 1998). If indeed sharp points can signify a wide range of symbolism, this diversity of meaning and potentiation of mood might explain the placement of visually striking bucrania (Fig. 12) in what appear to be shrine rooms in the 8000-year-old Neolithic village of Çatal Hüyük, possibly to facilitate the awe and mystique of ancestor worship (Mellaart 1967).

Fig. 10. Illustration of a marble version of *Bird in Space* by Constantin Brancusi, Australian National Gallery, Canberra (1931–1936). This long series of birds in marble and polished bronze exhibit subtle differences in their graceful lanceolate contours terminated by faceted tips. Photographic views typically show the pointed edges of these facets rather than their curving edges

Fig. 11. Detail of *Merline and Nimue* by Aubrey Beardsley from Thomas Malory's *Le Morte d'Arthur*, 1893

Fig. 12. Shrine VI.61 from the Neolithic town of Çatal Hüyük, Anatolia, Turkey (redrawn from Mellaart 1967). Note the salience of the horn-core bucrania that might have augmented arousal in a spiritual context

Not unlike what occurs in costume design for cinematic drama, digital graphic artists frequently adorn video game characters and those in role-playing games with pointed forms to increase their aggressive appearance. To illustrate this perceptual bias, the frequency of pointed forms on the body parts of 62 evil characters in the bestiary subsection of the web-based role-playing game *Everlore* (www.everlore.com) were analyzed as a function of the artists' attributions of character strength. Multinomial log-linear analysis revealed that significantly larger frequencies ($p<0.001$) of higher-ranking characters exhibited large teeth and sharp points radiating from their heads, ears, chests, and feet than did lower-ranking characters. Perhaps like these role-playing game characters, the ancient figures adorning temples and medieval cathedrals with gaping mouths and prominent teeth are thought to have had a protective (apotropaic) function of warding off evil (Sütterlin 1989). A similar application of pointed forms is found in so-called primitive art, notably that of African tribal masks (Coss 1968; Aronoff et al. 1988). Zigzag patterns are especially abundant (Uher 1991a, b) and they are sometimes combined with eye motifs (Fig. 13). On older works, zigzags were engraved frequently on Upper Paleolithic bone artifacts by European Cro-Magnon (Marshack 1972), often complemented by serrated forms resembling the overlapping seed heads in the spikelets of plants. An even earlier zigzag appears on an ox rib excavated at Pech de l'Azé, France, apparently engraved by Neanderthals (*H. neanderthalensis*)

Fig. 13. Plank mask from the Bobo tribe, Upper Volta region of Burkino Faso. These masks characterize entities of the wild that act as tribal envoys

Fig. 14. United States Air Force Academy Chapel (1956–1962), Colorado Springs, Colorado, designed by architect Walter Netsch. Students, visitors, and the American Institute of Architects have noted the conspicuousness of this building. (Photograph by Ted Jungblut)

approximately 130,000 years ago. Based on current knowledge of the perceptual effects of sharp points, architects can probably rely on serrated roof contours to catch attention on the skyline as exemplified by the Sydney Opera House designed by architect, Jörn Utzon, and by the United States Air Force Academy (Fig. 14).

Sources of Selection on Species Recognition

In this section I will focus on the visual aspects of species recognition, emphasizing the roles played by the distinctive features of humans and other species in artistic endeavors. Species recognition is defined as the ability of different species to distinguish members of their own kind from those of other species. It is not restricted to a single sensory modality and this ability is fundamentally useful for the appropriate segregation of species in mixed schools, flocks and herds. Because of its essential property for survival and reproduction, conspecific recognition might be based on an evolved species-typical archetype (cf. Mayr 1963, p. 95; Paterson 1993, p. 5) not unlike the perceptual organization subserving the awaiting states of predator species recognition (Coss 1993, 1999; Coss and Ramakrishnan 2000). As an example of this awaiting state under conditions of lack of experience, macaques separated from their mothers shortly after birth and cross-fostered with another species of macaque preferred to look at projected photographic slides of their own species, especially the head region, over those of the species they were reared with (Fujita 1993a). Human-reared pigtailed macaques (*Macaca nemestrina*) showed similar conspecific preferences (Fujita 1993b). Further evidence that the face is important in conspecific recognition is revealed by the provocative effects of threatening faces in isolation-reared macaques (Sackett 1966) and the effects of conspecific profiles on excitatory brain activity in macaque temporal cortex (Desimone 1991). Unique features of the head might play a role in conspecific recognition, especially if they are significant for fighting. Dalesbred sheep, a horned breed, show much greater neural activity in the temporal cortex when they view photographs and drawings of faces of horned sheep compared with those without horns (Kendrick and Baldwin 1987). Both conspecific and heterospecific recognition indubitably includes the differentiation of juvenile and adult forms; for humans, it can be considered the organizing substrate governing individual recognition (Coss and Schowengerdt 1998).

The Provocative Aspects of Eyespot Patterns and Aesthetic Expression

As mentioned previously, face perception relies on the distinctiveness of local facial features embedded in a generalized framework of the mouth and eyes. The importance of the eyes is illustrated by the repetitive scanning they received when the eye movements of subjects are measured as they look at schematic drawings of people (Noton and Stark 1971) and by the elevated physiological arousal that ensues with prolonged eye-to-eye contact (McBride et al. 1965; Nichols and Champness 1971; Gale et al. 1972, 1975). Although eye contact is generally provocative, it is much less so during nonthreatening social interactions (Wellens 1987); this context-mediated flexibility in the assessment of this facial feature permits the diversity of the aesthetic applications of eye-like schemata as both nonthreatening decorative patterns and as apotropaic devices.

Because detection and recognition of two facing eyes has been essential for survival in a variety of settings, the perceptual properties of this schema can be examined apart from other facial features. In experimental settings, visual fixation of schematic facing eyes depicted by two concentric circles in the horizontal plane causes an rapid increase in pupillary dilation (Coss 1965, 1970) in contrast to the effects of looking at other combinations of concentric circles (Fig. 15A).

Simplified even further as eyespot patterns, visual inspection of two black spots in the horizontal plane initiates rapid adjustments in cardio-

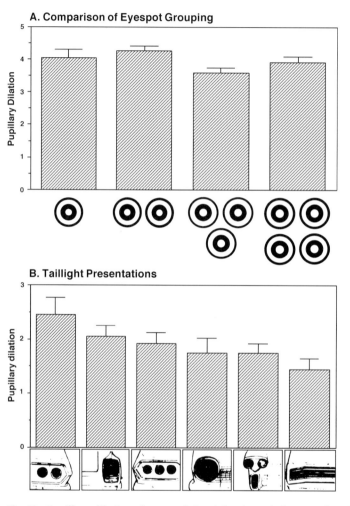

Fig. 15. Pupillary dilation after 5 s of visual fixation of photographic slides of two sets of concentric circle designs in different configurations and automotive taillight designs from the late 1950s to early 1960s: *left* to *right*; Chevrolet Biscane, Olsmobile F-85, Chevrolet Impala, Ford, Cadillac, Buick Skylark. Mean and standard error values are shown. In both studies, the averaged pupillary dilation levels elicited by the two sets of paired concentric circles and the taillight design, both of which present the schema of two facing eyes, was significantly greater than the combined average of the other respective circle configurations and taillight designs. A $n = 15$ men: $p<0.025$, Cohen's $d = 1.45$. B $n = 15$ men and 15 women: $p<0.025$, Cohen's $d = 1.31$

vascular tone as evinced by heart-rate deceleration followed by marked acceleration. This cardiovascular effect does not occur when other arrangements of two and three black spots are visually fixated (Aikens 1998b). Although these studies indicate that eyespots presented outside the context of a face were found to be arousing, it was not clear whether eyespots exhibited in a familiar configuration would retain their provocative properties. Gray-scale slides of five automotive taillights with equivalent luminance were shown to subjects and pupillary dilation was measured using a light-diffraction device (Coss 1965). Despite its presentation in a taillight configuration, two concentric disks appearing as facing eyes elicited the strongest physiological arousal among the series of taillights (Fig. 15B).

The surge of sympathetic nervous system arousal triggered by eye contact or the viewing of eyespot patterns in different contexts would seemingly account for the rapid onset of gaze aversion or visual avoidance that terminates eye contact. Gaze aversion is thought to function as both a nonthreatening signal of disengagement and as a way of modulating arousal by redirecting attention to less provocative sources of input (Coss and Goldthwaite 1995; Glenberg et al. 1998; Emery 2000). While this form of gaze behavior appears to be regulated automatically by arousal changes (Coss 1970; Kawashima et al. 1999; Emery 2000), it is recognition of the consequences of eye contact that promotes arousal and activates gaze aversion (Argyle and Cook 1976). Juvenile bonnet macaques (*Macaca radiata*), for example, quickly learn to suppress prolonged eye contact with dominant males that attack and chase them (Coss et al. 2002). Nevertheless, novel patterns with this schema are still provocative because two eye-like concentric circles in the horizontal plane induce much less visual inspection by mouse lemurs (*Microcebus murinus*) and normal and autistic children than other configurations of concentric circles (Coss 1978, 1979).

The ability of schematic images of eyes to modulate gaze behavior provides a partial explanation for the widespread belief that the display of eye-like patterns (Fig. 16) might have protective apotropaic properties (Eibl-Eibesfeldt 1989; Sütterlin 1989). For example, architect James Lennon was successful in reducing shop lifting, but not sales, in several small stores by suspending large banners near merchandise with a vertical array of schematic frowning eyes (Coss 1981). Women in South Yemen have been observed to wear arrays of large concentric eyespots on their chadors, which might reflect a common concern about the malevolent gaze of others. In southern India, women are known to smear lampblack on their eyelids to protect them from the evil glances of others as well as reminders not to cast such glances themselves. Belief in the "evil eye" occurs in many cultures, leading to the wearing of amulets with eye motifs designed to protect against someone who might be envious and likely to cast the evil eye (Schoeck 1981). Concern about the evil eye by some European Americans is manifested by their placement of charms or talismans on baby car-

Fig. 16. Celtic funerary pot with a putative apotropaic eye motif. Bronze age, Denmark

riages for protection that can include knives, scissors, keys, and crossed safety pins (Jones 1981). Dundes (1981) describes various cross-cultural contextual properties in which the threat of the evil eye and defense against it is manifested. He argues that male genitals are involved in some way with the evil eye complex because of the occurrence of iconographic representations of the eye with the phallus (also see Eibl-Eibesfeldt 1979).

Illusory Facial Expressions in Product Designs

The ability to accurately perceive object symmetry is greatest when the axis of symmetry is vertical (Evans et al. 2000). It is possible that sensitivity to changes in symmetry is a by-product of the adaptive value of determining head orientation, which involves identifying subtle changes in bilateral symmetry caused by the displacement of prominent vertical facial features like the nose (see Wilson et al. 2000). Johnson and Morton (1991) argue that newborn infants are innately attracted to the symmetrical configuration of two schematic eyes and a mouth. Follow-up research using similar schemata (cf. Simion et al. 1998; Easterbrook et al. 1999) indicates that the less complex schema of two eyes alone are sufficient to attract the newborn's attention. General face recognition abilities develop rapidly during the first months after birth, with infants exhibiting a considerable tolerance for facial inversion. By 5 years of age, recognition of individual faces of humans and monkeys, but not those of sheep, becomes hampered when these images are inverted (Pascalis et al. 2001). The lack of a face inversion effect for sheep faces suggests that the eyes can anchor relatively novel, but distinguishable facial configurations without the experiential constraint that restricts individual recognition of inverted primate faces.

Since the eye schema plays an important role in face recognition, and certain facial expressions such as frowning are consistent across cultures (see Eibl-Eibesfeldt 1989), it is not unreasonable to assume that the perceptual salience of frowning faces emerges from experiencing the wrath of others. The relevance of angry faces is manifested in the ability to detect schematic angry faces much faster than schematic happy faces among multiple faces shown in computer displays (Fox et al. 2000). Angry facial expressions are also the most provocative among a series of facial expressions in generating broad neocortical activation seen in fMRI neuroimaging (Kesler/West et al. 2001). Curiously, neuroimaging using fMRI and positron emission tomography (PET) does not indicate cortical input to the amygdala, an integrative subcortical structure that exhibits neural plasticity and does appear to contribute to neutral and sad facial expression processing (cf. Blair et al. 1999; Kesler/West et al. 2001). As an alternative view to social learning, the lack of amygdala involvement in the perception of angry facial expressions could indicate an innate predisposition for recognizing the aggressive intentions of individuals with frowning faces without the need for social conditioning via the amygdala (see Adolphs et al. 1998; Morris et al. 1998a,b). Although eye contact duration in nonhuman primates appears to be regulated by aversive social experience (Coss et al. 2002), and human assessment of the direction of gaze does involve the amygdala (Kawashima et al. 1999), a primary source of selection for the recognition of frowning faces historically would have been the low ambiguity of aggressive consequences following perceived threat. This reliable aspect of frowning faces would be quite different from the uncertain attribution of emotions given to a person who appears to be grieving (see Fridlund 1994), the latter of which would indubitably benefit from social learning.

The angle of viewing the face can distort assessments of facial expressions, especially when the face is viewed with the head tilted downward and the eyes are partially occluded by the brow, creating the illusion of frowning (Kappas et al. 1994). Similar facial expression illusions can be seen in consumer products with face-like appearances, such as contemporary portable stereo radios with housings exhibiting pairs of ovoid loudspeakers tilted in V-shaped angles. Illusions of angry and sad facial expressions are apparent in mock-ups with recessed controls or displays simulated by concentric circles partially occluded by V-shaped or inverted V-shaped housings (Fig. 17). Automotive stylists, corporate executives and journalists appear cognizant of the face-like property of the fronts of cars (Coss 1967) because the contours of fenders and headlights of numerous cars since the late 1960s have exhibited illusory facial expressions, notably aggressive frowning (see Fig. 17). The motivation for providing this form of aggressive appearance in automotive styling is consistent with the application of "muscular" fender lines and large wheels.

It is unclear, however, whether industrial designers intentionally create illusory facial expressions when they generate initial sketches, sculpt clay

Fig. 17. Illusory facial expressions of face-like objects. *Clockwise from top left*: relatively neutral, sad, and angry facial expressions can be seen in these design mock-ups and in the frowning, gaping appearance of a DaimlerChrysler model 2001 Sebring LX automobile

mock-ups of consumer products, or when images are created using computer software. Neurobiological studies of single neuron recordings of macaques suggest that facial expression recognition occurs after the initial categorization of the global properties of faces (Sugase et al. 1999). If relevant to humans, preattentive recognition of the face-like properties that emerge fortuitously during the conception phase of product design might bias further decisions to add illusory facial expressions. The results of such design decisions do seem to generate aesthetic appeal in automotive executives as reflected by the comments of Wayne Cherry, Vice President of design for General Motors: "Pontiac...will always have the two nostrils in front and lights arranged to give it a somewhat sinister shape" (Cook 1997). A similar statement about the 1995 Chrysler Cirrus LX by automotive reporter John Goepel (1995, p. 22) indicates clear recognition of a frowning facial expression: "The car's low, narrow-eyed front end strikes some as 'subtly aggressive,' overt aggression perhaps being out of place in what is, after all, a family car." Moreover, recent research on American tastes in automotive styling conducted by cultural anthropologist Clotaire Rapaille for DaimlerChrysler also reflects a not so subtle awareness of the physiognomic appearance of some sports utility vehicles. Rapaille states that: "vertical metal slats across the grilles give the appearance of a jungle cat's teeth and flared wheel wells and fenders that suggest the bulging muscles in a clenched jaw" (Bradsher 2000).

Reptilian Scales and Tessellated Graphics

Since Miocene times (Fig. 1), encounters with dangerous reptiles would have provided relatively consistent sources of selection operating on the perceptual ability to detect these animals in their appropriate microhabitats. Perhaps the most dangerous African predator to humans today is the Nile crocodile, which is known to become a relentless man-eater after it feeds on bodies dumped into rivers (Alderton 1991). As both stalking and sit-and-wait ambush predators, Nile crocodiles can launch themselves several body lengths onto land in pursuit of prey (Pooley et al. 1989), a style of attack that would have made these predators exceedingly dangerous to thirsty hominids. The weak aesthetic preference of humans for still, cloudy water in stagnant creeks and swamps, contrasted by their great appreciation of rushing and clear water in mountain streams and lakes (Herzog 1985), might reflect their knowledge of potential pathogens, but also the uncertainty of what could lurk in the water. Fear of crocodilians has been used as an argument by Yeager (1991) to explain the caution in crossing rivers exhibited by proboscis monkeys (*Nasalis larvatus*) in Borneo that are known to be killed by the False Gharial (*Tomistoma schlegelii*).

From the perspective of predator detectability, Nile crocodiles do exhibit distinctive features, such as grid-like scales and the zigzag pattern of spiky teeth that are still conspicuous when their mouths are closed. Another salient feature that might be useful for crocodile recognition is the serrated contour of bony osteoderms covered with skin on their dorsal scales (Coss 1965); these projecting scales become prominent when they glisten in surfacing crocodiles. Virtually no research has been conducted on crocodile recognition by prey species, and those few studies that have examined primate responses have done so mostly with the aim of investigating the preparedness construct and the role of observational conditioning. For example, a toy crocodile and snake can induce selective fear in naïve rhesus macaques (*Macaca mulatta*) after they observe on television other macaques reacting fearfully to these models; however, observational conditioning of fearfulness does not occur after naïve macaques observe these individuals appearing to behave fearfully to flowers or a toy rabbit (Cook and Mineka 1989).

A common feature shared by reptiles is the highly periodic crosshatch or grid pattern of scales, which includes the tessellated polygons on tortoise and turtle shells. The historical consistency of repetitive scale patterns that glisten in most species of snake would afford reliable predictors of hazards even when these scales are partially obscured by variegated textures of logs, vegetation and detritus. The innate snake-recognition system of California ground squirrels appears to capitalize on the long, cylindrical form of snakes and the repetitive properties of their contrasting scale patterns (Coss 1991a). Snake scale patterning can be important for differentiating venomous and nonvenomous snakes in birds that prey on snakes.

Smith (1977) describes how snake-naïve turquoise-browed motmots (*Emomota superciliosa*) show strong aversion to a pattern of wide yellow and narrow red rings characterizing a venomous coral snake (*Micrurus* spp.) compared with other patterns painted on wooden dowels.

A number of different African and Asian primates appear sensitive to moving animals with scale patterns as evinced by the fearfulness of young chimpanzees to snakes, lizards, tortoises and newly hatched alligators (Haselrud 1938). When shown a live brownish gopher tortoise (*Gopherus polyphemus*) and a green legless lizard, the glass snake (*Ophisaurus ventralis*), captive-born 1- to 2-year-old common chimpanzees (*Pan troglodytes*) tended to be cautious, exhibiting behavior ranging from hesitant investigation to complete avoidance (Yerkes and Yerkes 1936). The bulbous form of the tortoise might resemble a small portion of an African rock python (*Python sebea*), a killer of large-bodied primates, including humans that can achieve lengths of 5.2–6 m. Wild long-tailed macaques (*Macaca fascicularis*) will became excited and alarm call when they encounter a reticulated python (*Python reticulatus*) and will respond to a python model with enhanced vigilance and passive avoidance (van Schaik and Mitrasetia 1990; Palombit 1992). Our research on wild bonnet macaques from both forest and urban areas in southern India indicates that, among a series of realistic snake models, only the Indian python (*Python molurus*) elicits alarm calling and passive avoidance, whereas the Indian cobra (*Naja naja*) with conspicuous eyespots is much more likely to trigger startle and flight responses (Ramakrishnan et al. in prep.).

It is not clear how learning shapes the antisnake responses of primates. In captive-born rhesus macaques, watching the snake-induced fearfulness of experienced conspecifics appears to selectively potentiate the fear of snakes (Cook and Mineka 1989). Infant bonnet macaques in the wild have been observed to leave their mothers to approach small model snakes, causing their mothers to grab them in ways that might allow these infants to infer the danger posed by snakes (Ramakrishnan et al., in prep.). With the exclusion of pythons that are perceived as dangerous predators, wild bonnet macaques as they age show a reduction rather than an enhancement of fear of venomous and nonvenomous snakes, leading to appropriate monitoring of snake activity and maintenance of safe distance. The issue of learning is even murkier for chimpanzees. According to Kortlandt (1955, p. 229), chimpanzees do not become fearful of snakes until about 6 months of age after which their fear or aggression is expressed dramatically. As an example of the latter, Kortlandt and Kooij (1963) report incidences in which captive chimpanzees used sticks to club snakes that entered their enclosure. More analogous to our observations of adult bonnet macaques, Kortlandt (1967) observed rather casual responses by wild chimpanzees to a highly venomous Gaboon viper (*Bitis gabonica*). He notes: "The chimpanzees just quickened their pace and passed the snake, if necessary making a small detour, and looking shyly at it." In contrast to previous studies (e.g., Yerkes and Yerkes 1936), Schiller (1952) found that young, captive-reared chimpanzees under 3 years of age were not particularly fearful of a moving snake behind glass, inspecting it at close range. Older individuals were clearly more cautious, vocalizing and monitoring the snake from a safe distance. It is possible that developmental processes unrelated to snake experience, reflecting the effects of laboratory rearing on brain development, mediate this age-related change in snake response. For instance, Masataka (1993) reports that laboratory born squirrel monkeys (*Saimiri sciureus*) that acquire experience eating live insects developed a much greater fear of snakes than those without insect-eating experience did.

The role of diversified experience might be causally linked in a similar manner to the way humans respond to snakes. Like young chimpanzees, Prechtl (1949) reports that children under 1 year of age do not exhibit snake aversion. Jones and Jones (1928) found that children from 14 to 27 months of age were not frightened by a live snake released near the child or placed in a suitcase containing blocks, but fear became increasingly tentative or "guarded" as early as 26–79 months of age. Spindler (1959) also examined the responses of young children and showed that the onset of snake aversion begins about age two, consistent with the onset of a nighttime fear of scary things (Coss and Goldthwaite 1995). The onset of fearfulness after this age is coincident with a sharp decline after birth in the linear growth rate of the brain, prior to which gene expression mediating the connectivity of neural circuitry would predominate. As an alternative to this developmental neurobiology argument, Kortlandt (1955) suggests that the fear of snakes is coincident with a developmental period in which children begin to range beyond their mother's reach. He argues further about a possible specialized learning or "imprinting mechanism" leading to a restricted class of snake-like objects that has "evolved as a result of actual snake-bites in human ancestors."

A number of human studies show that adverse experiences with snakes and social referencing can potentiate the fear of snakes. In one study of snake-phobic individuals, frightening childhood experiences of unexpected encounters with snakes, or even witnessing a snake viciously beaten to death, appeared to contribute to excessive fear (Bandura et al. 1969). Careful presentation of snakes in the context of social modeling, relaxation during exposure, and snake-contact desensitization can reduce the fear of snakes (e.g., Bandura et al. 1969; Murphy and Bootzin 1973), indicating social learning processes akin to the natural context of how wild primates become more confident foraging near snakes. Neuroimaging using PET on individuals with snake phobia provides some insight into the neurobiological aspects of snake fear. Comparisons of regional cerebral blood flow during observation of snakes indicate that localized synaptic activity is much greater in the temporal lobes than in primary visual cortex (Wik et al. 1993). However, taken further, synaptic activity in the orbitofrontal cortex and prefrontal cortex decreases when snakes are viewed, suggesting a form of cognitive suppression congruent with anecdotal observations of suppression of motion or freezing to prevent further endangerment. Such a reduction of cognitive activity in these brain areas will be discussed further in the context of why snake images might be important in religious ceremonies and art.

Exposure to dangerous snakes was likely to have been a common occurrence in hominids, with inattention in trees leading to reaching for fruit near arboreal venomous snakes, such as the black mamba (*Dendroaspis polylepis*), or stepping on immobile vipers while foraging on the ground. Three cryptically patterned African vipers that would have posed serious threats (see Greene 1997) are the saw-scaled viper (*Echis pyramidum*) with a tessellated diamond scale pattern, the puff adder (*Bitis arietans*) with black and yellow crescents, and the large Gaboon viper with a complex crisscrossing pattern. The scale patterns of all three manifest a mixture of beauty and awe, with the Gaboon viper characterized by Ditmars (1931, p. 177) as the "world's most frightful-looking snake". He continues with: "Upon this awesome form the symmetrical pattern and really beautiful hues repel, rather than soften the picture." Although cryptically patterned, these snakes are relatively distinguishable when in motion, but difficult to distinguish from leaf litter and detritus when these snakes are coiled, a property that disrupts the uniformity of their scale patterns. To circumvent this crypticity, experience detecting these snakes while foraging would have undoubtedly enhanced recognition of where these snakes might be encountered as well as primed preattentive perceptual systems to detect and segregate any form of pattern regularity on detritus and leaf litter. Concordant with this argument of daytime detectability, the majority of bites by these vipers occur at night when people step on them accidentally (Nhachi and Kasilo 1994).

Because predatory pythons and venomous snakes might be outside the focus of attention in peripheral view as one moves about, the ability to detect complex contour information with peripheral vision would be essential to mitigate attacks or envenomation. As apparent in fMRI brain imaging, a checkerboard arrangement of textures presented 8° from the subject's point of visual fixation engenders more synaptic activity in the higher visual centers than a uniform texture (Kastner et al. 2000). Another imaging method, magnetoencephalography, revealed that a diamond shape on a randomly textured background produced larger neural responses in the higher visual centers than a random-dot pattern or, in preliminary tests, a circle or triangle on a randomly textured background (Okusa et al. 2000). Thus, some geometric patterns appear to have greater salience than others do when presented either outside the focus of attention or directly fixated. Related studies of macaques provide results consistent with the idea that, despite the infrequent occurrence of highly periodic patterns in nature, scale-like patterns portrayed by vertical and diagonal grids and checkerboards might undergo specialized processing in the visual stream by neurons in V4 and inferotemporal cortex (e.g., Tanaka et al. 1991; Kobatake and Tanaka 1994).

Snake scales have inspired awe and dread, ranging from disgust in phobic individuals who view them as "slimy" due to their glistening appearance (Bennett-Levy and Marteau 1984) to immediate startle responses in individuals who suddenly discover shed snake skins. The desire to approach and investigate snakes has probably biased their presentation as engravings or castings on numerous artifacts and their display in paintings and mosaics. Among some of the oldest artifacts of anatomically modern humans in Middle Stone Age sites are ochreous shales showing evidence of deliberate grinding and scraping or gouging (Henshilwood et al. 2001b). One example of ocher pigment was engraved with a crosshatch pattern with possible symbolic properties (see Henshilwood et al. 2002). While this design might be linked conceptually to reptiles, the interpretation is less ambiguous for pre-Columbian artifacts. Diamond-patterned crosshatching, zigzags, and step forms are predominant decorative motifs

Fig. 18. Similarities of geometric snake-scale pattern and brick work. *Left and right panels*, respectively, Gaboon viper (*Bitis gabonica*) and detail of the façade of Shir Dar funerary *madrasa*, Samarqand, 1636

in Mayan art which are thought to characterize rattlesnake scale patterns (Diaz-Bolio 1987). Snake scales are an important feature in snake imagery in studies of fear imagery and physiological arousal (e.g., Lang et al. 1983) and some of the tessellated mosaic designs in Islamic art (see Clévenot 2000) are strikingly similar to snake-scale patterns (Fig. 18).

There are three facets of snake perception, investigative attraction, group cohesiveness, and cognitive inhibition that might bias the choice of snakes for symbolic representations in religious art. In south Asia, Hindus worship the cobra in the hope of enhancing fertility and they associate the cobra with the sacred lingam, symbol of Lord Shiva. Dense crowds croon and pray to each cobra displayed in succession before the deity in the serpent ceremony (Miller 1970). Although entirely speculative, peripheral viewing of scale-like geometric patterns exhibited in Islamic mosques and on prayer rugs might enhance subcortical arousal while simultaneously inhibiting orbitofrontal and frontal cortex activation, possibly facilitating a meditative state and reverence. The aesthetic attraction to reptilian scales is clearly evident in the choice of skins from crocodiles and snakes for commercial use in consumer products. Current commercial exploitation of reptilian skin products (Greene 1997) has been substituted mostly by embossed leather prints and printed fabrics. In the last decade of the twentieth century, snake prints, notably python, were fashionable in women's apparel and accessories, often displayed in attention-getting contexts that possibly capitalized on the alluring interplay of human- and snake-species recognition. However, taken further, it is intriguing to speculate on whether perceptual sensitivity to tessellated patterns influenced the development of interlacing plant and animal materials, eventually leading to the weaving of baskets and fabrics and the natural emergence of geometric woven designs constrained by this technical process.

Leopard Spots and Spotted Textures

Another decorative texture, spots and polka dots, have attributes of pattern regularity that might be salient due to the texture-processing capabilities of "dot-pattern selective cells" discovered in macaque inferotemporal cortex (see Tanaka et al. 1991). Kruizinga and Petkov (2000) emphasize that dot-pattern-selective cells react most strongly to textures that consist of spots not unlike those occurring in nature as clusters of rounded leaves, trees, and flowers. Dot-pattern selective cells appear more tolerant of the irregularity of spot distribution than cells in striate cortex that are sensitive to the highly periodic textures found in gratings (von der Heydt et al. 1992). Located in different parts of the visual stream, these cell types together process a broad range of textures found in nature.

The flecks, spots and rosettes on the coats of modern felids are thought to have evolved to provide camouflage when viewed against the dappled light of leaf shadows in the forest. Both phylogenetic and developmental analyses of felid spots and rosettes suggest that these coat patterns are the ancestral trait in the felid lineage (Werdelin and Olsson 1997; Ortolani 1999). While these repetitive patterns might function as camouflage in certain circumstances for ungulate prey with dichromatic vision, primates with trichromatic vision have apparently capitalized on spots and rosettes on yellow coats as predator-recognition cues. Our initial research on bonnet macaques using a realistic leopard model to identify the perceptual cues for leopard-recognition indicates that spots and rosettes on the yellow coat contribute strongly to leopard recognition (Coss and Ramakrishnan 2000). Towels with a leopard print arranged like a resting leopard are also provocative to bonnet macaques, whereas troop members ignore an identical arrangement of towels with flower prints. Experience is not required for leopard recognition because urban monkeys that have never seen a leopard react as strongly by fleeing to trees and alarm calling as those that regularly encounter leopards in the forest. Other primates that appear to exhibit an evolved sensitivity to leopards are captive-born West Indian green monkeys (*Chlorocebus aethiops sabaeus*, Lisa Hollis-Brown pers. comm.), sooty mangabeys (*Cercocebus atys*), pigtailed macaques (*Macaca nemestrina*) and rhesus macaques (Jason Davis pers. comm.).

There are no equivalent studies of how humans respond to leopards and other large cats other than anecdotal accounts of initial freezing upon sighting these predators or the automaticity of climbing trees (e.g., Corbett 1948; Noronha 1992). Children and adults, especially parents protecting their children under attack, will mob tigers, leopards, and mountain lions, as evinced by their rushing these predators while yelling, throwing rocks, and hitting them repeatedly with sticks, rifles, shovels, a basket, and even a fish (Noronha 1992; Fitzhugh 2001). Such mobbing behavior would initially appear to be irrational, but aggressive mobbing can cause the predator to drop its victim and move away (Lee Fitzhugh pers. comm.). Like that engendered by reptilian scale patterns, humans do appear to be as fascinated with the spotted coats of the large cats, and aesthetic attraction to spotted leopard pelts might explain their first appearance as apparel in the early Neolithic on hunters and on individuals of high stature in the village of Çatal Hüyük (Mellaart 1967). Current interest in the coats of spotted cats has led to the endangerment of many species by illegal poaching, in part, because some societies still associate the wearing of spotted pelts with high social status (Wolfe and Sleeper 1995).

The rationale for wearing animal prints for social display awaits detailed empirical investigation because of its importance to animal conservation. Because of the long history of hominid contact with felid predators in Africa, there might be an evolved basis for the interest in spotted pelts and even the desire for domestic cats as companion animals. Crawford (1934) notes the complex attributes underlying ailurophobia that he characterized as representing an emotional conflict of love and fear of cats. A similar pattern of conflict has been photographed in urban bonnet macaques as they initially cornered a young domestic cat with a spotted gray coat, gathered around it cautiously, and peered in awe as the dominant male proceeded to groom its fur. While fearful of leopards, the context of viewing a miniature leopard-like animal engendered attraction mixed with ambivalence.

The presence of humans in the experimental context of leopard-model presentations has no noticeable impact on the level of fear triggered in bonnet macaques by these models. In essence, humans who have not mistreated troop members are generally irrelevant in the context of viewing a truly dangerous adversary. A similar focused attention occurs during snake perception as when a male long-tailed macaque alarm called repeatedly while he stared at the python-skin shoes of a laboratory visitor (Carolyn Crockett, pers. comm.). Uncoupling of a human who poses no threat and a dangerous-looking pattern might explain why a woman wearing a leopard print dress was mobbed excitedly by a wild common chimpanzee who was habituated to humans. For the tourists who observed this event, the woman was the significant object and her dress meaningful only to the extent that it was a fashionable outfit with a mildly provocative wildlife theme. Described another way, the women did not lose her significance to other tourists because she was a member of the same species, whereas, for the chimpanzee, she was a different species and thus much less relevant than the leopard print which was the presumed target of harassment. However, apart from the owner, this dress could acquire strong significance if gathered into a lump and placed under a bush, in which case the context of discovery by experienced forest personnel might indeed be frightening. In a context analogous to this chimpanzee anecdote, villagers in southern India captured a bonnet macaque and painted it with yellow spots to resemble a leopard to frighten troop members who were entering their homes and causing damage. When released, the painted macaque was immediately attacked by troop members who, contrary to villager expectations, were not fearful because they apparently recognized the macaque as a member of their own species. An abnormal pattern of yellow spots in conjunction with conspecific recognition severely altered the motivation of troop members who might normally be fearful of a spotted texture if it appears suddenly, but might mob it if it appears less threatening. Perhaps the aesthetic attraction of humans to leopard pelts and printed fabrics engenders a similar admixture of recognition cues in which a spotted texture imparts a subjective impression of potency to the individual wearing it.

While there is little empirical information on human recognition of predator configurations, fascination with leopard rosettes and python scales can be detected in infants as young as 7 months of age, possibly as a precursor to pattern-recognition processes that emerge later in development as animal fears. In follow-up research on the aforementioned mouthing of glossy and dull plates by infants and toddlers, 14 of 23 children in

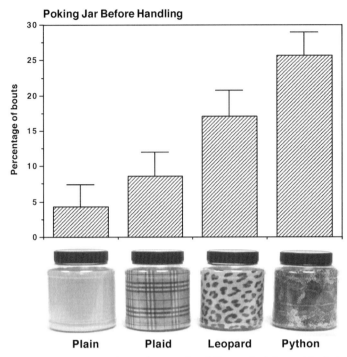

Fig. 19. Frequencies that infants and toddlers poked jars with their forefingers prior to handling them for five episodes ($n = 14$; mean age = 12 months; range 7–15 months). Mean and standard error values are shown. Jar weight and size: 136 g; 15 cm height × 14.5 cm diam. In planned comparisons, the jars with the python and leopard patterns elicited poking at significantly higher frequencies with large effect sizes than the jar with the plaid print (respectively: $p<0.0025$; Cohen's $d = 2.2$; $p<0.05$, Cohen's $d = 1.4$)

3 day-care facilities handled each of 4 lightweight plastic jars a minimum of 5 times in a repeated-measures design. Jars were presented as toys in random order during playtime for 5-min observation periods over multiple sessions. The jars contained yellowish-orange paper pressed against their interior side and bottom surfaces and three jars displayed the following full-size patterns: plaid fabric print, leopard rosettes, and rock-python patches and scales. Prior to handling the jars, the children often engaged in careful prodding of the jar patterns typically with the outstretched index finger of the right hand. The jars with the python and leopard patterns engendered significantly larger frequencies of poking than the plain and plaid jars (Fig. 19). Poking with the forefinger was interpreted as an investigative probing action (see Blake et al. 1994) because children immediately suspended further interaction with the jars if they rolled when poked, a behavior especially noticeable for the jars with the leopard and python patterns.

Conclusion

The making of aesthetically interesting artifacts involves admixtures of imagination and contemplation organized in ways that guide the production of artistic behavior. Intervening assessment that unfolds during the process of creative decision making might be subject to perceptual biases either at the conceptual level of visual imagery or during the evaluative process of inspecting the graphical result (cf. Johnson 1993; Verstijnen et al. 1998). Since the primary visual cortex has been shown to be involved in the formation of mental images (e.g., Kosslyn et al. 1995), it is conceivable that preattentive processes might engender different degrees of image vividness or coherence, possibly constrained by image configuration. Mental-image manipulation, however, is more likely to occur in the same cortical areas where the scanning and recognition of real images occur (see Cocude et al. 1999; O'Craven and Kanwisher 2000), so that any evolved specializations of neural circuitry might facilitate the construction of specific types of imagery and hence their manifestations in works of art. In addition to the contemplative process of conceptualization, some geometric designs in artwork are similar to those that emerge as phosphenes and hallucinatory images after the ingestion of psychotropic substances (Kellogg et al. 1965; Lewis-Williams and Dowson 1993). Irrespective of whether mental images are generated consciously (e.g., Coss 1969) or during altered states of consciousness, the pattern of neural organization subserving such image generation would have been subject to a long history of natural selection for solving essential visual tasks.

My initial conception of how natural selection might configure specific visual pattern-recognition abilities that bias artistic expression (Coss 1965, 1968) was influenced by early ethological conceptions of sign stimuli and innate releasing mechanisms (e.g., Lorenz 1950; Tinbergen 1951). This view characterized the automaticity of motor responses triggered by key visual features of an object, with additional key features promoting stronger responses (see Curio 1993, p. 157). As apparent in my review of innate pattern-recognition processes, I have since emphasized the complementary roles of cognition and learning in the assessment of the pattern meaning and concomitant organization of behavior.

In any given circumstance, the source of selection on a pattern-recognition ability is the entirety of the chain of events in which the perceiver fails to detect, recognize, and execute appropriate behavior that impacts fitness (see Coss and Owings 1985; Coss and Goldthwaite 1995). For static images, successive failure at the detection phase over thousands of generations is assumed to result in progressive neural specialization for extracting contour and texture information relevant for texture-background segmentation. Recognition of pattern meaning, yielding its pop-out from background, is coupled inextricably to action patterns that promote fitness. Selection would operate on the outcome of passive interactions with abiotic materials and dangerous sessile organisms controlled entirely by the perceiver and on the outcome of dynamic interactions, such as avoidance of conspecific aggression or evading predatory attacks. In the case of sighting a motionless predator, the innate information entailed in the predator-recognition process includes cognitive inferences about the predator's cursorial capabilities. As inferred from the startling, investigative and refuge-seeking behavior of predator-naïve primates, a spotted texture on a yellow background signifies something that can chase and climb rapidly, whereas an elongated patch with a glistening crosshatch and brown and black blotches signifies something that can be investigated reasonably close, but with the potential of lunging suddenly. Both textures signify a dreadful context to these primates, with subsequent action patterns consistent with the nature of the threat. It is not known whether humans wearing apparel with these animal textures convey to perceivers similar differences in approachability other than a general sense of the dangerousness that these animal pose in nature.

Because of their link to ambushing predators and aggressive conspecifics, eyespot patterns would more likely connote a form of uncertainty or hazard analogous to that of detecting real facing eyes behind unoccluded patches of vegetation; this attribution alone might explain the historical application of eyespot patterns as apotropaic devices. As a powerful component of face perception, subsumed by face-sensitive neural circuitry, eye-like features can be construed from the incidental and perhaps deliberate arrangements of controls and displays by product designers and these can include expressive face-like contours.

Physical and aesthetic attraction to sharp, piercing forms would vary as a function of the perceived context in which they appear, with a hazardous affordance relevant for only three-dimensional shapes in specific configurations or microhabitats. Any efficacy in visualizing sharp, piercing forms that influences their representation in artwork probably reflects a generalized vigilance towards shapes that can inflict injury. An increase in sympathetic nervous system arousal following perception of pointed forms would not explain the immediate freezing or sudden evasive action that occurs when spiny plants are encountered unexpectedly. This slower, more diffused state of arousal, however, might enhance the perceived awe and aesthetic appreciation of objects with strong cultural significance as apparent for the aforementioned sculptures and buildings. As graphical images, pointed, serrated and zigzagging contours would likely attract attention repeatedly, thus delaying visual habituation useful for fostering the power and strength of iconic symbolism or individuals wearing apparel and accessories displaying these patterns.

Several of the provocative patterns and textures selected as exemplars have glossy surfaces, a property that appears to be universally valued because of their wet connotation. The glittery quality of the scales of moving snakes might explain their positive association with undulating streams of flowing water by preliterate societies; albeit, snake-scale glossiness also increases the dreadful appearance of snakes to phobic individuals. Nevertheless, the glossy and sparkling luminance of still and rippling water conveys information about the availability of this vital resource, which is apparently recognized early in development by infants and toddlers that selectively suck and mouth glossy surfaces. Perhaps one of the oldest forms of aesthetic appreciation is manifested by the hand production of glistening flakes and facets of flint and obsidian tools followed by the polishing of artifacts to achieve a lustrous surface finish. Modern computer software and manufacturing methods yield the ubiquitous presence of glossy and sparkling visual effects in electronic media and on the surface finishes of apparel and consumer products.

As suggested by Orians (1998), humans might have an innate predisposition for extracting information about habitat quality using various visual cues that include the appearance of trees adapted to seasonal variation in moisture. The cross-cultural appeal of trees with wide crowns (Sommer 1997) might be relevant to this issue because trees with wider crowns afford better shade to prevent dehydration than trees with narrower crowns. Preschool children were especially cognizant of this attribute as evinced by their predominant choice of an Umbrella Thorn with a wide thick crown to stay under to keep cool on a hot day (Coss and Moore 1994, 2002). Yet, their choice of this tree for shade was not associated with their choice of this tree for refuge from a lion or as the prettiest tree (Coss and Moore 2002). Preschool children also differed markedly from adults (cf. Summit and Sommer 1999) in their choice of a pine tree with tall, conical crown as the prettiest tree (see Fig. 2A), a finding that does not support the idea that trees with wide crowns are innately appealing for purely aesthetic reasons.

The aesthetic appeal or perceived beauty of objects and microhabitats can be considered a higher-order abstraction or emergent property that results from complex assessment by perceptual, cognitive and emotional processes at lower levels of organization. Cultural experiences and evolved predispositions are manifested at these lower levels of organization, the interaction of which is little understood. Because of the entanglement of these effects, future study of evolved perceptual biases in art and design should consider the developmental aspects of creativity as a heuristic component of the quest to discover fundamental principles in artistic expression.

Summary

Since Darwin's initial development of the theory of evolution by natural selection, anthropologists, biologists, and psychologists have been inspired to consider this construct as a possible explanation for many facets of human behavior, including creative expression in art and design. Consistent with this view, this chapter focuses on the visual pattern-recognition abilities shaped by natural selection that might explain cross-cultural similarities of certain aesthetic elements selected by artists and designers. During the conception phase of creative decision making, pre-attentive or unconscious perceptual processes configured to detect and recognize ecologically important shapes and textures are likely to influence artworks by affecting the readiness of some visual images to emerge into consciousness. More overt biases occur during the production phase in which subtle and even unintended visual images are recognized and modified to augment their prominence. As a consequence, some patterns are represented in artwork or designs at frequencies well above the likelihood of encountering them in natural settings. Choice of patterns for experimental study of their aesthetic appeal and provocative properties was guided by assessment of the probable sources of natural selection engendering pattern recognition. Research on infants and toddlers indicated that sensitivity to some patterns emerged early in development and might be the childhood precursors of adult aesthetic preferences; other patterns might acquire their significance via evolutionarily prepared associative learning. Aesthetic appreciation of trees with wide canopies and landscapes with water appears to characterize the assessment of safe habitats that afford refuge from predators, shade, and drinking water. Recognition cues for water are seemingly reflected by the glossy and glistening surface finishes of some of the earliest polished tools of human ancestors and they are ubiquitously displayed by modern artwork, shining cloth, and consumer products. Other patterns engender awe or caution and yet retain their aesthetic appeal. Examples include eye-like schemata and sharp, piercing shapes displayed by contemporary sculpture and graphic design. Adult fascination with the coat patterns and skins of historical predators, such as leopard spots and python scales, is evident in their application as animal prints on apparel. Infants and toddlers prod these textures with their forefingers in an exploratory fashion that might be the precursor of adult recognition of the potency of these patterns for social display.

References

Adolphs R, Tranel D, Damasio AR (1998) The human amygdala in social judgment. Nature 393:470–474
Aikens NE (1998a) The biological origins of art. Praeger, Westport, CT
Aikens NE (1998b) Human cardiovascular response to the eye spot threat stimulus. Evol Cognition 4:1–12
Alderton D (1991) Crocodiles and alligators of the world: a synopsis. Blandford, London
An X, Bandler R, Ongur D, Price JL (1998) Prefrontal cortical projections to longitudinal columns in the midbrain periaqueductal gray in macaque monkeys. J Comp Neurol 401:455–479
Appleton J (1984) Prospects and refuges revisited. Landscape J 3:91–103
Appleton J (1996) The experience of landscape, revised ed. Wiley, New York
Argyle M, Cook M (1976) Gaze and mutual gaze. Cambridge University Press, Cambridge
Aronoff J, Barclay AM, Stevenson LA (1988) The recognition of threatening facial stimuli. J Pers Soc Psychol 54:647–655
Asfaw B, White T, Lovejoy O, Latimer B, Simpson S, Suwa G (1999) *Australopithecus garhi*: a new species of early hominid from Ethiopia. Science 284:629–635
Balling JD, Falk JH (1982) Development of visual preferences for natural environments. Environ Behav 14:5–28
Bandura A, Blanchard EB, Ritter B (1969) Relative efficacy of desensitization and modeling approaches for inducing behavioral, affective, and attitudinal changes. J Pers Soc Psychol 13:173–199
Barry JC (1987) Large carnivores (Canidae, Hyaenidae, Felidae) from Laetoli. In: Leakey MD, Harris JM (eds) Laetoli a Pliocene site in northern Tanzania. Clarendon Press, Oxford, pp 235–258
Bennett-Levy J, Marteau T (1984) Fear of animals: what is prepared? Br J Psychol 75:37–42
Berggren U (1992) General and specific fears in referred and self-referred adult patients with extreme dental anxiety. Behav Res Therapy 30:395–401
Bernáldez F, Gallardo D, Abelló RP (1987) Children's landscape preferences: from rejection to attraction. J Environ Psychol 7:169–176
Bilotta J, Lindauer MS (1980) Artistic and nonartistic backgrounds as determinants of the cognitive response to the arts. Bull Psychonomic Soc 15:354–356
Black JE, Greenough WT (1986) Developmental approaches to the memory process. In: Martinez JL Jr, Kesner RP (eds) Learning and memory, a biological view. Academic Press, Orlando, pp 55–81
Blair RJ, Morris JS, Frith CD, Perrett DI, Dolan RJ (1999) Dissociable neural responses to facial expressions of sadness and anger. Brain 122:883–893
Blake J, O'Rourke P, Borzellino G (1994) Form and function in the development of pointing and reaching gestures. Infant Behav Dev 17:195–203
Blumstein DT, Daniel JC, Griffin AS, Evans CS (2000) Insular tammar wallabies (*Macropus eugenii*) respond to visual but not acoustic cues from predators. Behav Ecol 11:528–535
Bradsher K (2000) Was Freud a minivan or s.u.v. kind of guy? New York Times, July 17
Brain CK (1970) New finds at the Swartkrans australopithecine site. Nature 225:1112–1119
Brain CK (1981) The hunters or the hunted? University of Chicago Press, Chicago
Brandon JG, Coss RG (1982) Rapid dendritic spine stem shortening during one-trial learning: the honeybee's first orientation flight. Brain Res 252:51–61
Brown CF, Kulik J (1977) Flashbulb memories. Cognition 5:73–99

Brown F, Harris J, Leakey R, Walker A (1985) Early *Homo erectus* skeleton from west Lake Turkana, Kenya. Nature 316:788–792

Bruce C, Desimone R, Gross CG (1981) Visual properties of neurons in a polysensory area in superior temporal sulcus of the macaque. J Neurophysiol 46:369–384

Buhyoff GJ, Wellman JD, Daniel TC (1982) Predicting scenic quality for mountain pine beetle and western spruce budworm damaged forest vistas. For Sci 28:827–838

Buschbeck EK, Strausfeld NJ (1997) The relevance of neural architecture to visual performance: phylogenetic conservation and variation in Dipteran visual systems. J Comp Neurol 383:282–304

Clearwater YA, Coss RG (1991) Functional aesthetics to enhance well-being in isolated and confined settings. In: Harrison AA, Clearwater YA, McKay C (eds) Human experience in Antarctica: applications to life in space. Springer, Berlin Heidelberg New York, pp 331–348

Clévenot D (2000) Splendors of Islam – architecture, decoration and design. Vendome Press, New York

Cocude M, Mellet E, Denis M (1999) Visual and mental exploration of visuo-spatial configurations: behavioral and neuroimaging approaches. Psychol Res 62:93–106

Cook M, Mineka S (1989) Observational conditioning of fear to fear-relevant versus fear-irrelevant stimuli in rhesus monkeys. J Abnormal Psychol 98:448–459

Cook WJ (1997) GM is getting looks. US News World Rep 122:48–53

Corbett JE (1948) The man-eating leopard of Rudraprayag. Oxford University Press, New York

Cosmides L, Tooby J (1987) From evolution to behavior: evolutionary psychology as the missing link. In: Dupré J (ed) The latest on the best: essays on evolution and optimality. The MIT Press, Cambridge, MA, pp 277–306

Coss RG (1965) Mood provoking visual stimuli: their origins and applications. University of California Press, Los Angeles

Coss RG (1967) De nouveaux concepts esthétiques et le comportement du consommateur. Design Industrie No 84–85, July

Coss RG (1968) The ethological command in art. Leonardo 1:273–287

Coss RG (1969) Electro-oculography: drawing with the eye. Leonardo 2:399–401

Coss RG (1970) The perceptual aspects of eye-spot patterns and their relevance to gaze behaviour. In: Hutt C, Hutt SJ (eds) Behaviour studies in psychiatry. Pergamon Press, London, pp 121–147

Coss RG (1977) Constraints on innovation: the role of pattern recognition in the graphic arts. Proc 8th Western Symposium on Learning: Creative Thinking, pp 24–44

Coss RG (1978) Perceptual determinants of gaze aversion by the lesser mouse lemur (*Microcebus murinus*), the role of two facing eyes. Behaviour 64:248–270

Coss RG (1979) Perceptual determinants of gaze aversion by normal and psychotic children: the role of two facing eyes. Behaviour 69:228–254

Coss RG (1981) Reflections on the evil eye. In: Dundes A (ed) The evil eye. The University of Wisconsin Press, Madison, WI, pp 181–191

Coss RG (1985) Evolutionary restraints on learning: phylogenetic and synaptic interpretations. In: Weinberger NM, McGaugh JL, Lynch G (eds) Memory systems of the brain: animal and human cognitive processes. Guilford Publications, New York, pp 253–278

Coss RG (1991a) Context and animal behavior III: the relationship between early development and evolutionary persistence of ground squirrel antisnake behavior. Ecol Psychol 3:277–315

Coss RG (1991b) Evolutionary persistence of memory-like processes. Concepts Neurosci 2:29–168

Coss RG (1993). Evolutionary persistence of ground squirrel antisnake behavior: reflections on Burton's commentary. Ecol Psychol 5:171–194

Coss RG (1999) Effects of relaxed natural selection on the evolution of behavior. In: Foster SA, Endler JA (eds) Geographic variation in behavior: perspectives on evolutionary mechanisms. Oxford University Press, New York, pp 180–208

Coss RG, Biardi JE (1997) Individual variation in the antisnake defenses of California ground squirrels (*Spermophilus beecheyi*). J Mammal 78:294–310

Coss RG, Goldthwaite RO (1995) The persistence of old designs for perception. In: Thompson NS (ed) Perspectives in ethology 11: behavioral design. Plenum Press, New York, pp 83–148

Coss RG, Moore M (1990) All that glistens: water connotations in surface finishes. Ecol Psychol 2:367–380

Coss RG, Moore M (1994) Preschool children recognize the utility of differently shaped trees: a cross-cultural evaluation of aesthetics and risk perception. In: Francis M, Lindsey P, Rice JS (eds) The healing dimensions of people-plant relations: proceedings of a research symposium. Center for Design Research, University of California, Davis, California, pp 407–423

Coss RG, Moore M (2002) Precocious knowledge of trees as antipredator refuge in preschool children: An examination of aesthetics, attributive judgments, and relic sexual dinichism. Ecol Psychol 14:181–222

Coss RG, Owings DH (1985) Restraints on ground squirrel antipredator behavior: adjustments over multiple time scales. In: Johnston TD, Pietrewicz AT (eds) Issues in the ecological study of learning. Lawrence Erlbaum Associates, Hillsdale, NJ, pp 167–200

Coss RG, Perkel DH (1985) The function of dendritic spines: a review of theoretical issues. Behav Neural Biol 44:151–185

Coss RG, Ramakrishnan U (2000) Perceptual aspects of leopard recognition by wild bonnet macaques (*Macaca radiata*). Behaviour 137:315–335

Coss RG, Schowengerdt BT (1998) Evolution of the modern human face: aesthetic and attributive judgments of a female profile warped along a continuum of paedomorphic to late archaic craniofacial structure. Ecol Psychol 10:1–24

Coss RG, Towers SR (1990) Provocative aspects of pictures of animals in confined settings. Anthrozoös 3:162–170

Coss RG, Clearwater YA, Barbour CG, Towers SR (1989) Functional decor in the international space station: body orientation cues and picture perception. NASA Tech Memorandum 102242

Coss RG, Marks S, Ramakrishnan U (2002) Early environment shapes the development of gaze aversion by wild bonnet macaques (*Macaca radiata*). Primates 43 (3):217–222

Crawford NA (1934) Cats holy and profane. Psychoanal Rev 21:168–179

Curio E (1975) The functional organization of anti-predator behaviour in the pied flycatcher: a study of avian visual perception. Anim Behav 23:1–115

Curio E (1993) Proximate and developmental aspects of antipredator behavior. Adv Stud Behav 22:135–238

Dailey ME, Smith SJ (1996) The dynamics of dendritic structure in developing hippocampal slices. J Neurosci 16:2983–2994

Darwin C (1885) The descent of man and selection in relation to sex, 2nd edn. John Murray, London

Davey GCL (1995) Preparedness and phobias: specific evolved associations or a generalized expectancy bias? Behav Brain Sci 18:289–325
de Heinzelin J, Clark JD, White T, Hart W, Renne P, Wolde Gabriel G, Beyene Y, Vrba E (1999) Environment and behavior of 2.5-million-year-old Bouri hominids. Science 284:625–629
Dennell R (1986) Needles and spear-throwers. Nat Hist 95:70–78
Desimone R (1991) Face-selective cells in the temporal cortex of monkeys. J Cognitive Neurosci 3:1–8
Desmarais JH (1998) The Beardsley industry. Ashgate Publishing, Brookfield, Vermont
Diaz-Bolio J (1987) The geometry of the Maya and their rattlesnake art. Area Mayan-Mayan Area, Merida, Yucatan, Mexico
Dissanayake E (1974) A hypothesis of the evolution of art from play. Leonardo 7:211–217
Dissanayake E (1979) An ethological view of ritual and art in human evolutionary history. Leonardo 12:27–31
Dissanayake E (1992) Homo aestheticus: where art comes from and why. Free Press, New York
Ditmars RL (1931) Snakes of the world. Macmillan, New York
Dundes A (1981) Wet and dry, the evil eye: an essay in Indo-European and Semitic worldview. In: Dundes A (ed) The evil eye. University of Wisconsin Press, Madison, WI, pp 257–298
Easterbrook MA, Kisilevsky BS, Hains SMJ, Muir DW (1999) Faceness or complexity: evidence from newborn visual tracking of facelike stimuli. Infant Behav Dev 22:17–35
Eibl-Eibesfeldt I (1979) Human ethology: concepts and implications for the sciences of man. Behav Brain Sci 1:1–26
Eibl-Eibesfeldt I (1989) Human ethology. Aldine de Gruyter, New York
Eimer M (2000) Effects of face inversion on the structural encoding and recognition of faces: evidence from event-related brain potentials. Cognitive Brain Res 10:145–158
Elston GN, Rosa MG (1998) Morphological variation of layer III pyramidal neurones in the occipitotemporal pathway of the macaque monkey visual cortex. Cerebral Cortex 8:278–294
Emery NJ (2000) The eyes have it: the neuroethology, function and evolution of social gaze. Neurosci Biobehav Rev 24:581–604
Engert F, Bonhoeffer T (1999) Dendritic spine changes associated with hippocampal long-term synaptic plasticity. Nature 399:66–70
Evans CS, Wenderoth P, Cheng K (2000) Detection of bilateral symmetry in complex biological images. Perception 29:31–42
Fischer M, Kaech S, Knutti D, Matus A (1998) Rapid actin-based plasticity in dendritic spines. Neuron 20:847–854
Fitzhugh L (2001) Update to mountain lion attack information. Department of Wildlife, Fish, and Conservation Biology, University of California, Davis
Fox E, Lester V, Russo R, Bowles RJ, Pichler A, Dutton K (2000) Facial expressions of emotions: are angry faces detected more efficiently? Cognition Emotion 14:61–92
Fridlund AJ (1994) Human facial expression: an evolutionary view. Academic Press, San Diego
Fujita I, Fujita T (1996) Intrinsic connections in the macaque inferior temporal cortex. J Comp Neurol 368:467–486
Fujita K (1993a) Development of visual preference for closely related species by infant and juvenile macaques with restricted social experience. Primates 34:141–150

Fujita K (1993b) Role of some physical characteristics in species recognition by pigtailed macaques. Primates 34:133–140

Gale A, Lucas B, Nissim R, Harpham B (1972). Some EEG correlates of face-to-face contact. Br J Soc Clin Psychol 11:326–332

Gale A, Spratt G, Chapman AJ, Smallbone A (1975). EEG correlates of eye contact and interpersonal distance. Biol Psychol 3:237–245

Gibson JJ (1979) The ecological approach to visual perception. Houghton Mifflin, Boston

Glenberg AM, Schroeder JL, Robertson DA (1998) Averting the gaze disengages the environment and facilitates remembering. Mem Cognition 26:651–658

Goepel J (1995) Driving impressions of the Geo Metro Lsi, Chevrolet Cavalier LS, Chrysler Cirrus LX, Ford Explorer. Motorland 116:20–63

Gold PE, McCarty RC (1995) Stress regulation of memory processes: role of peripheral catecholamines and glucose. In: Friedman MJ, Charney DS, Deutch AY (eds) Neurobiological and clinical consequences of stress: from normal adaptation to post-traumatic stress disorder. Lippincott Raven, Philadelphia, pp 151–162

Goren CC, Sarty M, Wu PYK (1975) Visual following and pattern discrimination of face-like stimuli by newborn infants. Pediatrics 56:544–549

Greene HW (1997) Snakes: the evolution of mystery in nature. University of California Press, Berkeley

Grosof DH, Shapley RM, Hawken MJ (1993) Macaque V1 neurons can signal 'illusory' contours. Nature 365:550–552

Guthrie RD, Petocz RG (1970) Weapon automimicry among mammals. Am Nat 104:585–588

Haile-Selassie Y (2001) Late Miocene hominids from the Middle Awash, Ethiopia. Nature 412:178–181

Hall GS (1897) A study of fears. Am J Psychol 8:147–249

Hamilton WJ III, Bulger JB (1990) Natal male baboon rank rises and successful challenges to resident alpha males. Behav Ecol Sociobiol 26:357–362

Haselrud GM (1938) The effect of movement of stimulus objects upon avoidance reactions in chimpanzees. J Comp Psychol 25:507–528

Hendey QB (1974) The late Cenozoic carnivora of the south-western cape province. Ann S Afr Mus 63:1–369

Henshilwood CS, d'Errico F, Marean CW, Milo RG, Yates R (2001a) An early bone tool industry from the Middle Stone Age at Blombos Cave, South Africa: implications for the origins of modern human behaviour, symbolism and language. J Human Evol 41:631–678

Henshilwood CS, Sealy JC, Yates R, Cruz-Uribe K, Goldberg P, Grine FE, Klein RG, Poggenpoel C, van Niekerk, K, Watts I (2001b) Blombos cave, Southern Cape, South Africa: preliminary report on the1992–1999 excavations of the Middle Stone Age levels. J Archaeol Sci 28:421–448

Henshilwood CS, d'Errico F, Yates R, Jacobs Z, Tribolo C, Duller GAT, Mercier N, Sealy JC, Valladas H, Watts I, Wintle AG (2002) Emergence of modern human behavior: Middle Stone Age engravings from South Africa. Science 295:1278–1280

Herzog TR (1985) A cognitive analysis of preference for waterscapes. J Environ Psychol 5:225–241

Hess EH (1975) The tell-tale eye. Van Nostrand Reinhold, New York

Hiss T (1991) The experience of place. Vintage Books, New York

Hooff Jaram van (1967) The facial displays of the catarrhine monkeys and apes. In: Morris D (ed) Primate ethology. Weidenfeld and Nicolson, London, pp 7–68

Howell FC, Petter G (1975) Carnivora from Omo Group Formations, southern Ethiopia. In: Coppens Y, Howell FC, Isaac GL, Leakey REF (eds) Earliest man and environments in the Lake Rudolf Basin. University of Chicago Press, Chicago, pp 314–360

Hull BR IV, Buhyoff GJ (1983) Distance and scenic beauty, a nonmonotonic relationship. Environ Behav 15:77–91

Hunt KD (1992) Social rank and body size as determinants of positional behavior in *Pan troglodytes*. Primates 33:347–357

Hunt RM (1996) Basicranial anatomy of the giant viverrid from 'E' Quarry, Langebaanweg, South Africa. In: Stewart KM, Seymour KL (eds) Paleoecology and paleoenvironments of late Cenozoic mammals. University of Toronto Press, Toronto, pp 588–597

Jacobs BF, Winkler DA (1992) Taphonomy of a Middle Miocene autochthonous forest assemblage, Ngorora Formation, central Kenya. Palaeogeogr Palaeoclimatol Palaeoecol 99: 31–40

Jacobs BF, Deino AL (1996) Test of climate-leaf physiognomy regression models, their application to two Miocene floras from Kenya, and 40Ar/39Ar dating of the Late Miocene Kapturo site. Palaeogeogr Palaeoclimatol Palaeoecol 123:259–271

Jersild AT, Holmes FB (1935) Children's fears. Bureau of Publications, Teachers College, Columbia University, New York

Jirari CG (1971) Form perception, innate form preferences and visually mediated head-turning in the human neonate. PhD Dissertation, University of Chicago. University Microfilms Order No AAC T-22258

Johnson M (1993) A cognitive model for the perception and translation of a three-dimensional object/array onto a two-dimensional surface. Visual Arts Res 19:85–99

Johnson MH, Morton J (1991) Biology and cognitive development: the case of face recognition. Blackwell, Cambridge

Johnson MH, Dziurawiec S, Ellis H, Morton J (1991) Newborns' preferential tracking of face-like stimuli and its subsequent decline. Cognition 40:1–19

Jones HE, Jones MC (1928) A study of fear. Childhood Educ 5:136–143

Jones LC (1981) The evil eye among European-Americans. In: Dundes A (ed) The evil eye. The University of Wisconsin Press, Madison, WI, pp 150–168

Jung CG (1916) Psychology of the unconscious. Moffat Yard, New York

Jung CG (1972) Man and his symbols. Doubleday, New York

Kappas A, Hess U, Barr, CL, Kleck RE (1994) Angle of regard: the effect of vertical viewing angle on the perception of facial expressions. J Nonverbal Behav 18:263–280

Kastner S, De Weerd P, Ungerleider LG (2000) Texture segregation in the human visual cortex: a functional MRI study. J Neurophysiol 83:2453–2457

Kawashima R, Sugiura M, Kato T, Nakamura A, Hatano K, Ito K, Fukuda H, Kojima S, Nakamura K (1999) The human amygdala plays an important role in gaze monitoring. A PET study. Brain 122:779–783

Kellogg R, Knoll M, Kugles J (1965) Form-similarity between phosphenes of adults and pre-school children's scribblings. Nature 208:1129–1130

Kendler KS, Neale MC, Kessler RC, Heath AC, Eaves LJ (1992) The genetic epidemiology of phobias in women. The interrelationship of agoraphobia, social phobia, situational phobia, and simple phobia. Arch Gen Psychiatry 49:273–281

Kendler KS, Karkowski LM, Prescott CA (1999) Fears and phobias: reliability and heritability. Psychol Med 29:539–553

Kendrick KM, Baldwin BA (1987) Cells in temporal cortex of conscious sheep can respond preferentially to the sight of faces. Science 236:448–450

Kesler/West ML, Andersen AH, Smith CD, Avison MJ, Davis CE, Kryscio RJ, Blonder LX (2001) Neural substrates of facial emotion processing using fMRI. Cognitive Brain Res 11:213–226

Kibunjia M (1994) Pliocene archaeological occurrences in the Lake Turkana basin. J Human Evol 27:159–171

Kobatake E, Tanaka K (1994) Neuronal selectivities to complex object features in the ventral visual pathway of the macaque cerebral cortex. J Neurophysiol 71:856–867

Koenderink J (2000) Trieste in the mirror. Perception 29:127–133

Kortlandt A (1955) Aspects and prospects of the concept of instinct. Archives néerlandaises de zoologie XI. EJ Brill, Leiden, pp 1–284

Kortlandt A (1967) Experimentation with chimpanzees in the wild. In: Starck D, Schneider R, Kuhn H-J (eds) Neue Ergebnisse der Primatologie – progress in primatology. Gustav Fischer Verlag, Stuttgart, pp 208–224

Kortlandt A (1980) How might early hominids have defended themselves against large predators and food competitors? J Human Evol 9:79–112

Kortlandt A, Kooij M (1963) Protohominid behaviour in primates (preliminary communication). Symp Zool Soc Lond 10:61–88

Kosslyn SM, Thompson WL, Kim IJ, Alpert NM (1995) Topographical representations of mental images in primary visual cortex. Nature 378:496–498

Kovacs G, Vogels R, Orban GA (1995a) Selectivity of macaque inferior temporal neurons for partially occluded shapes. J Neurosci 15:1984–1997

Kovacs G, Vogels R, Orban GA (1995b) Cortical correlate of pattern backward masking. Proc Natl Acad Sci USA 92:5587–5591

Kruizinga P, Petkov N (2000) Computational model of dot-pattern selective cells. Biol Cybernetics 83:313–325

Lang PJ, Levin DN, Miller GA, Kozak MJ (1983) Fear behavior, fear imagery, and the psychophysiology of emotion: the problem of affective response integration. J Abnormal Psychol 92:276–306

Lazarus RS (1981) A cognitivist's reply to Zajonc on emotion and cognition. Am Psychol 36:222–223

Leakey MG, Feibel CS, McDougall I, Walker A (1995) New four-million-year-old hominid species from Kanapoi and Allia Bay, Kenya. Nature 376:565–571

Leakey MG, Feibel CS, Bernor RL, Harris JM, Cerling TE, Stewart KM, Storrs GW, Walker A, Werdelin L, Winkler AJ (1996) Lothagam: a record of faunal change in the Late Miocene of East Africa. J Vertebrate Paleontol 16:556–570

Leakey MG, Feibel CS, McDougall I, Ward C, Walker A (1998) New specimens and confirmation of an early age for *Australopithecus anamensis*. Nature 393:62–66

Leakey MG, Spoor F, Brown FH, Gathogo PN, Klarle C, Leakey LN, McDougall I (2001) New hominin genus from eastern Africa shows diverse Middle Pliocene lineages. Nature 410:433–440

Lee-Thorp J, Thackeray JF, van der Merwe N (2000) The hunters and the hunted revisited. J Human Evol 39:565–576

Levitt JB, Lewis DA, Yoshioka T, Lund JS (1993) Topography of pyramidal neuron intrinsic connections in macaque monkey prefrontal cortex (areas 9 and 46). J Comp Neurol 338:360–376

Lewis-Williams JD, Dowson TA (1993) On vision and power in the Neolithic: evidence from the decorated monuments. Curr Anthropol 34:55–65

Lichtenstein P, Annas P (2000) Heritability and prevalence of specific fears and phobias in childhood. J Child Psychol Psychiatry Allied Disciplines 41:927–937

Logothetis NK, Pauls J, Augath M, Trinath T, Oeltermann A (2001) Neurophysiological investigation of the basis of the fMRI signal. Nature 412:150–157
Lorenz KZ (1950) The comparative method in studying innate behaviour patterns. Symp Soc Exp Biol 4:221–268
Lyons E (1983) Demographic correlates of landscape preference. Environ Behav 15:487–511
Maletic-Savatic M, Malinow R, Svoboda K (1999) Rapid dendritic morphogenesis in CA1 hippocampal dendrites induced by synaptic activity. Science 283:1923–1927
Marks IM, Gelder MG (1966) Different ages of onset in varieties of phobia. Am J Psychiatry 123:218–221
Marshack A (1972) The roots of civilization. McGraw-Hill, New York
Masataka N (1993) Effects of experience with live insects on the development of fear of snakes in squirrel monkeys, *Saimiri sciureus*. Anim Behav 46:741–746
Masello R (1986) Deathscapes. Omni 8:80–85
Matus A (2000) Actin-based plasticity in dendritic spines. Science 290:754–758
Mayr E (1963) Animal species and evolution. Harvard University Press, Cambridge, MA
McBride G, King MG, James JW (1965) Social proximity effects on galvanic skin responses in adult humans. J Psychol 61:153–157
McHenry HM (1992a) Body size and proportions in early hominids. Am J Phys Anthropol 87:407–431
McHenry HM (1992b). How big were early hominids? Evol Anthropol 1:15–19
McPherson EG, Rowntree RA (1993) Energy conservation potential of urban tree planting. J Arboricult 19:321–331
McWhinnie HJ (1968) A review of research on aesthetic measure. Acta Psychol 28:363–375
Mellaart J (1967) Çatal Hüyük: a Neolithic town in Anatolia. Thames and Hudson, London
Meltzoff AN, Moore MK (1977) Imitation of facial and manual gestures by human neonates. Science 198:75–78
Miller EK, Erickson CA, Desimone R (1996) Neural mechanisms of visual working memory in prefrontal cortex of the macaque. J Neurosci 16:5154–5167
Miller H (1970) The cobra, India's "good snake". Natl Geogr 138:393–408
Mitchell G (1970) Abnormal behavior in primates. In: Rosenblum LA (ed) Primate behavior: development in field and laboratory research. Academic Press, New York, pp 195–249
Morris JS, Friston KJ, Buchel C, Frith CD, Young AW, Calder AJ, Dolan RJ (1998a) A neuromodulatory role for the human amygdala in processing emotional facial expressions. Brain 121:47–57
Morris JS, Öhman A, Dolan RJ (1998b) Conscious and unconscious emotional learning in the human amygdala. Nature 393:467–470
Murphy CM, Bootzin RR (1973) Active and passive participation in the contact desensitization of snake fear in children. Behav Therapy 4:203–211
Nelson CA (2001) The development and neural bases of face recognition. Infant Child Dev 10:3–18
Newman RW (1970) Why man is such a sweaty and thirsty naked animal: a speculative review. Human Biol 42:12–27
Nhachi CF, Kasilo OM (1994) Snake poisoning in rural Zimbabwe–a prospective study. J Appl Toxicol 14:191–193
Nichols KA, Champness BG (1971) Eye gaze and the GSR. J Exp Soc Psychol 7:623–626
Nietzsche F (1909) Human all-too-human. TN Foulin, Edinburgh
Noronha RP (1992) Animals and other animals. Sanchar Publishing House, Delhi
Noton D, Stark L (1971) Eye movements and visual perception. Sci Am 224:35–43

O'Craven KM, Kanwisher N (2000) Mental imagery of faces and places activates corresponding stimulus-specific brain regions. J Cognitive Neurosci 12:1013–1023

O'Craven KM, Downing PE, Kanwisher N (1999) fMRI evidence for objects as the units of attentional selection. Nature 401:584–587

Öhman A (1986) Face the beast and fear the face: animal and social fears as prototypes for evolutionary analyses of emotion. Psychophysiology 23:123–145

Öhman A (1994) The psychophysiology of emotion: evolutionary and non-conscious origins. In: d'Ydewalle G, Eelen P, Bertelson P (eds) International perspectives on psychological science, volume 2: the state of the art. Lawrence Erlbaum, Hillsdale, NJ, pp 197–227

Öhman A, Soares JJF (1994) "Unconscious anxiety": phobic responses to masked stimuli. J Abnormal Psychol 103:231–240

Okusa T, Kakigi R, Osaka N (2000) Cortical activity related to cue-invariant shape perception in humans. Neuroscience 98:615–624

Orians GH (1980) Habitat selection. In: Lockard JS (ed) The evolution of human social behavior. Elsevier, New York, pp 49–66

Orians GH (1986). An ecological and evolutionary approach to landscape aesthetics. In: Penning-Rowsell EC, Lowenthal D (eds) Landscape meanings and values. Allen and Unwin, London, pp 3–22

Orians GH (1998) Human behavioral ecology: 140 years without Darwin is too long. Bull Ecol Soc Am 79:15–28

Orians GH, Heerwagen JH (1992) Evolved responses to landscapes. In: Barkow JH, Cosmides L, Tooby J (eds) The adapted mind. Oxford University Press, New York, pp 555–579

Ortolani A (1999) Spots, stripes, tail tips and dark eyes: predicting the function of carnivore colour patterns using the comparative method. Biol J Linn Soc 67:1–71

Packer C (1979) Male dominance and reproductive activity in *Papio anubis*. Anim Behav 27:37–45

Palombit RA (1992) A preliminary study of vocal communication in wild long-tailed macaques (*Macaca fascicularis*), I. Vocal repertoire and call emission. Int J Primatol 13:143–182

Papa M, Bundman MC, Greenberger V, Segal M (1995) Morphological analysis of dendritic spine development in primary cultures of hippocampal neurons. J Neurosci 15:1–11

Parks TE, Coss RG, Coss CS (1985) Thatcher and the Cheshire Cat: context and the processing of facial features. Perception 14:747–754

Patel SN, Stewart MG (1988) Changes in the number and structure of dendritic spines 25 hours after passive avoidance training in the domestic chick, *Gallus domesticus*. Brain Res 449:34–46

Paterson HEH (1993) Evolution and the recognition concept of species: collected writings. Johns Hopkins University Press, Baltimore

Patsfall MR, Feimer NR, Buhyoff GJ, Wellman JD (1984) The prediction of scenic beauty from landscape content and composition. J Environ Psychol 4:7–26

Patterson JH (1986) The man-eaters of Tsavo. St Martin's Press, New York

Perrett DI, Rolls ET, Caan W (1982) Visual neurones responsive to faces in the monkey temporal cortex. Exp Brain Res 47:329–342

Peters A, Kaiserman-Abramof IR (1970) The small pyramidal neuron of the rat cerebral cortex. The perikaryon, dendrites and spines. Am J Anatomy 127:321–356

Pickford M, Senut B (2001) The geological and faunal context of Late Miocene hominid remains from Lukeino, Kenya. Comptes rendus de l'Académie des Sciences, Série 2, sciences de la terre et des planètes 332:145–152

Pillemer DP (1984) Flashbulb memories of the assassination attempt on President Reagan. Cognition 16:63–80
Pooley AC, Hines TC, Shield J (1989) Attacks of humans. In: Ross CA (ed) Crocodiles and alligators. Facts on File Inc, New York, pp 172–187
Potts R, Behrensmeyer AK, Ditchfield P (1999) Paleolandscape variation and early Pleistocene hominid activities: Members 1 and 7, Olorgesailie Formation, Kenya. J Human Evol 37:747–788
Prechtl HFR (1949) Das Verhalten von Kleinkindern gegenüber Schlangen. Wiener Z Philos Psychol Pädagogik 2:68–70
Puce A, Smith A, Allison T (2000) ERPs evoked by viewing facial movement. Cognitive Neuropsychol 17:221–239
Quade J, Cerling TE, Bowman JR (1989) Development of Asian monsoon revealed by marked ecological shift during the latest Miocene in northern Pakistan. Nature 342:163–166
Richardson WB (1942) Reaction toward snakes as shown by the wood rat (*Neotoma albigula*). J Comp Psychol 34:1–10
Rivers VZ (1995) Shining cloth. Surface Design J 20:4–6
Rivers VZ (1999) The shining cloth: Dress and adornment that glitter. Thames and Hudson, New York
Robinson MD (1998) Running from William Jame's Bear: a review of preattentive mechanisms and their contribution to emotional experience. Cognition Emotion 12:667–696
Rollenhagen A, Bischof HJ (1994) Spine morphology of neurons in the avian forebrain is affected by rearing conditions. Behav Neural Biol 62:83–89
Romanes GJ (1886) Animal intelligence, 4th edn. K Paul, Trench and Company, London
Rusakov DA, Stewart MG, Sojka M, Richter-Levin G, Bliss TV (1995) Dendritic spines form 'collars' in hippocampal granule cells. Neuroreport 6:1557–1561
Sackett GP (1966) Monkeys reared in isolation with pictures as visual input: evidence for an innate releasing mechanism. Science 154:1468–1473
Schiller PH (1952) Innate constituents of complex responses in primates. Psychol Rev 59:177–191
Schlosser K (1952) Der Signalismus in der Kunst der Naturvölker. Kommissionsverlag Walter G Mühlau, Kiel
Schoeck H (1981) The evil eye: forms and dynamics of a universal superstition. In: Dundes A (ed) The evil eye. The University of Wisconsin Press, Madison, WI, pp 192–200
Schroeder HW (1986) Estimating park tree densities to maximize landscape esthetics. J Environ Manage 23:325–333
Seligman MEP (1970) On the generality of the laws of learning. Psychol Rev 77:406–418
Seligman MEP (1971) Phobias and preparedness. Behav Therapy 2:307–320
Semaw S, Renne P, Harris JWK, Feibel CS, Bernor RL, Fesseha N, Mowbray K (1997) 2.5-million-year-old stone tools from Gona, Ethiopia. Nature 385:333–336
Senut B, Pickford M, Gommery D, Mein P, Cheboi K, Coppens Y (2001) First hominid from the Miocene (Lukeino Formation, Kenya). Comptes rendus de l'Académie des Sciences, Série 2, sciences de la terre et des planètes 332:137–144
Shipman P, Walker A (1989) The costs of becoming a predator. J Human Evol 18:373–392
Simion F, Valenza E, Umilta C, Barba BD (1998) Preferential orienting to faces in newborns: a temporal-nasal asymmetry. J Exp Psychol Human Perception Performance 24:1399–1405
Simonton DK (1999) Origins of genius: Darwinian perspectives on creativity. Oxford University Press, New York

Singer R, Wymer J (1982) The Middle Stone Age at Klasies River Mouth in South Africa. University of Chicago Press, Chicago

Skre I, Onstad S, Torgersen S, Lygren S, Kringlen E (2000) The heritability of common phobic fear: a twin study of a clinical sample. J Anxiety Disorders 14:549–562

Smart FM, Halpain S (2000) Regulation of dendritic spine stability. Hippocampus 10:542–554

Smith SM (1977) Coral-snake pattern recognition and stimulus generalization by naïve great kiskadees (Aves: Tyrannidae). Nature 265:535–536

Sommer R (1997) Further cross-national studies of tree-form preference. Ecol Psychol 9:153–160

Sommer R, Summit J (1995) An exploratory study of preferred tree form. Environ Behav 27:540–557

Sommer R, Summit J (1996) Cross-national rankings of tree shape. Ecol Psychol 8:327–341

Sommer R, Guenther H, Cecchettini CL (1992) A user-based method for rating street trees. Landscape Res 17:100–107

Sommer R, Summit J, Clements A (1993) Slide ratings of street tree attributes: some methodological issues and answers. Landscape J 12:17–22

Spencer H (1888) The principles of psychology, 3rd edn. D Appleton and Company, New York

Spindler P (1959) Studien zur Vererbung von Verhaltensweisen 2. Verhalten gegenüber Schlangen. Anthropol Anz 23:187–218

Stern JT Jr, Susman RL (1983) The locomotor anatomy of *Australopithecus afarensis*. Am J Phys Anthropol 60:279–317

Sugase Y, Yamane S, Ueno S, Kawano K (1999). Global and fine information coded by single neurons in the temporal visual cortex. Nature 400:869–872

Summit J, Sommer R (1999) Further studies of preferred tree shapes. Environ Behav 31:550–576

Susman RL, Stern JT Jr, Jungers WL (1984) Arboreality and bipedality in the Hadar hominids. Folia Primatol 43:113–156

Sütterlin C (1989) Universals in apotropaic symbolism: a behavioral and comparative approach to some medieval sculptures. Leonardo 22:65–74

Tanaka K (1996) Representation of visual features of objects in the inferotemporal cortex. Neural Networks 9:1459–1475

Tanaka K, Saito H-A, Fukada Y, Moriya M (1991) Coding visual images of objects in the inferotemporal cortex of the macaque monkey. J Neurophysiol 66:170–189

Taylor MJ, Itier RJ, Allison T, Edmonds GE (2001) Direction of gaze effects on early face processing: eyes-only versus full faces. Cognitive Brain Res 10:333–340

Thaker HM, Kankel DR (1992) Mosaic analysis gives an estimate of the extent of genomic involvement in the development of the visual system in *Drosophila melanogaster*. Genetics 131:883–894

Thomas EM (1990) Reflections the old ways. The New Yorker, October 15, pp 78–109

Thompson P (1980) Margeret Thatcher: a new illusion. Perception 9:483–484

Tinbergen N (1951) The study of instinct. Clarendon Press, Oxford

Toni N, Buchs PA, Nikonenko I, Bron CR, Muller D (1999) LTP promotes formation of multiple spine synapses between a single axon terminal and a dendrite. Nature 402:421–425

Tooby J, De Vore I (1987) The reconstruction of hominid behavioral evolution through strategic modeling. In: Kinzey WG (ed) The evolution of human behavior: primate models. State University of New York Press, New York, pp 183–237

Topál J, Csányi V (1994) The effect of eye-like schema on shuttling activity of wild house mice (*Mus musculus domesticus*): context-dependent threatening aspects of the eyespot patterns. Anim Learning Behav 22:96–102

Tuan Y-F (1978) Children and the natural environment. In: Altman I, Wohlwill JF (eds) Human behavior and environment, vol 3. Plenum Press, New York, pp 5–32

Turner A (1990) The evolution of the guild of larger terrestrial carnivores during the Plio-Pleistocene in Africa. Geobios 23:349–368

Turner A (1997) The big cats and their fossil relatives. Columbia University Press, New York

Turner A, Wood B (1993) Taxonomic and geographic diversity in robust austrolopithecines and other African Plio-Pleistocene larger mammals. J Human Evol 24:147–168

Tyler CJ, Dunlop SA, Lund RD, Harman AM, Dann JF, Beazley LD, Lund JS (1998) Anatomical comparison of the macaque and marsupial visual cortex: common features that may reflect retention of essential cortical elements. J Comp Neurol 400:449–468

Uher J (1991a) Die Ästhetik von Zick-Zack und Welle, ethologische Aspekte der Wirkung linearer Muster. PhD Dissertation, University of Munich

Uher J (1991b) On zigzag designs: three levels of meaning. Curr Anthropol 32:437–439

Ulrich RS (1977) Visual landscape preference: a model and application. Man Environ Syst 7:279–293

Ulrich RS (1981) Nature versus urban scenes, some psychophysiological effects. Environ Behav 13:523–556

Ulrich RS (1983) Aesthetic and affective response to natural environment. In: Altman I, Wohlwill JF (eds) Human behavior and environment, vol 6. Plenum Press, New York, pp 85–125

Ulrich RS (1984) View through a window may influence recovery from surgery. Science 224:420–421

van den Hout MA, de Jong P, Kindt M (2000) Masked fear words produce increased SCRs: An anomaly for Öhman's theory of pre-attentive processing in anxiety. Psychophysiology 37:283–288

van Schaik CP, Mitrasetia T (1990) Changes in the behaviour of wild long-tailed macaques (*Macaca fascicularis*) after encounters with a model python. Folia Primatol 55:104–108

van Rossum D, Hanisch UK (1999) Cytoskeletal dynamics in dendritic spines: direct modulation by glutamate receptors? Trends Neurosci 22:290–295

Varia R (1986) Brancusi. Universe, New York

Verstijnen IM, van Leeuwen C, Goldschmidt G, Hamel R, Hennessey JM (1998) Creative discovery in imagery and perception: combining is relatively easy, restructuring takes a sketch. Acta Psychol 99:177–200

von der Heydt R, Peterhans E, Dursteler MR (1992) Periodic-pattern-selective cells in monkey visual cortex. J Neurosci 12:1416–1434

Wallhäusser E, Scheich H (1987) Auditory imprinting leads to differential 2-deoxyglucose uptake and dendritic spine loss in the chick rostral forebrain. Brain Res 428:29–44

Wellens AR (1987) Heart-rate changes in response to shifts in interpersonal gaze from liked and disliked others. Perceptual and Motor Skills 64:595–598

Werdelin L, Lewis ME (2000) Carnivora from the South Turkwel hominid site, northern Kenya. J Paleontol 74:1173–1180

Werdelin L, Olsson L (1997) How the leopard got its spots: a phylogenetic view of the evolution of felid coat patterns. Biol J Linn Soc 62:383–400

White TD, Suwa G, Asfaw B (1994) *Australopithecus ramidus*, a new species of early hominid from Aramis, Ethiopia. Nature 375:88

Wik G, Fredrikson M, Ericson K, Eriksson L, Stone-Elander S, Greitz T (1993) A functional cerebral response to frightening visual stimulation. Psychiatry Res Neuroimaging 50:15–24
Willats J (1987) Marr and pictures: an information-processing account of children's drawings. Arch Psychol 55:105–125
Willats J (1992) Seeing lumps, sticks, and slabs in silhouettes. Perception 21:481–496
Wilson HR, Wilkinson F, Lin L-M, Castillo M (2000) Perception of head orientation. Vision Res 40:459–472
Wolde Gabriel G, Haile-Selassie Y, Renne PR, Hart WK, Ambrose SH, Asfaw B, Heiken G, White T (2001) Geology and palaeontology of the Late Miocene Middle Awash valley, Afar rift, Ethiopia. Nature 412:175–178
Wolfe A, Sleeper B (1995) Wild cats of the world. Crown, New York
Wood B (1992) Origin and evolution of the genus *Homo*. Nature 355:783–790
Woodcock DM (1982) A functionalist approach to environmental preference. PhD Dissertation, The University of Michigan, University Microfilms International No: 8215108
Yang K-H, Blackwell KT (2000) Analogue pattern matching in a dendritic spine model based on phosphorylation of potassium channels. Network Computation in Neural Systems 11:281–297
Yeager CP (1991) Possible antipredator behavior associated with river crossings by proboscis monkeys (*Nasalis larvatus*). Am J Primatol 24:61–66
Yerkes RM, Yerkes AW (1936) Nature and conditions of avoidance (fear) response in chimpanzee. J Comp Psychol 21:53–66
Yuste R, Bonhoeffer T (2001) Morphological changes in dendritic spines associated with long-term synaptic plasticity. Annu Rev Neurosci 24:1071–1089
Zajonc RB (1980) Feeling and thinking: preferences need no inferences. Am Psychol 35:151–175
Zapfe H (1981) Ein Schädel von *Mesopithecus* mit Biß-Spuren (A skull of *Mesopithecus* with bite marks). Folia Primatol 35:248–258
Zube EH (1974) Cross-disciplinary and intermode agreement on the description of evaluation of landscape resources. Environ Behav 6:69–89
Zube EH, Pitt DG, Evans GW (1983). A lifespan developmental study of landscape assessment. J Environ Psychol 3:115–128

From Sign and Schema to Iconic Representation. Evolutionary Aesthetics of Pictorial Art

CHRISTA SÜTTERLIN

Introduction

The question of artificial structures being an issue of evolutionary thinking follows an indirect way of approach. What is biological in a political system or in an architectural environment? Are symbols natural structures? We tend to argue that all this has been *made* and not *evolved*. It seems indeed quite complicated to declare such highly complex phenomena as universals. However, if they are *made* and not *evolved* structures, they are made by natural beings, i.e., evolved humans (Salter 1995).

Signs of artistic performance are evident since the earliest days of human evolution. The fact that most of our ancestors spent so much time applying signs, signals and symbols in their environment – for communication or for decoration – at a point of time when they were surely concerned with much more urgent problems of food seeking and self defense, allows reconsideration of this phenomenon from an evolutionary perspective. Art must have served functions – in both proximate and in ultimate terms.

The aesthetics of physical or natural beauty being an issue of different contributions in this volume (see Cunningham and Shamblen, Grammer et al., and Thornhill and Gangestad, this Vol.), the question here is quite a different one: how does it look when humans start to create their own aesthetic ambient? Any probable answer of this question would have to deal with much more than art alone, but also with any human attempt to interfere with his environment in a way that is more symbolic than functional.

Towards a Platonic View of Aesthetics and Art

The world of the human imagination – as the term implies – seems to be a *visual* world made of images, and so are most human ideas about beauty, even when conveyed verbally. The earliest such ideas found in different writings of Pre-platonic philosophy refer to form, coloring, ornamentation and decoration of bodies or objects, and they go together with terms like "tall", "shining", "rich" and "ornamented" (Grassi 1980). Beauty was at the same time a kind of attribute visible to the beholder and something inherent to the object. A somehow more advanced and powerful topic defended by Pythagoras and his school as "Unity within Diversity" (*Einheit im Mannigfaltigen*), pointed to a discrete cohesion or correlation within the different parts of a given structure, artistic or real. Heraklit confirmed: "The invisible harmony is more powerful than the visible" (Capelle 1953). If this quality is potentially inherent to all sort of natural things, the artist finds his role and dignity in imitating nature (mimesis)!

Plato referred to many of these ideas, but was the first to insist on a synthesis of more general principles underlying the phenomenon, as already proposed by Pythagoras. What Plato calls "proportion" (harmony) and "symmetry", "order" and "regularity" refer to more structural properties of beautiful things and give evidence of a divine law – the Platonic "Idea" as a kind of ideal prototype or archetype – working within them.

However, there is a problem: as both "ideas" and archetypes do not belong primarily to the visible world, the question arises, how should an artist be able to perceive what is an *invisible harmony*? Seeing it is a first prerequisite for reproducing it. It would seem that only "copying" nature would bring about beautiful and ugly subjects at random, as distributed in our physical environment. Plato's answer was quite rigid: ideas – as invisible structures – are accessible only through intellectual insight.

Aristotle was probably the first to see the role of the artist in a critical and at the same time more respectable light. It is only the artist's eye that is able to perceive the divine laws of harmony underlying visible nature and to sort out the nonfitting parts. By this special gift, he is even able to achieve and optimize what nature left at random and deficient. He completes the incomplete. However, Aristotle did not precisely say, *how*.

This view of an early aesthetic seems to guide our opinion and attitude up to today. What most people associate with the topic aesthetics is something "highly selective" in the way we perceive or shape our environment. We think of exclusive taste, all kinds of designs and ornaments seen in our life, of nicely arranged furnishings, of decorated clothes and elaborate objects in our everyday world[1].

Hence, we think – at least partially – of art.

This connotation had not changed during my own education as an art historian, except that I learned that (and how) "selectivity" had been modeled through different cultures and the sequence of centuries. I went on with the naive presumption that art (and aesthetics) are *doing* something with everyday reality – something better of course – while our perception is concerned with constructing a rough and unrefined picture of its physical platitudes. My view of life could be represented as follows: somewhere down at the bottom is our everyday reality, condensed in the "blueprints"

---

[1] This partly coincides with Dissanayakes definition of "making" something "special" (1992).

of our perception. A bit above, there is "aesthetic perception", picking out different aspects of blueprint perception according to their inferences with appeal and pleasure, and at the top there is art as a product and a further distillation of aesthetic perception.

My own interest in perceptual psychology started when I learned that the question of aesthetics begins much earlier than in our elevated emotions and taste, much earlier than in art, and much further down at the bottom. In my post-doctoral years spent at the Institute of Medical Optics and Medical Psychology, University of Munich, I learned that already our "everyday-perception" is *doing* something with reality, and the way it functions is highly selective. *All* perception is biased. It is just how our sensory system works. Hence, the question arose: what links could be between what we call "natural perception" on the one side and "aesthetic perception" on the other?

Or is there no link at all, but graduation within one and the same function?

In other words: does our natural perception work in a similar way to aesthetic perception?

Sensory Aesthetics

If we rely on the semantics of the word "aesthetics" tracing back to its original Greek etymology – meaning "evaluative perception" and "sensory cognition" – it comes close to the meaning it revealed in Alexander Baumgarten's first textbook called *Aesthetica* in 1750–58, i.e., "cognitio inferior" or the intuitive and *sensory* pathway to cognition. In this case, we should be able as well to trace some attitudes called "aesthetic" back to some attitudes inherent to our perception in general.

How should we understand this fact?

Looking at a simple arrangement of lines as presented by Wolfgang Metzger in the early 1950s of the last century, we feel a curious constraint to organize these lines within our visual field: lines of a certain closeness are perceived as contours of a figure, whereas new figures arise if some of these lines are interconnected (Metzger 1953; Fig. 1). The same holds true for similarity of form or brightness: similar elements within a random assortment are comprised as a figure (Fig. 2). Sometimes similarity and closeness are competitive.

The fact that we highly appreciate *symmetry* and *order* in a given visual situation is due to their simplifying, i.e., complexity reducing effect. If we face a complex stimulus with no conceivable order, we feel uncomfortable. Max J. Kobbert (1986) invited ten art students for a design test with the task of creating ten aesthetically appealing versus ten aesthetically nonappealing figures. The results show an evident aesthetic preference for sim-

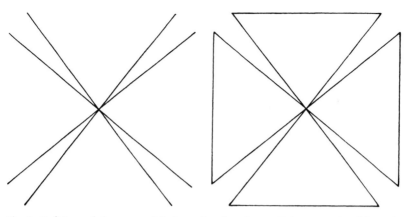

Fig. 1. *Left* Law of closeness; *right* law of enclosedness. (Schuster and Beisl 1978)

ple versus confused figures, whereby "simple" turns out to be symmetric and regular, while "confused" the opposite. The experiment confirmed earlier results found by Pierce (1894), Puffer (1903), Eysenck (1942) and Wertheimer (1945).

The same happened to be the case in an early experiment with long-tail and capuchin monkeys and carrion crows, done by Bernhard Rensch in the late 1950s. The animals were presented a series of paired plates with either regular or irregular patterns and their preference was observed when they picked one pattern up by hand or beak. All individuals of all species showed a significant preference for regular and not for irregular forms (Rensch 1958).

The reason for these effects may be much less an "aesthetic" taste of our perception than a clear perceptional bias of our aesthetics. The experiments of Rensch made evident that the preferred patterns were the ones

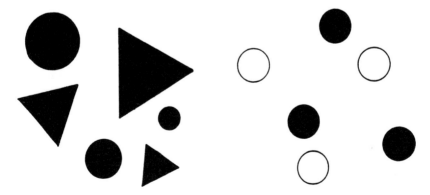

Fig. 2. *Left* similarity of form; *right* similarity of brightness. (Arnheim 1965)

that were easier to *memorize* in an independent test. Simplicity and regularity save expense in storing information since the capacity of memory for new information is limited. Attempts to quantify the amount of informational reduction realized by human perception are given by Franke (1994): from an input of about 10 million *bits* per second as retinal information, only 160 bits per second reach human conscience, while 1000–2000 bits stay in the short-term memory. In comparison: up to 100 million bits form in the long-term memory stock. The enormous reduction of incoming "new" and complex information into an already acquainted one that contains generalized features of all stored patterns seems to be an aim of sensory processing, but not the only one.

The way we assimilate new random information into more standardized stored models (or abstracts/templates) is showed by an early experiment of Rudolf Arnheim (1965). He projected simple geometrical figures with a slightly distorted symmetry and invited subjects to reproduce the figure after projection from their memory. The subjects produced two types of specific errors, but never the original figure (Fig. 3) Either they equalized the slight distortion into perfect symmetry, or they accentuated asymmetry. It's just how our perception works. If something doesn't fit into a familiar model, it needs distinct deviating features to catch our attention for "new" information. This categorization into known/unknown patterns seems to be a basic binary criterion for perceptual processing and predictive for artistic performance as well. Standardization (Vereinheitlichung) and/or accentuation (Prägnanz-Tendenz) are two main phenomena investigated by Gestalt psychology, being at the basis of aesthetic experience.

It seems, however that aesthetic satisfaction of perceiving patterns that *allow* for the formation of regular optical fields is higher compared to patterns *offering* regularity. This is explained by the experiments of Dörner

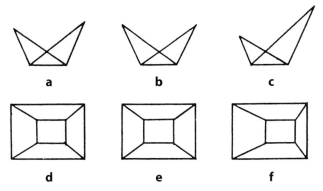

Fig. 3. Experiment by Arnheim (1965). **a** and **d** are the originally projected figures with a minimal distortion of their correct geometry. Subjects would either annul (**b** and **e**) or exaggerate (**a** and **f**) distortion during reproduction of the original figure by memory

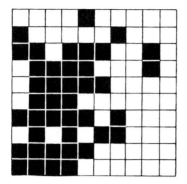 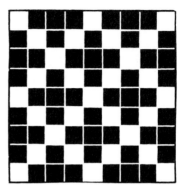

Fig. 4. Patterns with unpleasant (*left*) and pleasant (*right*) organization of elements. The pleasant pattern allows for the formation of super signs, i.e., of secondary geometrical order. (Schuster and Beisl 1978, after Dörner and Vehrs 1975)

and Vehrs (1975). Subjects were given an equal amount of (60) colored plates, which they should arrange on a designed grid of (160) blank squares into (a) an aesthetically appealing and (b) an aesthetically nonappealing pattern. Results showed a group of arrangements with similar features for the aesthetic and two subgroups of arrangements for the nonaesthetic configurations. All appealing patterns were rather complex structures, but structures that shared one property: they were easy to subdivide into more simple figures that again could be integrated into an overall regular structure (Fig. 4). Evidently, our eyes are somehow able to see through the superficial pattern and discover the secret order behind, and exactly these "cognitive" properties of our vision seem to be at the origin of aesthetic pleasure. We are tuned to recognize regularities – even hidden ones – because our memory is based on standardized experiences (templates). The process of secondary sign-formation is also called *super-sign*-formation and has to do with reducing complexity into order.

The two groups of nonappealing structures were either too complex for any possible reduction into more simple figures, or they were too simple and regular, offering structures of primary order only, like for example a big triangle, a square, or a chess-board pattern. Both were not suitable for any further super-sign-formation. The first group was called "confused", the second "boring". Therefore, aesthetic taste seems to hold the balance exactly between the two extreme possibilities for missing human pleasure.

The findings of Dörner and Vehrs coincide with all we said before, when talking of the optical laws and constraints of Gestalt psychology. To see "figures" against a background of random information, to see "similarities", means to find regularities which means reducing complexity – and all this adds to our comfort and satisfaction. Again, it seems to be exactly what Aristotle called the very specific gift of the artist's eye! In fact, as we know today, every one's eye is doing this every moment of the day, and

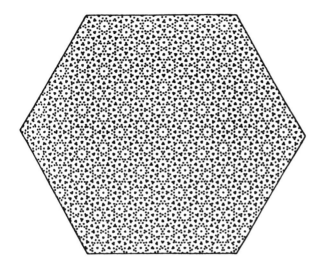

Fig. 5. Illustration of the autonomous organizing process in the visual field. (Marr 1982)

thus behaves like the artist's eye. It filters out regularities which could be the common features of a given visual situation and in this very function it even behaves like the human mind when "seeing" the hidden laws within diversity, which Plato said is only accessible for human thinking. Titles of modern monograph's like *The Intelligent Eye, The Artful Eye* etc. (Gregory 1970, 1995) give evidence of a new evaluation of human and nonhuman perception as an integrated part of the brain.

We may resume that stimuli allowing for the formation of super signs – like hidden polygons, squares and other regular figures – are likely to elicit higher aesthetic satisfaction than patterns with the given order of a chessboard or, on the other hand, with *no* conceivable order. Aesthetic preference seems to follow a curve of an inverse U-turn between the axes order and complexity and culminate at its apex, as proposed earlier by Walker (1973). Figure 5 shows the process of autonomous perceptual super-sign-formation within a regular primary pattern, involving playful effects.

In art however, the aesthetics of lines, circles, triangles and squares, i.e., simple figures offering reduced complexity as an attained effect, proved to be a reliable guideline through history up to our days. The tradition beginning with early ornaments and ending with Op and Concrete Art in the last century provided a rich selection of geometrical aesthetics. The beauty of a simple figure answers both to our perceptual as well as mental needs for orientation and perfection (Fig. 6).

Traditional cultures of our days, many of them living on a Neolithic or post-Neolithic level – and some of them still without contact to civilized societies – cultivate ornaments of primary order in objects of special value. The G/wi for example, a San group from the Central Kalahari (Botswana) and one of the last hunter-gatherer groups living today, decorate

Fig. 6. The aesthetics of geometry performed in a composition of Piet Mondrian (1929)

Fig. 7. Beaded head-band and purse decoration of the !Ko San, Central Kalahari, Botswana. Private Collection of I. Eibl-Eibesfeldt, Starnberg

their wooden bracelets with bands of fine incised zigzag patterns. Elaborated zigzag ornaments are also the ground-motive of beaded headbands and leather purses of the !Ko (Botswana; Fig. 7). Here, we know that the variations within the patterns are connected with a rather strict symbolism conveying information about the personal and social identity of the maker or bearer (Wiessner 1984; Fig. 8).

Comparable performances of geometrical patterns are found in the Ojibwa and Potawatomi Indians of northwestern America as variations of local styles (Boas 1955), whereas the Himba, traditional pastoralists of northern Namibia, manufacture geometrically decorated belts for young women promised to a husband (Fig. 9). Except for their baskets with simple regular line ornaments, they are not familiar with any alternative style and decoration within their living memory. Furthermore, others like the Mbole in Upper Congo use highly sophisticated geometrical patterns in straw mats as emblems of different clans or of status within the ritual community. More complex configurations are thus based on the variation of simple geometrical elements like the square, the triangle or polygon, following the rules of an evolved aesthetics based on super signs of a higher order.

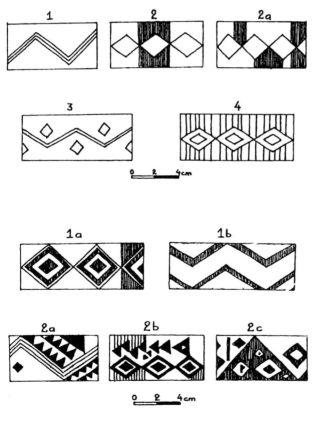

Fig. 8. Headband patterns of different clans in the G/wi San, Central Kalahari, Botswana.
(P. Wiessner 1984)

Fig. 9. Young Himba women (Namibia) with their characteristic belts when promised to a husband. (Photo: I. Eibl-Eibesfeldt)

Ndebele house decoration in South Africa is famous for its similar aesthetics. Franz Boas (1955) points out that "elementary aesthetic forms like symmetry and rhythm are not entirely dependent upon technical activities, but these are common to all art styles; they are not specifically characteristic of any particular region". (p. 11).

If we translate "rhythm" into "repetition" as a unifying and invariable element, we reach a kind of ground principle of artistic performance: the play with regularities and similarities! Plato's vision of beauty seems inherent to perceptual needs.

This seems to be exceedingly true for the earliest signs of art-like performance in man. They were found in Central Germany (Bilzingsleben), dating back about 300,000 years, and are assigned to *Homo erectus* (Fig. 10).

The pattern on a mammoth bone is characterized by a bunch of regular parallel lines of nearly equal length, easily set off against the irregular background of the material. Impossible to say what the producer intended to communicate by this marking, but "readable" as an aesthetic attempt up to our time.

This hitherto "first found document of human artistic manifestation" – as Rudolf Feustel (1987) puts it carefully – tells a lot about aesthetic vision relying on natural perception. Regularity and repetition – actually *redundancy* – as a visible "figure" against a random background seems to be the first step needed for discriminating a conscious marking from accidental

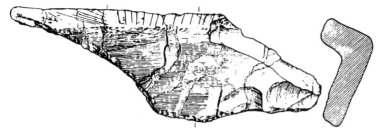

Fig. 10. Engraved elephant tibia of Bilzingsleben (Thuringia, Germany) from about 300,000 B.P. Total length 45 cm. (Herrmann and Ullrich 1991)

graining in the bone or stone. Organization and regularization as signs of intentional control over the unpredictable nature – the nature of own muscle activity included – begins in our perception, since perception evolved in the service of predictability. Each view of life, even that of a frog, is based on the hypothesis of finding regularities, i.e., repetition. Without predictability, no organism is able to survive. Even chimpanzees, when given materials for painting, avoid pure chance. Congo, a subject tested by Desmond Morris (1963), tried to systematize consequently a figure called "fan" by Morris, which from its base avoids the superimposing of contours and colors. Furthermore, when I. Eibl-Eibesfeldt (1966) made the same experiment with chimps in Hellabrunn Zoo near Munich, he observed a similar behavior with traced "rainbow structures" carefully set side by side over the whole sheet of paper (Fig. 11).

Structures of high regularity and self-reliance seem to be appealing and to catch human – and pre-human – attention already at this early stage of production, since they are signaling intentionality, i.e., some nonaccidental power working within. Humans thus are easily ready to assign such structures a message, or at least some importance. We shouldn't forget: regular primary patterns like spots, stripes and lines are the structural basis of signal properties in *all* species, whether on hide, skin or feathers, and encoding a message on the grounds of such "well-readable" structures belongs to the oldest inventions in evolution.

A comparison with children's drawings seems obvious. We know of the first so-called scribbling phase, which evidently demonstrates nothing but the pleasure of producing something that leaves marks and traces. Kellogg (1970), however, found regularities that compounded a repertoire of about 12 prototypes of scribbling patterns in children aged from 1 to 3 years. In addition, the drawings proved a remarkable sense of balance in filling the sheet and a tendency for serial arrangement of lines in equal distances (Al-Shamma-Gerber 1984).

In art, this kind of early pattern remains paradigmatic for further evolution. Symmetry and regularity are even more elaborated and completed. The next step after making a series of paralleled line bundles is the series

Fig. 11. Chimpanzee painting of a high ranking female. The different colors are set carefully one after the other, like in a fan. Sense of balance. (I. Eibl-Eibesfeldt 1966)

of paralleled bundled curves or zigzags, which proves to be highly decorative and corresponds to secondary super-sign-formation. Examples from Pavlov (Moravia), dated 25,000 B.P. probably served as adornment (Fig. 12). Two examples of traditional cultures of today show the persistence of this basic aesthetics, taken up in many different material contexts: Tukano Indian drawings as well as bark drawings of Mbuti women of the

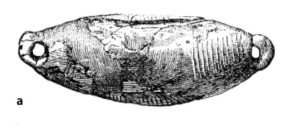

Fig. 12. a Decorated appendage from Pavlov (Moravia), Gravettian age, 27,000–20,000 B.P. **b** Decorated flat brooch (ebony) with parallel line bundles, Pavlov (Moravia) ca. 25,000 B.P. (Müller-Beck and Albrecht 1987)

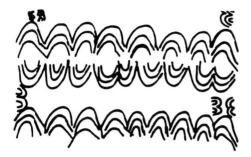

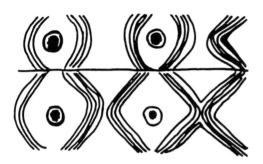

Fig. 13. Tukano Indian drawings. (Reichel-Dolmatoff 1978)

African rain forest (Fig. 13). The hypothesis of these somehow hypnotically insistent patterns being the result of hallucinatory drugs taken during rituals was brought up by different authors (Reichel-Dolmatoff 1978; Lewis-Williams and Dowson 1988, 1989). The fact of drug inference, however hard to prove for all such examples, would not actually displace the perceptual argument, since "entoptic phenomena", as claimed by Lewis-Williams and Dowson, are in fact nothing but the results of a hyperstimulated perception (Eibl-Eibesfeldt and Sütterlin 1997). Moreover, it does not explain the occurrences in children and animal drawing (Aiken 1998).

Regularity or redundancy being a hypothesis built into our perception becomes evident as well in brain pathology: what is not seen because of a lesion in the area of visual representation is completed by central brain mechanisms on the basis of a hypothetical homogeneity of the surrounding pattern. If this pattern is a wallpaper with regular motives, the patient does not see any hole, but a continuous structure. Contextual structures are interpreted as being isomorphic and redundant (Pöppel 1982). The same is true if a geometrical figure with incomplete contours is projected to healthy subjects within milliseconds, they will still see the whole figure, since lacking parts in the optical structure are virtually completed in the direction of redundancy. (Fig. 14; Kanisza 1976; Metzger 1975, pp. 430 ff.; Pöppel 1982).

The fact that we expect regularity is one thing, that we actively seek it is another.

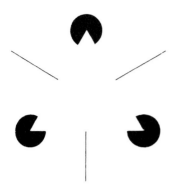

Fig. 14. Virtual trigon after Kanisza (1976)

If we tend to transform incoming information into the structure of information already stored – which is always based on an average – there is only one chance for really new information to be received: to be as clear and pronounced as possible. Recalling the Arnheim experiment (Fig. 6): either a figure is randomized or highly sharpened and this exaggeration is not in the stimulus, but in our perception. *Prägnanztendenz* is a term created by Gestalt psychologists who found that we tend to see the best possible version of a given prototype, especially when the stimulus is weak or damaged. Already babies learn very easily and without any further reward to adapt a blurred stimulus projection into a well-focused one by a mechanism working in the comforter when they suckle (Kalnins and Bruner 1973). Evidently, to see a clear picture is rewarding enough.

Line perception, for instance, with its different mechanisms of lateral inhibition – another well studied phenomenon of Gestalt psychology – can be called instrumental in enhancing contrasts along the edge of two different surfaces. This facilitates discrimination and categorization of our optical environment.

Moreover, the same is true for the different mechanisms that guarantee the visual constancy of our changing environment, like size, color, brightness and so on.

For instance, the fact that we perceive a tree under any circumstances as a tree, although its size on our retina is only the one of a fly on the window if the tree is seen in a far distance, is basically due to cerebral functions that account size with distance before we try to guess ourselves. Guessing is not good enough and takes too much time in case of danger. We have to be *sure* about what we see. When we see a piece of charcoal in the sunlight, which is brighter than a snowy hill in the shade, we perceive the charcoal as being black and the snowy hill as being white, even if we are instructed about the objective brightness values. All these phenomena, known as Optical Illusions, give evidence of the stringent perceptual mechanisms working within our sensory experiences, mostly against our better judgement.

Species-Specific Adaptations

Selective Sensitivity to Semantic Information: A Cognitive Pathway

The phenomena mentioned remind us of another important issue as well: the fact that our perception is not a blueprint of our physical environment, but a reduction, interpretation and evaluation of (and into) aspects relevant for our survival. However, no confusion should result in what we otherwise call a subjective attitude in the new sense of a philosophical "Constructivism". We do not refer to individual or socially induced judgements. All optical illusions as well as many other aspects of human perception may be called subjective as opposed to the objective or physical state of our environment, but they function reliably for humans of any kind and culture. Working with groups in Melanesia and Namibia, I obtained equally significant results in testing various "illusions" as they were substantial for European studies (Sütterlin, in prep.). This means that judgements of this kind do not vary intra- and inter-individually nor cross-culturally, but they are valuable for all members of the species and represent what we call species-specific adaptations (Morinaga 1933; Obonai 1935).

Saying that our perception is no blueprint of our environment means also that there is no cerebral point-to-point-representation of visual space. Instead, all information is analyzed into very special features (Merkmale) or categories. Some of it is always suppressed while other information is enhanced. This is exactly what we generally attribute to art only.

We just have to answer the simple question: what would we actually "see", if we could register indifferently the full amount of possible visual information looking at a particular scenery? Billions of pixels and dots, but no shapes, no colors, no "trees" and no "houses," no "humans" and "animals" etc.! And certainly no "beauty".

As we know from the research of Hubel and Wiesel (1959), there are special neurons responding only to the orientation of lines, others, as found by Semir Zéki (1980) to a special color, and again others to motion etc. This means – in the category color – that we do not respond to the physical wavelength of a stimulus, but code its color as we perceive it (Pöppel 1982, p. 181).

Selective sensitivity to certain features implies generalization. Categories – and even more complex prototypes – are always more average than the single event, since they comprehend and encode the common features of many different individual events, which again reduces complexity and saves memory in our cerebral hardware. Long before they become an issue of philosophical reflection, simplification as well as schematization are functional properties of categorical perception first sensory cognition.

However, there are also more complex configurations that underlie categorical perception, like for instance the human face. Evidence for this was found in patients who lost their competence of identifying individual faces after a stroke or occipital brain injuries (Damasio 1985). The existence of such brain structures explains the importance of recognizing people by an unmistakable indicator of their personality.

Selective sensitivity for complex stimuli was probably already evolved in higher apes, as we may conclude from experiments by G.P. Sackett (1966) and C. Groß (Groß et al. 1981).

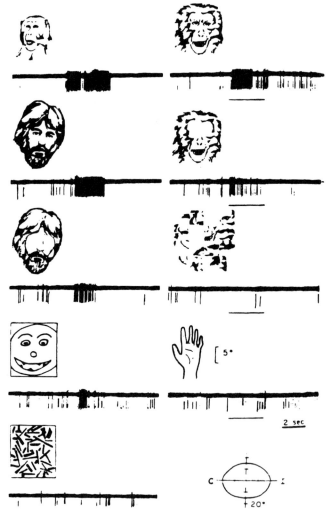

Fig. 15. Responses of single neurons from the temporal cortex of a macaque monkey to different stimuli. (Gross et al. 1981)

Charles Groß and his colleagues tested the neuronal activity of single neurons of a macaque monkey exposed to different stimuli like a conspecific's face, a human face, a human face without eyes, a schematized human face and a random pattern of about the same informational influx (Fig. 15). What we learn from his score is the selectivity of responses, especially the high activation in the case of a conspecific's face as well as of an individual human face – of the experimenter in this case – and the reduced response in the case of nonsemantic information. Interestingly the schematized human face shows about the same response as the eyeless human face.

What we further learn from such scores is that sensory processing serving reduction of informational complexity does not simply work in quantitative – or physical – terms as a random process of "making less" of what has been too much. Perception does not function like a dust-bin that just discharges the overflow when full. This would be a indifferent choice. Rather, selective sensitivity is a process tuned to semantic qualities and thus, highly discriminative and undemocratic. Analyzing and categorizing information means transforming it into meaningful information, whereby a quite rigid hierarchy of importancies is followed. This hierarchy is mirrored in the score of Fig. 15. The monkey's sensitivity shows a clear preference for important things like faces, first conspecific and second human faces, later face rudiments and face schema, hands, and so on.

Actually, categorization serves a clear discrimination and identification of perceived matter in terms of a valid interpretation. In order for us to recognize immediately whether an impending object is simply a figure or an animal, a human being, a plant or a tree, again internalized reference patterns – i.e., prototypes, templates – are required that contain general information of all comparable events. These templates are generally acquired by statistical learning[2]. They encode all information necessary for identification, retaining the congruent features of all stored figures according the hypothesis of similarity. Against this prototype incoming information is compared.

We all know that in the process of template formation children generalize: once they have learned to call a dog "wauwau" they also address other mammals as "wauwau".

The same is true for children's drawings which are still interesting to look at. The phase mentioned earlier as the "scribbling phase" is replaced at the age of about three by the so-called schema-phase that predominates until school age.

---

[2] A computer experiment by Hans Daucher (1967) that resumes the earlier experiments of Francis Galton (Forrest 1974) illustrates the process of statistical learning by superimposing 20 portrait photo of female students. The result shows a pretty standard face that has absorbed and wiped out all individual features.

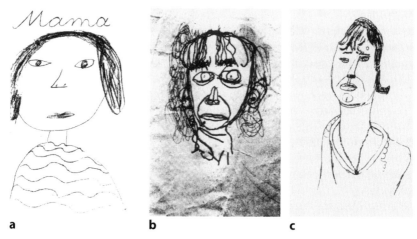

Fig. 16. Three drawings of a 7-year-old girl of her mother. **a** as a free drawing (schema phase); **b** drawing after model; **c** further elaboration after model 2 weeks later. (Nguyen-Clausen 1987)

Furthermore, these schemas or prototypes are not really influenced by learning. As Poek and Orgass (1964) could demonstrate, most of the 4-year-old children who are able to discriminate verbally neck and human torso draw the so-called *Kopffüssler*, i.e., the human figure consisting of a head with feet annexed – without the neck and torso. This means that the prototypes are more conservative than is possible for a child to know.

They represent a significant generalization on essential features of homo: round-faced and two-legged. All other features they share more or less with other species and the reduction exactly follows the generalization of essentials. Hence, the process of schematizing and generalizing seems more basic than the acquisition of details that fit – or do not fit – into these schemata[3]. I would like to mention the example of a 7-year-old girl who was invited to draw her mother. What she first drew was a completely schematized view of a woman in general: round face, nose in side view, hair like a cap. Only after her mother asked her to draw her as she *really* looked did the portrait become closer to the model (Fig. 16). Later on, she would even add dressing details. This means that it needs much more effort and scrutiny to concentrate on individual features – actually requiring a learning and observing effort – than to project an internalized general pattern, since individual features are perceived as deviations from the rule. The same is true when children draw a bird with four feet after they

---

[3] This is in agreement with Jean Piaget (1975). His concept of "assimilation" meant that children used to first resolve a drawing problem with an established schema as long as this is able to integrate the new subject. Only when the subject definitely does not fit in, the *schema* is adapted – which is called "accommodation".

have observed that most animals are four-footed. They generalize a learned pattern. Exceptions from the rule and specialization generally are subsumed later.

In art history as well we find figuration much later than early abstraction, around 35,000 years ago, in the so-called Gravettian Age. When it appears, it begins again with a formula of schematization or high stylization (Fig. 17). The so-called Woman of Predmost (Moravia, dated about 27,000 B.P.) shows a female representation on a tusk composed of abstract parallel line bundles, curve clusters and circular patterns, clearly readable as a female body. The accentuation of breasts, hips and pelvis highlight the most important characteristics, whereas the head and face remain geometrically encoded.

What we see is a perfect female "prototype" and what we normally expect is to find such a high degree of stylization at the end of a long evolution in realistic representation, where detailed information about a subject has been already internalized. Here, abstraction precedes realism in a way that is only explained by perceptional processing via prototypical

Fig. 17. "Woman of Predmost" (Moravia). Gravettian engraving on a tusk, ca. 27,000 B.P., Museum Brno. (Herrmann and Ullrich 1991)

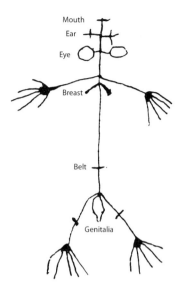

Fig. 18a. First drawing of a female figure done by an unexperienced Eipo man (Papua New Guinea). Private collection of I. Eibl-Eibesfeldt, Starnberg (Germany)

Fig. 18b. Bronce age rock painting from South of France (Grotte Chelo) Female body Schema with exaggerated vulva (Glory 1948)

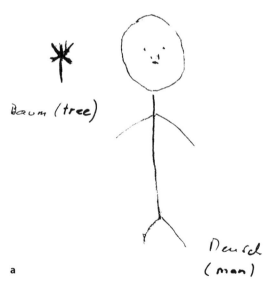

Fig. 19.a First drawing of a human figure done by an unexperienced Himba woman (Namibia). Strong schematization. Collection Ch. Sütterlin

Fig. 19.b Himba Woman while producing her first drawing: a human figure. Photo: Ch. Sütterlin

templates. It is easier to project internal models than to delineate precisely what we see in concrete. In 1975, when Eibl-Eibesfeldt invited an Eipo man in Papua New Guinea to draw a human figure, he got the outline seen in Fig. 18.a.

I need to add that this happened in the first contact period when the Eipo had no tradition in drawing at all except for the abstract ornamentation of their shields. The question is about the first attempt at sketching done by an Eipo.

What we find is a highly abstract projection of all information necessary for the recognition of a human figure compiled on the vertical axe with some characteristics of a female added, like the breasts and the sexual organ. Frontal presentation facilitates the full visual prospect of the human ground plan and when I did the same experiment with young Himba women in 1999, I got a similar abstract (Fig. 19)

Again, the subjects were absolutely nonexperienced and accustomed only to the geometrical symbolism of their iron belts. I had to show them how to hold the pencil. In fact, the most astonishing part was that none of the subjects faced with this rather difficult task tried to look at and sketch one of their fellows standing around, mostly children, who could have served as models. The subjects hesitated in the beginning, seemed to meditate about how to start, and after a while began to draw very concisely (Fig. 19). In fact, the same again happened in a first experiment with Yanomami Indians (Venezuela) in 1975 done again by Eibl-Eibesfeldt.

There is no doubt about a rudimentary symmetrical concept underlying our perception of a human being. It contains every information necessary to distinguish the figure reliably from an antelope or a salamander. All extremities are spread apart in order to be clearly discriminated.

Further examples of art history show only small variations of the schema which can be called a human prototype.

A Bronze Age rock art painting of the Grotte Chelo (Dept. of Var, France) adds what we could call the signalism of female sexual organs that runs in the line of the Eipo female (Fig. 18.b).

The comparison of rock art representations of the Bronze Age in northern Italy, southern France, Spain, Sweden and Russia reveal the same high redundancy of the human model: always the reduction into the symmetrical enumeration of paired organs along the vertical axe, and all individual markers (hair, clothes etc.) oppressed (Fig. 20).

Finally, there is not much more to learn about examples from cultures worldwide.

Early rock paintings from Venezuela (dated about 1500–6000 years), Hawaii and Tahiti and the South Pacific confirm the hypothesis of a highly consistent template on the morphological concept of man (Lommel 1962; Sütterlin 1994).

If there is variation and innovation, sexual organs are involved or the expressive exaggeration of other relevant parts like the hands, as in some

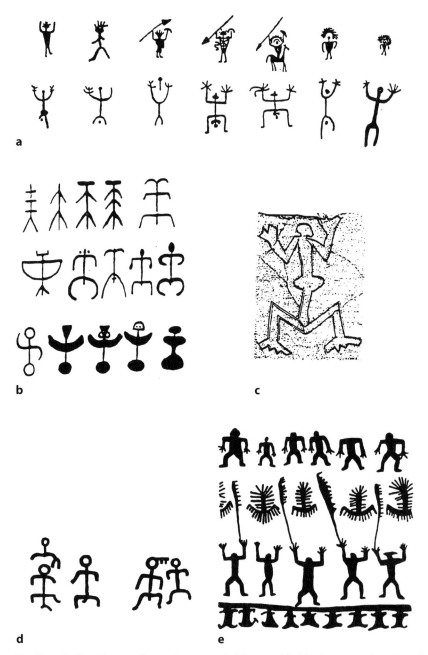

Fig. 20. a Stylized human figures from a rock (Naquane) in Val Camonica (northern Italy), Bronze age (Anati 1960). **b** Bronze Age rock painting from Spain (2000–1000 B.C.) (Glory 1948). **c** Human figure scheme from Neolithic ceramics, found in Kolesovice (Czech. Rep.; Müller-Karpe 1968). **d** Rock paintings from Hawaii, ca. 1600 B.C. (Lommel 1962). **e** Rock painting from Tahiti, ca. 1500 B.C. (Lommel 1962)

From Sign and Schema to Iconic Representation   153

Fig. 21. Figures with exaggerated, upraised hands. **a** Val Camonica (northern Italy), Bronze Age (Anati 1960). **b** Different rock paintings from the Upper Paleolithic with exaggerated hands from southern Russia, Oman and the Sinai Peninsula. *1* Wadi Adai (Oman); *2* Bilal seid Exoties (Oman); *3* Ghubra Tanuf (Oman); *4* Umm Hagag (Negev); *5* Gegam Mountains (Armenia); *6* Ramet Matred (Negev). (Ozols 1988)

rock paintings of the Negev (Sinai) and Oman (South Arabia). They introduce what has been called the "socio-sexual signalism" (Wickler 1967) and other releasers like the hands, and they are the only expressive clues observable. Both are part of a distinct male display behavior, which sometimes is completed by the presence of slim hand weapons. Actually, frontal presentation plays an additional part in enhancing visibility from the relevant large side, and belongs to male display behavior as well (Fig. 21).

Selective Sensitivity to Expressional Clues: The Emotional and Communicative Pathways

The pictograms of the human figure can be regarded as a kind of rune or label that shows the same basic meaning cross-culturally, like the one of Coca Cola that is known in every corner of this world under its special swinging graph. With the significant difference that the logo of the Coca

Fig. 22. **a** Protective entrance figure from an Indonesian temple, Pura Dalem, Sidan (Java), showing different kinds of hand, tongue and toothdisplay (Photo: Ch. Sütterlin). **b** Magic figure in northwestern Australian rock art. "Large side" as well as signal values are combined. Copied by Katharina Lommel 1955 (Lommel 1989)

Cola drink was spread all over from one definable source, whereas the human rune evolved, at least in most of the prehistoric cases, in an independent, i.e., convergent, process, as a universal artistic code (Sütterlin 1994).

If its rudimentary symmetric ground plan satisfies our cognitive and aesthetic concept of homo, its further occasional attributes, like the enlarged or upraised hands, spread fingers, the sexual organs, etc. hit quite a different, but equally immediate understanding. The occurrence of such figures in a magic or ritual context illustrates their signal-like value as display figures (Fig. 22; Eibl-Eibesfeldt and Sütterlin 1992).

The frontal position not only offers the typical, but also the "large side" of the figure as well – especially with arms and legs spread – and is easily "read" as an impressive display. Children still like to take this position when they want to scare others. Since early days, animals are usually represented from their lateral side, which shows more of their species-specific characteristics as well as their impressive "large side". Beautiful examples are found in early rock art of Tennes, Balsford, Sweden (Kühn 1952, p. 82). This refers to relevant positions in wildlife, as we know from fishes, squirrels and buffaloes, etc. Others display their most powerful features –

horns, teeth, large wings or feathers, eye spots, etc. – in order to impress their conspecifics or even their enemies when threatened, and we are reminded once more of the very early capacity in many species to assign meaning to minimal signs and signals. Communication is one of the earliest manifestations of life. Single-celled organisms already communicate with chemical and tactile signals, higher animals with acoustical and visual signs. The blue stripes on the flanks of a pine lizard proclaim "I am a male", females, on the other hand, are a uniform gray. If one paints blue stripes on their flanks, they will be attacked by males as if they were rivals. Conversely, if the stripes on a male are painted gray, the lizard will be courted by other males. Signals of this sort that control social interaction are known as "social releasers", or "releasers" for short (Eibl-Eibesfeldt 1987). To fulfill their communicative function, these signals need specific visual qualities that make them memorable and unambiguous. The striped bug (*Graphosoma lineatum* L.), for example, wears a flag-like colored pattern which evolved for warding off enemies. In a behavioral context, we observe a further evolution into expressive movements and ritualized behavior, like the parental feeding as friendly behavior becomes ritualized as courtship feeding, thus providing a condition for developing bonding signals (Huxley 1966).

For the human understanding of social situations, elementary and simplified codes are sufficient like the different gestures (chin flick, head toss, nose thumb, etc.; Morris et al. 1979), and mimic signs, which are easily interpreted and persuading, even on highly schematized faces. The happy or sad Smiley was eagerly taken up as a symbol in advertisement. We are especially sensitive to the eye, not only in a behavioral context (Coss 1972; Koenig 1975). We all know that a staring look is perceived as intrusive, and the simple application of an eye sign on a surface or facade makes us feel "watched" and slightly uncomfortable. This human attitude has been ingeniously called "susceptibility to dummies" ("Attrappensichtigkeit") by Otto Koenig (1975). Eye motives and eye-like patterns persist in the magic (evil-avoiding) context and art as well in all cultures (Eibl-Eibesfeldt and Sütterlin 1992). This could explain the presence of such signals or expressional clues even in highly abstract performances of indigenous as well as early rock art (Fig. 23).

Moreover, similar effects are true for the perception of specific colors, like red. Red is the color of many eatable berries and fruits, in addition to the color of blood. Its appearance always has something alarming and the evolution of specific receptors for just this color seems quite adaptive (Zéki 1980). Special care in the use of the color red is observed in the clothing and art of traditional as well as modern cultures. Even in the very pragmatic realm of road signs, it just needs a spot of well-placed red color to attract our special attention.

In this context of visual communication, simplicity serves efficiency and nonambiguity since a more complex stimulus would distract and

Fig. 23. Greek drinking bowls with eye-motive (530–520 B.C.). Munich Antiques Collection

weaken the message. Art performs differently in the sense that aesthetic pleasure consists in discovering messages within a more subtle texture of patterns (Fig. 24).

If we said earlier that the category "face" as a complex innate template or learning program is an exception (p. 145), we should correct the general outline at this very remarkable point: the ethological concept of species-

Fig. 24. Aesthetic signalism: Joan Miró "La chanteuse mélancolique" (1955). Private collection

specific and social releasers gives evidence about the stringent and consistent interpretation of complex social situations in the context of nonverbal communication. Facial mimics, expressive movements, ritualized behavior etc. are highly complicated visual sets that underlie clear interpretations cross-culturally and are evolved in the service of an unambiguous mutual understanding. Evidently, visual communication is so important that its codes and functions had become hard-wired in our perceptual outfit. It is not only an advantage if we can identify relevant events of our physical environment, but even more, if we are able to interpret its intentions and meaning so as to respond in an optimal way to survive.

This is the case when a 2-week-old baby reacts with eye blinking and avoidance behavior in front of a symmetrically extending projected black spot – what she or he evidently interprets as an object in impending approach. The behavior is not observed when the extension of the black spot proceeds asymmetrically (Ball and Tronick 1971). It would be hard to prove that this interpretation is learned from experience. With experiments of this kind – and I include the one with the visual cliff by Gibson and Walk (1960) – we are not far from the evidence G. Sackett presented with his study on the recognition of a threatening conspecific by socially nonexperienced rhesus monkeys (1966)[4].

We know of human babies' early competence to respond differentially to the expression of their care person, mostly the mother (Harris 1994), and of the similar facial expressions human adults produce and read cross-culturally (Ekman 1973; Eibl-Eibesfeldt 1995). We know also of preschool children being able to discriminate male body contour schemas reliably from female body schemas (Skrzipek 1978; Conolly et al., submitted 2000). There seems to be a clear prejudice about what sexual dimorphism looks like – and which is preferred! – the male schema being characterized by wide shoulders and narrow hips and the female by a differential hip-to-waist ratio. In general, younger children prefer their own sex in body contours – probably in terms of gender identification – juveniles and adults prefer the body contours of the opposite sex.

Other special sensitivities, as we know, concern babies' features in all kinds of natural and artificial phenomena. This means that single and proportional characteristics extracted from natural babies – big head compared to the rest of the body, big eyes, small suckling mouth, round cheeks and facial predominance of forehead, etc. – are preferred in children, animals and even rudimentary designs such as in comics (Fig. 25). Preferences of this kind can be called perceptual adaptations to relevant events of our lives, and it is certainly adaptive to develop responsiveness to the special features of small children and the two different sexes with respect

---

[4] The Sackett experiment testing the recognition of conspecifics with unexperienced juvenile rhesus monkeys hints at the possiblity of more complex innate templates in nonhuman primates.

Fig. 25. "Baby schema" (Kindchen-Schema) from a prospect of an international animal protection agency

to establishing relative emotional bonds. All such adaptations or preferences work as perceptual biases or prejudices which means that they are judgments that actually precede experience, as shown in the babies' reaction to impending collision (p. 145), since evidently they are encoded as neuronal templates. Our perception seems full of such prejudices both positive and negative. Let me add in this context our positive attitude to nature prospects. Phytophilia, the love of flowers and green plants (Wilson 1984; Eibl-Eibesfeldt 1995), appears to be an adaptive preference in general, since flowers and plants are indicators of a healthy and fertile environment – and the one for Savanna-like habitat features in particular (Orians 1980). Both play an important role in indoor-decoration (curtains, tableware) and art.

Now, if all these adaptations work as preferences, they are predestinated for aesthetic experience, since aesthetic experience is *preferential* by nature, and what we have found hitherto are human artifacts that answer nicely to questions concerning properties and necessities of our natural perception. However, where is the beauty, variety, exuberance and finally, meaning of all this together – the history of art? What people most associate with art are cultural events of high sophistication, like the late work of Rembrandt or Titian, the School of Barbizon, Expressionism, Art Nouveau, and so on. How would all perceptual adaptations be apt to explain the whole richness of artistic elaboration in this world?

It seems, indeed that art for a very long time runs its course in areas that satisfy (a) a very elementary, sensory and (b) a species-specific level of perception.

Moreover, we may have learned how humans used this cultural instrument to create iconic codes, symbols or messages and an environment that optimizes communication with and stimulation of many specific biases of their sensual system.

Fig. 26. Lascaux cave art (Dordogne, France), Magdalenian (ca. 15,000 B.P.) protorealism of representation. Postcard

However, there is of course a culture-specific level as well, otherwise we would not study art history. I did not mention the first conceivable – and successful – attempt of artistic realism in Magdalenian cave art that historically lies between the Woman of Predmost and Bronze Age rock art. In the depth of western European ritual centers, a first gleam of environmental mirroring of iconic representation comes into being, at least in style. Up to today, we do not know precisely what supplementary message is encoded in the herds and single horses, bulls and stags, but we can do without any message in front of such a blooming vision of Paleolithic fauna. Prototypes are still present, but they meet incarnation, that is: the realism of a refined observation of details, positions, diversity and behavior (Fig. 26).

Here, we assist in the beginning of a purely local tradition as part of cultural diversity and style in art. What has been attributed to Neolithic revolution as a process of increasing settlement shows its signs at the end of the Upper Paleolithic: the development of local traditions that become markers of a specific culture. Human beings start to live in communities of varying size which demarcate themselves by cultural differences including the style of ornamentation, the shape of pottery, the way people dress and new inventions in architecture. The hypothesis of artistic evolution being enhanced and accelerated by growing group concurrence is shared meanwhile even by fractions within classical art history.

The fact shows two sides: group- or culture-specific styles bring about separation from other groups as well as cohesion *within* the same group. Styles have imprinting effects and function as symbolic agreement on a shared environment. People identify not only with customs and constitutions, but also with common symbols and styles on more sensual and emotional grounds. In particular when groups take up larger sizes, rational insight alone is not sufficient to establish a bond to a cultural or national identity, but an emotional tie has to be established based on the primary affective bond of family (Eibl-Eibesfeldt 1994; Sütterlin 1998). Art can contribute to this cultural elaboration of political goals by transferring perceptual prejudices onto prejudices defined by the social system. Emotional and perceptual codes are apt to bind our attention in a way that triggers our response to the message enclosed. Art, by its immediate access to our affective life, is able to create emotional engagement and the readiness to act.

The Cultural Elaboration of Style. Art and the Perception of Nature

Cultural diversification seems to be the most adaptive instrument for humans to deal with the more rapid trail of cultural evolution.

For the question of how our more basic adaptations in perceiving our natural and social environment are transferred to the level of cultural adaptations, let me focus first on some few examples of recent art history. From this point of view, I may also try to find an explanation for why it took so long for art to evolve to a bloom that answers more to our perceptual needs for realism, and why art for such a long time persisted in more conceptual roots, which actually are the roots of our perception.

We could give special attention to the evolution of the human portrait which allows for humans' specific visual adaptation to individual faces, to the painted still lifes with fruits and flowers, or landscapes which evolved in the seventeenth century that satisfy our positive bias for stimuli of cultivated nature. We could also look to the long tradition of body representation in both sexes, which mostly follow our sex-specific prejudices, and we could even include the full vision of the babies schema especially in a religious context, where the appeal of lovely angels or putti had to court potential members of the Christian movement, along with the highly touching Mother–Child complex of Mary with the Child Jesus present in so many altar pieces, sculptures and icons (Bedaux 1999; Figs. 27, 28).

We would find that not only the species-specific biases are fully respected, but also the need for standardization and prototype-formation in given natural subjects.

Fig. 27. G.L.Bernini. Angels face from the high altar in St. Peter, Rome. Photo: I. Eibl-Eibesfeldt

The human figure in sculptural art, for instance, first deals with rudimentary idol-like representation (Babylonian or Minoan art for instance), and an archaic schematism is observed in Greek sculpture before it gives way to the Classic style with its idealized typified Olympus of figures (Boardman 1978, 1995). What we could call a portraying realism, only begins in Roman sculpture. Painted portraits are mostly esteemed – even by the portrayed persons and family members – when slightly "idealized", which means, shifted towards symmetry and transparent for discovering generalized human features approaching the type. Flower still lifes and landscape painting not only meet our need for nature, but they persuade the most when arranged for a certain balance. Actually, a more realistic concept of "nature" takes exceedingly long to evolve, natural motives like plants or flowers being an issue of ornament and marginal stylization for centuries. It seems that nature becomes an artistic topic in particular with the urbanization of cultures, as a kind of substitute. "Landscape", for example, as an autonomous artistic motive in European painting, appears only in the fifteenth century with Giorgione, Dürer and Aldtorfer. When the "ideal landscape painting" arose around 1600 in Italy, with paintings from Claude Lorrain and Nicolas Poussin, it showed an interesting fusion with the ethological concept of Savanna habitat preference, characterized by open plains with scattered trees, rivers or lakes, some animals and scarce signs of human residence, all typical aspects of biodiversity

Fig. 28. Pablo Picasso: "The Harem" (1906). Cleveland Museum of Art, Hanna Collection

Fig. 29. Caspar David Friedrich "Summer" (1803). Adaptations of earlier types of "ideal landscape" are still present, but slightly adapted to a very local type of mid-German landscape. München, Neue Pinakothek

(Fig. 29). The view from a moderate hill often implemented in such paintings further corresponds perfectly to the conditions of "prospect and refuge" assigned to a most comfortable site, since it guarantees a covered back and full overview into hiding places of possible enemies (Appleton 1975). This ideal of artistic landscape remained a model for further diversification first into various types of sea and mountain painting, before "portraits" of concrete landscapes in the painter's environment and even "national landscapes" were produced again, this probably in terms of an adaptive bonding with home nature and landscape (Sütterlin 2002). This means that even nature as an aesthetic and artistic topic is first submitted to concepts evolved for our natural perception, before it arrives at a tentative "naturalism".

There would be still more to discover in the run through different centuries.

Finally, we would learn to understand why there is evolution (and history) in art. Bonding with culture is very old, but from the very first attempts to establish exclusive messages for one's own group, the different codes follow a line that leads from a general sensory aesthetics (geometrical as well as signalistic) to schema formation and further standardization up to modes that incorporate concrete details.

The people from Pavlov may have identified with their lines, grids and spirals, Babylonians with their highly schematized idols, and Greeks certainly identified with their typified Apollos and Aphrodites. The repertoire has been enormously enriched since then, but in a way that the basic guiding rules of encoding visual information (regularity, repetition, redundancy) are still present within the more elaborated ones. By variations and cultural innovation into style and moderate complexity, they may appear even more precious and powerful. Therefore, we can read guiding structures of hues, shades and green or brown pixels in a portrait painted by Cézanne, indicating the color-accords underlying and connecting the different parts. Or we can find the principles of order and regularity projected into political architecture, insinuating the reliability and power of the system in discussion (Sütterlin 1998). Cézanne's portrait will certainly evoke higher aesthetic pleasure than a photo-portrait or landscape without the artistic elaboration of readable meta-structures and the semantics of order and regularity becomes much more powerful when integrated in artifacts of political and material context than in the performance of a simple square.

Above all, the impact of *composition*, i.e., the selective restriction to meaningful parts – counteracting diversity – and the appeal of timeless prototypes within modern incarnation and modern context prove the persistence of "Platonic ideas" and visual archetypes in the inevitable heritage of human sensory outfit (Fig. 30).

However, why did it take so long to evolve and suddenly accelerate?

One could argue with good reasons that without increasing competition amongst different groups and cultures – which promoted cultural evolution and diversification – humans probably would go on performing what seems to be much easier: the projection of innate templates and schemata in art that satisfy our primary outfit.

It is, as I tried to make clear, much more time-consuming and a greater strain to learn and observe what things really look like, and to adapt inner rules to both specific and changing environments than to simply obey these rules. It takes education, effort, and special attention. The adaptive force to do so is its reward: an evolved cultural bonding on the basis of common roots.

Culture still follows nature, up to our days. Only the intentional refusal of perceptional guidelines, as we assist in some tendencies of art nowadays, bears the risk of losing potentials of visual communication. However, these risks are carried and explored mostly by individuals that function as a spear-point of further cultural evolution. Selection will be made by new recipient groups to arise.

Fig. 30. Oskar Schlemmer "Staircase of Women" ("Frauentreppe") (1924). Kunstmuseum, Basel. (Photo: Ch. Sütterlin)

Summary

This paper reconsiders the question why man's artistic activities originated in symbolic abstraction and not in the naturalistic approximation of reality. We propose a new answer to this question by incorporating findings from developmental psychology, neuropsychology and human ethology.

Human mental representation of the external world is mainly a visual one, and the same is true for its creative representation. If we look at the overall picture of artistic manifestations from a historic point of view, we see a scenery of nearly immeasurable abundance and variety. With art, we generally associate something like the overwhelming flower-bouquets of Breughel and Rubens, the subtle portraits of Holbein and Titian, or the nostalgic landscapes of Turner and the masters of Barbizon.

The very early beginnings of artistic production, however, give evidence of a clear priority of highly abstract sign-like and signalistic encodings instead of a more elaborate and realistic description of reality and history shows that art followed rather "conceptual" paths for a very long time.

One main reason evidently is that visual arts and representation to a major extent, recur to a biological system which has evolved primarily for other than "aesthetic" reasons. Visual perception, like any sensory system that serves the reception of external information, is tuned to "design" a reliable description of the environment in which the organism is able to move, to find and avoid, to recognize and to orient himself, in short: to survive.

Thus, perception does not mirror the environment in a mathematically correct copy, point to point, but scans it first for what is regular and repeated, i.e., statistically probable to occur. This saves memory in the brain and helps to organize the overflow of information in the visual field. The analysis of information is primarily concerned with clearness and distinctness (Gestalt), the elimination of accidentals and the accentuation of regularities, and all this in the service of identification and recognition of relevant environmental data.

Plato (fifth to fourth century B.C.) was one of the first thinkers who assigned this very faculty of discriminating law and order within the ephemeral flow of appearances essential to human mental capacity – but only on the level of intelligent insights. Kant and Baumgarten (eighteenth century), who both dealt with aesthetic phenomena, assigned some of these capacities to the human senses, talking of their lower or instinctive intelligence ("gnoseologia inferior"). Modern sciences showed that important concepts and hypotheses in this world are wired into any sensory process, animal and human (Eibl-Eibesfeldt). They work as apriori insights ("prejudices") and as part of our phylogenetic adaptation, taking influence from aesthetic judgments as well, as can be seen in the cross-cultural comparison of art and ornament.

Coherence and regularity are the most general hypothesis organisms have to live with.

However, there are more specific concepts, tuned to the level of the species and which guide the expectations of what a favorable or dangerous situation, allies and enemies or a sexual partner, etc. should look like. Modern ethology and human ethology gave evidence of varieties of such "semantic" concepts working reliably within particular species including man, especially on the level of communication and ritualization. As on the purely formal level of Gestalt, these kinds of prejudices act as visual codes or guidelines in channeling information towards a certain prototype/archetype of inner representation and influence the formation of aesthetic preferences as well. There is a large layer of shared experiences and aesthetic agreement in art worldwide – otherwise international exhibitions on the Gold of the Pharaohs or the late work of Titian would definitely be a flop.

The need to channel, reduce and "encode" information from our surroundings immediately upon its arrival – i.e., the need to create neuronal templates that enable us to detect recurrent patterns of features – stands at the outset of any process of cognition involving perception, *re*-cognition and evaluation. The formation of abstracts and schemata, the "encoding of reality", is itself a basic achievement of human perception. Thus, it may well be that this fundamental achievement of our visual faculty constitutes the neuropsychological basis of sign and schema formation in the arts. This casts doubt on the assumption that processes of abstraction must proceed from a faithful replication of real objects, and are thus ex post facto. It follows that our mental image of the outside world is primarily governed by notions and concepts only partly related to the realistic reproduction of that world. Apparently, our comprehension of everything that is individual and unique must be "acquired." This concerns the more cultural and individual level of preferences, and also seems to be the reason for why the approximation to a more portraying "realism" takes much longer to develop, in children's drawings as well as in art. To incorporate – or "assimilate" (Piaget) – the very particular and unmistakable features of a person or an event into the general outlines of perceptual prejudice (schema or type) by time-consuming observation is much more difficult and strenuous than expressing pure templates. However, these learning processes are important for further evolution and help in identifying with one's own group and environment – the same as the learning of culture-specific symbols and styles. In all likelihood, phylogenetic pre-adaptations codetermine our concepts of reality from the level of sense perception all the way to their external representation.

Acknowledgments. This work was produced with support from the Max-Planck Gesellschaft, the Krupp-Stiftung, Essen, and Ms. Traudl Engelhorn-Vecchiatto, Brent-Fontanivent.

References

Aiken NE (1998) The biological origins of art. Praeger, Westport, CT
Al-Shamma-Gerber N (1984) Der formale Bestand der Schemazeichnung und der Aufbau von Bedeutungsträgern aus definierten Grundmustertypen. Eine empirische Längsschnittstudie zur Entwicklung der Zeichenfähigkeit im Kindergartenalter. Diss LMU München, München
Anati E (1960) La civilisation du Val Camonica. Arthaud, Paris
Appleton J (1975) The experience of landscape. Wiley, London
Arnheim R (1965) Kunst und Sehen. Eine Psychologie des schöpferischen Auges. De Gruyter, Berlin
Ball W, Tronick F (1971) Infant responses to impending collision, optical and real. Science 171:818–820
Baumgarten AG (1750–58) Aesthetica. Reprint 1983. Felix Meiner, Frankfurt am Main
Bedaux JB (1999) From normal to supranormal: observations on realism and idealism from a biological perspective. In: Bedaux JB, Cooke B (eds) Sociobiology and the arts. Editions Rodopi, Amsterdam
Boardman J (1978) Greek sculpture: the archaic period. Thames and Hudson, London
Boardman J (1995) Greek sculpture: the late classical period. Thames and Hudson, London
Boas F (1955) Primitive art. Dover, New York
Capelle W (ed) (1953) Die Fragmente der Vorsokratiker. Kröner Verlag, Stuttgart
Changeux JP (1994) Art and neuroscience. Leonardo 27(3):189–201
Coss R (1972) Eye-like schemata: their effect on behavior. Thesis, University of Reading
Damasio A (1985) Prosopagnosia. Trends Neurosci 8(3):132–135
Daucher H (1967) Künstlerisches und rationalisiertes Sehen. Gesetze des Wahrnehmens und Gestaltens. Schriften der pädagogischen Hochschulen Bayerns. Ehrenwirth Verlag, München
Dissanayake E (1992) Homo aestheticus. Where art comes from and why. The Free Press, Maxwell Macmillan, New York
Dörner D, Vehrs W (1975) Ästhetische Befriedigung und Unbestimmtheitsreduktion. Psychol Rev 37:321–334
Eibl-Eibesfeldt I (1966) Ethologie. Die Biologie des Verhaltens. Akademische Verlagsgesellschaft Athenaion, Frankfurt am Main
Eibl-Eibesfeldt I (1987) Grundriß der vergleichenden Verhaltensforschung, 7th edn. Piper, München
Eibl-Eibesfeldt I (1994) Kultur im Dienste der Wertevermittlung. In: Liedtke M (ed) Kulturethologie. Realis, München, pp 168–180
Eibl-Eibesfeldt I (1995) Die Biologie menschlichen Verhaltens. Grundriß der Humanethologie. Piper, München
Eibl-Eibesfeldt I, Sütterlin C (1992) Im Banne der Angst. Zur Natur- und Kunstgeschichte menschlicher Abwehrsymbolik. Piper, München
Eibl-Eibesfeldt I, Sütterlin C (1997) Zeichen und Abbild. Zu den Ursprüngen künstlerischen Gestaltens. Kultur Notizen 20:29–36

Ekman P (1973) Cross-cultural studies of facial expression. In: Ekman P (ed) Darwin and facial expression. Academic Press, New York, pp 169–222
Eysenck HJ (1942) The experimental study of the "Good Gestalt". A new approach. Psychol Rev 49:344–364
Feustel R (1987) Eiszeitkunst in Thüringen. In: Müller-Beck H, Albrecht G (eds) Die Anfänge der Kunst vor 30,000 Jahren. Theiss, Stuttgart, pp 60–63
Forrest DW (1974) Francis Galton, the life and work of a Victorian genius. Elek, London
Franke HW (1994) Die neuen Bildwelten. Paper presented at the Symposium on "Naturwissenschaft und Kunst – Kunst und Naturwissenschaft" 1–3.12.1994 Universität Leipzig, Klingenberg, Leipzig
Gibson EJ, Walk RD (1960) The visual cliff. Sci Am 202:64–71
Glory A (1948) Les peintures de l'age du métal en France méridionale. Préhistoire:7–135
Grassi E (1980) Die Theorie des Schönen in der Antike. Dumont, Köln
Gregory RL (1970) The intelligent eye. Weidenfeld and Nicolson, London
Gregory RL (1995) The artful eye. Oxford University Press, Oxford
Groß CG, Bruce CJ, Desimone R, Fleming J, Gattas R (1981) Cortical visual areas of the temporal lobe. In: Woolsey CN (ed) Cortical sensory organization 2. Humana Press, Totowa, NJ, pp 187–216
Harris P (1994) The child's understanding of emotion: developmental change and the family environment. J Child Psychol Psychiatriy 33 (1):3–28
Herrmann J, Ullrich H (1991) Menschwerdung. Akademie, Berlin
Hubel D, Wiesel T (1959) Receptive fields of single neurons in the cat's striate cortex. J Physiol 148:547–591
Huxley J (1966) A discussion on ritualization of behaviour in animals and in man. Philos Trans R Soc Lond Ser B 251:772
Kalnins IV, Bruner JS (1973) The coordination of visual observation and instrumental behavior in early infancy. Perception 2(3):307–314
Kanisza G (1976) Subjective contours. Sci Am 234(4):48–52
Kellogg R (1970) Analyzing children's art. The national press book, Palo Alto
Kobbert MJ (1986) Kunstpsychologie: Kunstwerk, Künstler und Betrachter. Wissenschaftliche Buchgesellschaft, Darmstadt
Koenig O (1975) Urmotiv Auge. Piper, München
Kühn H (1952) Felsbilder Europas. Kohlhammer, Stuttgart
Lewis-Williams JD, Dowson TA (1988) Sign of all times: entoptic phenomena in Upper Paleolithic art. Curr Anthropol 29:201–245
Lewis-Williams JD, Dowson TA (1989) Images of power: understanding Bushman rock art. Southern Book Publishers, Johannesburg
Lommel A (1962) Motiv und Variation in der Kunst des zirkumpazifischen Raumes. Staatliches Völkerkundemuseum, München
Lommel A (1989) Die Kunst des Alten Australien. Prestel, München
Marr D (1982) Vision. WH Freeman, New York
Metzger W (1953) Gesetze des Sehens, 2nd edn. Kramer, Frankfurt am Main
Metzger W (1975) Gesetze des Sehens, 3rd edn. Kramer, Frankfurt am Main
Morinaga S (1933) Untersuchungen über die Zöllnerische Täuschung. Jpn J Psychol 8:195–242
Morris D (1963) Die Biologie der Kunst. Rauch, Düsseldorf
Morris D, Collett P, Marsh P, O'Shaughnessy M (1979) Gestures. Their origins and distribution. Jonathan Cape, London

Müller-Beck H, Albrecht G (1987) Die Anfänge der Kunst vor 30,000 Jahren. Theiss, Stuttgart
Müller-Karpe H (1968) Handbuch der Vorgeschichte, vol 2. (Neolithicum). Beck, München
Nguyen-Clausen A (1987) Ausdruck und Beeinflußbarkeit kindlicher Bildnerei. In: von Hohenzollern JG, Liedtke M (eds) Vom Kritzeln zur Kunst. Stammes- und individualgeschichtliche Komponenten der künstlerischen Fähigkeiten. Klinkhardt, Bad Heilbrunn
Obonai T (1935) Contributions to the study of psychophysical induction, VI. Experiments on the Müller-Lyer illusion. Jpn J Psychol 10:37–39
Orians GH (1980) Habitat selection: general theory and applications to human behavior. In: Lockard SJ (ed) The evolution of human social behavior. Elsevier, New York
Ozols J (1988) Zur Ikonographie der eiszeitlichen Handdarstellungen. Antike Welt 19:46–52
Piaget J (1975) Das Erwachen der Intelligenz beim Kinde. Klett, Stuttgart
Pierce E (1894) Aesthetics of simple forms. Psychol Rev 1:483–495
Poek K, Orgass B (1964) Die Entwicklung des Körperschemas bei Kindern im Alter von 4–10 Jahren. Neuropsychologia 2:109–130
Pöppel E (1982) Lust und Schmerz. Grundlagen menschlichen Erlebens und Verhaltens. Severin und Siedler, Berlin
Puffer ED (1903) Studies in symmetry. Psychol Rev Monogr Suppl 4:467–539
Reichel-Dolmatoff G (1978) Beyond the Milky Way: hallucinatory imagery of the Tukano Indians. UCLA Latin America Center, Los Angeles
Rensch B (1958) Die Wirksamkeit ästhetischer Faktoren bei Wirbeltieren. Z Tierpsychol 15:447–461
Sackett GP (1966) Monkeys reared in isolation with pictures as visual input: evidence of an innate releasing mechanism. Science 154:1468–1473
Salter F (1995) Emotions in command. A naturalistic study of institutional dominance. Oxford University Press, Oxford
Schuster M, Beisl H (1978) Kunst-Psychologie. DuMont, Köln
Skrzipek KH (1978) Menschliche "Auslösermerkmale" beider Geschlechter I. Attrappenwahluntersuchungen des geschlechtsspezifischen Erkennens bei Kindern und Erwachsenen. Homo 29:75–88
Sütterlin C (1994) Körperschemata im universellen Verständnis. In: Michel P (ed) Die biologischen und kulturellen Wurzeln des Symbolgebrauchs beim Menschen. Schriften zur Symbolforschung. Lang, Bern
Sütterlin C (1998) Art and indoctrination. From the Biblia Pauperum to the Third Reich. In: Eibl-Eibesfeldt I, Salter F (eds) Indoctrinability, ideology and warfare. Berghahn Books, Oxford
Sütterlin C (2003) Wandel der Natur-Rezeption in der europäischen Malerei. In: Liedtke M (ed) Naturrezeption. Matreier Gespräche. Schriftenreihe der Otto-Koenig-Gesellschaft, Wien. Austria Medien Service, Graz
Walker EL (1973) Psychological complexity and preference: a hedgehog theory of behavior. In: Berlyne DE, Madsen KB (eds) Pleasure, reward, preference: their nature, determinants, and role in behavior. Academic Press, New York
Wertheimer M (1945) Productive thinking, 1st edn. Harper and Brothers, New York
Wickler W (1967) Socio-sexual signals and their intraspecific imitation among primates. In: Morris D (ed) Primate ethology. Weidenfeld and Nicholson, London, pp 69–147
Wiessner P (1984) Reconsidering the behavioral basis for style: a case study among the Kalahari San. J Anthropol Archeol 3:190–234
Wilson EO (1984) Biophilia. Harvard University Press, Cambridge, MA
Zéki S (1980) The representation of colours in the cerebral cortex. Nature 284:412–418

III The Fitness of Beauty – Aesthetics and Adaptation

Beauty and Sex Appeal: Sexual Selection of Aesthetic Preferences

UTA SKAMEL

"Imagine a world in which people had no mate preferences."
David Buss 1998b, p. 405

Introduction: An Evolutionary Perspective on Sexual Aesthetics

When female barn swallows search for a prospective father of their future offspring, they are far from accepting every male they meet. Instead, they strongly prefer males with long and symmetrical tail feathers bearing large white spots. This is, according to modern evolutionary theory, a wise choice since these white spots are supposedly reliable signals of the males' phenotypic quality, such as their resistance against parasites (Kose and Møller 1999). Likewise, from an evolutionary point of view, a human male makes a wise choice when he falls in love with a young, attractive woman, because with such a woman he will potentially be able to sire a large number of healthy offspring. Thus, the mate choice "decisions" of both the female barn swallow and the human male appear to be biologically adaptive, i.e., they are likely to maximize the genetic fitness of the respective individual. At the same time, however, their choice is also an aesthetic one, whereby members of one sex value certain traits shown by members of the opposite sex as attractive.

For the understanding of the evolution of the aesthetics of human sexually attractive traits, Darwin's (1871) theory of sexual selection provides a powerful framework. Darwin focused on sexually dimorphic, often ornamental and highly exaggerated traits such as the peacock's tail, which abound throughout the animal kingdom. In Darwin's view, such traits were difficult to explain by his theory of natural selection because they did not appear to increase, but rather to decrease the individual's survival prospects. His solution to this paradox was the theory of sexual selection, which he defined as "the advantage which certain individuals have over others of the same sex and species solely in respect of reproduction" (Darwin [1871] 1998, p. 216). This advantage could arise from two 'kinds of sexual struggle': "In the one it is between individuals of the same sex, generally the males, in order to drive away or kill their rivals, the females remaining passive; whilst in the other, the struggle is likewise between the individuals of the same sex, in order to excite or charm those of the opposite sex, generally the females, which no longer remain passive, but select the more agreeable partners" (p. 638).

Not least because Darwin ascribed female animals a "taste of beauty" and an "aesthetic sense", this latter part of his theory – female choice – quickly became highly controversial (e.g., Wallace 1889) and was then largely ignored (see Cronin 1992 for review). Only a few authorities such as Huxley (1938) and Fisher (1958) tried to resolve this central problem of female choice. They argued that mate choice does not necessarily reflect any aesthetic sense or preference on behalf of the choosy individual, but that neural mechanisms enabling individuals to differentiate and assess potential mates by certain physical attributes would suffice (cf. Wiley and Poston 1996). Nevertheless, the whole subject lay dormant until Trivers (1972) published his seminal paper on "sexual selection and parental investment", where he argued that the "key variable controlling the operation of sexual selection" is the relative parental investment of the sexes in their young.

In evolutionary biology, the term 'choice' has a rather broad meaning. It refers to any behavior that influences an individual's chances of mating with certain individuals of the opposite sex. Complex cognitive processes or a special internal state need not be involved (Kirkpatrick and Ryan 1991; see also Bateson 1983; Maynard Smith 1987; Wiley and Poston 1996; Barrett and Warner 1997). In contrast to Darwin's much criticized assumption, mate choice does not rely on any kind of 'aesthetic sense'. Aesthetic preferences may be and probably are often involved, but there are some contexts – e.g., postcopulatory or 'cryptic' female choice (Eberhard 1996; see Birkhead 2000, for a discussion) – where potential aesthetic preferences are irrelevant. From an evolutionary point of view, the term 'choice' does not require that animals must be able to discriminate between potential mates in the classic sense, as implied by the concept of indirect mate choice (Wiley and Poston 1996). Inciting male competition, mating with multiple males or mating with the first available male (e.g., Alatalo et al. 1998; Cunningham and Birkhead 1998) are primarily indiscriminative, but nevertheless evolutionary effective methods of choosing a mate. Choice does not even require behavior. Like Darwin, some recent authors proposed to differentiate between active choice and passive attraction (Parker 1983; Searcy and Andersson 1986). However, the concepts of indirect, postcopulatory and indiscriminate choice, as well as mechanisms of sexual selection in plants (Willson 1990) suggest that the discrimination between active and passive females might be no longer useful.

Searching for aesthetics in mate choice while at the same time using such a broad definition of 'choice' seems fruitless at first. However, the term 'aesthetics' might be more useful for the understanding of the evolution of sexually selected traits in its more restricted, conventional definition such as "the science of beauty". Taking this approach, aesthetics would imply both the production of beauty and its perception, evaluation and preference within the framework of evolutionary adaptedness. The strength of this approach lies in its conceptual closeness to modern biological signal theory, which analyzes the production of communicative units and their effects. According to signal theory, a signal is a trait whose adaptive value for the receiver lies in its informative function, while the sender benefits from evoking a certain response (Zahavi 1991; Maynard Smith and Harper 1995). Assuming that beauty is a biological signal (or a signal complex), in the context of mate selection, the following predictions could be derived:

1. Their sensory and cognitive equipment should enable individuals to perceive fitness-relevant aesthetic traits. At the same time, only beauty should evolve that is adapted to the 'psychological landscape' of the choosy individual, which also means that traits should be of high detectability, discriminability and memorability (Guilford and Dawkins 1991).

2. Preferences for beauty are evolutionary significant only if the preference leads to a response, and if this response has reproductive consequences. Ornamental signals should, therefore, not only attract the choosy sex (as has been suggested by Wallace 1889), but also enhance his reproductive success and that of the chosen individual.

3. Receivers should be sensitive to aesthetic traits of a sender whenever such traits are 'self reporting signals' (Maynard Smith and Harper 1995) which might advertise potentially useful information about the sender's quality. However, conflicts of interests quickly arise in interactive systems, and conflicts of interests between mates are almost universal (Trivers 1972). Therefore, receivers should be especially interested in the reliability of a signal (Bradbury and Vehrenkamp 2000).

4. In many interactions, senders also benefit from emitting reliable signals. Beauty must not be honest, however. According to modern biological signal theory, deception can occur (Bradbury and Vehrenkamp 2000). By using 'manipulative' (Krebs and Dawkins 1984) or 'cheating' signals (Hauser 1997), the sender changes the informational content of the message for his own benefit (Maynard Smith and Harper 1995; Hasson 1997; Bradbury and Vehrenkamp 2000).

The functional significance of beauty remained obscure for a long time. In Darwin's (1871) view, beauty existed only to serve reproduction, whereas for Wallace (1889), beauty was a mere expression of strong vitality, and some contemporary authors even regarded sexual selection as "selection without adaptation" (e.g., Futuyama 1990; see Cronin 1992 for the misunderstandings perpetrated by this view), just as most nonbiologists regard beauty as nonfunctional. From a sexual selectionist point of view, physical beauty appears to be functional and by interpreting beauty as a guarantee for quality, the long-standing antithesis between nonfunctional beauty and functional biology becomes blurred (Symons 1995). Nevertheless, just how many human and nonhuman sexually attractive traits evolved remains a matter of debate (see Cronin 1992). Similarly, there appears as yet little agreement about whether human sexual signals are always honest or sometimes also deceptive (see below).

How Sex Appeal Might Have Come into Our Minds

Since Darwin's times several formal models have been developed explaining how females might benefit from mating with males carrying attractive, but otherwise apparently useless traits, which sometimes even lower the viability of their bearers (reviewed in Kirkpatrick and Ryan 1991; Andersson 1994). Some of these models may also shed light on the evolution of the human aesthetic sense and may explain why humans consider certain traits as beautiful and whether they are capable of evaluating the fitness of a prospective mate through beauty signals.

When a female values a certain trait as attractive, this may be simply an effect of an innate 'taste' (cf. Ridley 1994), being the result of natural selection. If, for example, the visual system of a female is adapted to finding high-quality food through its distinct coloring, she may also value a similarly colored male as attractive. Thus, when exploited by males, such a 'sensory bias' (e.g.,Ryan and Keddy-Hector 1992; Endler and Basolo 1998) may secondarily influence female mate choice and, thereby, male reproductive success. These signals are important starting points for the evolution of signals relevant for mate choice, but it seems unlikely that they would survive in the long run unless the preference for this trait also improves the fitness of the female (Miller 2000).

Reproductive benefits of 'good taste' would be more apparent when females choosing attractive males also produce attractive sons, as was first proposed by Fisher (1958); suppose a population where males vary in some trait, say in the coloration of their plumage. The majority of the females, no matter how small, prefer males with brightly colored feathers – for whatever reason. If both the trait and the females' preference are heritable, the son will inherit the signal of the father, and the daughter the preference for this signal, leading to an association of the alleles of preference and trait in the offspring. In the next generation, more brightly colored males will sire more offspring due to higher mating success, and females choosing attractive males will produce attractive sons and choosy daughters. In this 'runaway' process, each successive generation of males will become more extravagant than the last one, until some other selective forces, such as a high mortality through an increased predation risk, stops the process. Genetic models have shown that the Fisherian self-reinforcing runaway selection is theoretically possible (O'Donald 1962, 1980; Lande

1980, 1981; Kirkpatrick 1982; Pomiankowski et al. 1991; Jones et al. 1998), and empirical data have confirmed several of its predictions (Bakker 1993; Gilburn and Day 1994; Wilkinson and Reillo 1994; Whittier and Kaneshiro 1995; McLain 1998; Presgraves et al. 1999). Most notably, some studies found that attractive males sire attractive sons without increased viability (Etges 1996; Jones et al. 1998; Wedell and Tregenza 1999; (Brooks 2000). In a Fisherian runaway process, beauty is defined by the majority and its only biological function is that the majority prefers beauty. Signaling beauty is a promise here: choose me, and your offspring will be chosen in turn.

Unlike this 'good taste' perspective of mate selection (Cronin 1992), the 'good-genes' hypothesis considers conspicuous traits to be indicators for general viability (Williams 1966; Trivers 1972). Here, beauty is neither arbitrary nor defined by the taste of the majority, but highly specific; Females prefer traits which are reliable indicators of certain fitness components. Assuming that quality is heritable, a female choosing a male advertising his quality can pass these genes on to her offspring. In the host–parasite coevolution version of the good-genes hypothesis (Hamilton and Zuk 1982; Ebert and Hamilton 1996), the ability to adapt to a changing environment is a prime attribute for individual fitness; only those resistant against parasites and pathogens can develop such expensive adornments, and since parasites also adapt to the resistances of their hosts, this results in coadaptive evolutionary cycles of hosts and parasites. According to good-genes models, beauty is a guarantee for well-adapted offspring with good survival chances. This hypothesis received substantial empirical support from a wide range of taxa (see Møller and Alatalo 1999, for a metaanalysis of these studies), although the magnitude of viability effects appears to be relatively small in most cases.

The 'handicap' hypothesis (Zahavi 1975, 1977, 1993; Zahavi and Zahavi 1997) is another version of a good-genes model and explains how such costly traits could evolve through mate choice; the more a male invests in a trait, the more reliable it is for the female as an indicator for the male's genetic quality. Only a high quality male can survive wearing exaggerated handicaps, and this makes him attractive for females. Cheating would not pay because the cheater's investment would not be outweighed by his potential benefits. Beauty, therefore, is a reliable fitness indicator because of its high cost. A beautiful individual advertises, 'I am so full of heritable quality that I can also afford to be beautiful'. The handicap model was criticized at first (Davis and O'Donald 1976; Maynard Smith 1976; Kirkpatrick 1986), but was later confirmed through mathematical modeling (e.g., Grafen 1990a, b; Maynard Smith 1991). Moreover, the fact that most sexually selected traits are costly appears rather plausible; but whether these costs exist, as the hypothesis assumes, to make the signal more reliable, will be much more difficult to decide.

While the handicap model assumes that the potential conflict between the sexes must result in honest signals, the 'chase-away' model (Holland

and Rice 1998; Getty 1999; Rice and Holland 1999) adopts the notion that this conflict will result in antagonistic coevolutionary cycles; an already existing sensory bias of the females will lead to the evolution of males that exploit this bias with certain 'attractive' traits. Choosing these males as mates would not only confer no benefit; females would suffer a fitness loss by mating too frequently or at the wrong times. As a result, female resistance against attractive males evolves. The males, then, will develop more excessive traits to negate the increased threshold and the females will in turn raise their threshold. In any case, from the view of the chase-away model, choosing an attractive male would be maladaptive for the female.

Unlike common models of mate choice, the 'heterozygosity' hypothesis (Brown 1997; Penn and Potts 1999) claims that the genetic quality of a mate is strongly dependent on the genetic makeup of the choosing female. Females should choose males whose genotype is as different from their own as possible, so that future offspring are as heterozygous as possible. High levels of heterozygosity are associated with increased growth, developmental stability, and enhanced resistance against pathogens (reviewed by Brown 1997), while low levels of heterozygosity, which result from matings between close genetic relatives, strongly increases the risk of inbreeding depression (Crnokrak and Roff 1999). Thus, the central message of the heterozygosity hypothesis is, 'what is beauty for one female might be ugly for the other' (Brown 1997).

Determinants of Human Physical Beauty

Mating System, Mutual Mate Choice, and Parental Roles

As sexually selected signalers, humans show at least two peculiarities (Low 1999): first, humans modify and intensify their sexual signals, and second, in comparison to many other species, their sexually selected traits appear rather paradox (see Low 1999). Several characteristics such as the sexual dimorphism in body size and muscular development (e.g., Dixson 1998; Geary 1998), and in the number of (actual or desired) sexual partners (Symons 1979; Buss 1994), and, not at least, the predominance of facultatively polygynous human societies (Daly and Wilson 1983; Murdock 1981) appear to support the idea that the ancient human mating system was polygynous. Moreover, both the lack of sexual swellings and the rather small testes and the small number of sperm per ejaculation (cf. Harcourt et al. 1981; Dixson 1998) strongly suggest that the ancient human mating system was not promiscuous (as is the case in our nearest living relatives, the chimps and bonobos). One would expect, therefore, that human males, as males in other polygynous species, should be the ornamented sex; instead, it is female attractiveness (and its artificial display) which is much more apparent (Jones 1995; Low 1999) and which is offered on the mating market, while men make themselves attractive through money and power (e.g., Thiessen et al. 1992; Wiederman 1993).

The fact that men and women use different mate choice criteria is a result of their different reproductive strategies; human mothers, like all other mammalian mothers, invest much more in their offspring than fathers do. Consequently, womens' potential reproductive rate is much lower than that of men, making women more the object of male competition than vice versa (Trivers 1972; Clutton-Brock and Vincent 1991). This holds true even if human fathers, as it is often argued, invest much more into their offspring than most other mammalian fathers. Thus, if women potentially benefit from the help of males, they should not only look at their prospective mates' genetic quality, but also at their ability and readiness to provide resources. In fact, women in different cultures prefer partners with a high social status who can provide protection and material resources (Symons 1979; Buss 1989), even under conditions where (long-term) accumulation of resources is impossible (Betzig and Turke 1992; Hames 1996). However, if the role of males as providers is greatly overemphasized as some recent authors assume (e.g., Hrdy 1999; Hawkes et al. 2001), women should be especially sensitive to the males' genetic quality, and, as it appears, they are. At least they are sensitive to aesthetic traits in males, such as attractive, i.e., symmetrical and masculine, faces (Thornhill and Gangestad 1999), an above-average height (Pawlowski et al. 2000), and a moderately athletic body (Jackson 1992; Barber 1995).

In contrast to females, males can increase their reproductive success by impregnating as many eggs as possible without investing much in their offspring. So why should human males be choosy at all? Principally, the question of choosiness is a question of a cost-benefit-calculation (e.g., Alatalo et al. 1998); if the benefits relative to the costs are either very high or very low, mate selectivity may result. A male should be choosy, for example, when he seeks a long-term partner, because the choice of the right female can be crucial for his reproductive success. She should be young, because with increasing age, her reproductive value rapidly declines, she should be healthy, and she should have maternal qualities, because good and healthy mothers are likely to successfully rear healthy offspring. Such a 'high-benefit scenario' corresponds to male preferences for very young women (e.g., Voland and Engel 1990), whose fecundity may be lower than among women in their mid-twenties, but whose residual reproductive value is higher (Symons 1979; Buss 1989). Moreover, in the 'environment of evolutionary adaptedness' (Bowlby 1969), women in their twenties may actually have been less fecund than younger women, because the latter's fecundity was less constrained by previous births and periods of lactational amenorrhea (Profet, quoted in Symons 1995). Thus, it would not be surprising if a male who had the choice between a very young woman or a slightly older, but lactating one, would choose the younger, and in this case also the more fecund woman. In both cases, however, the male also incurs a cost, because he has to compete with other males over access to females, and he may lose. One might also argue, however, that the costs for a male, at least for a somewhat attractive one, might be so small that the choice is worthwhile anyway, even if the benefits are small. The fact that, under certain ecological conditions, women compete for men (Gaulin and Boster 1990; Cashdan 1995) can make male mate choice decidedly easier.

Yet why is it adaptive for women to make themselves attractive and perhaps even compete actively for men? At least in cultures where resources are controlled by a few, powerful males it is necessary for women to compete for males as 'resource objects', just as other primate females compete for access to food (Hrdy 1997). In fact, women seem to compete mainly for men who can guarantee them long-term supply, and far less for men with a low income or when they seek a short-term sexual adventure (Schmitt and Buss 1996; Buss and Shackelford 1997). Since the female's physical beauty constitutes an important factor of female mate competition, the question of the cost-benefit ratio must be posed in this case as well. In general, sexually selected signals are assumed to be costly, but whether this is also true for female physical beauty remains to be shown (but see Zahavi and Zahavi 1997). As a matter of fact, how much additional energetic and metabolic costs a woman signaling broad hips and large breasts might incur is still unknown. Moreover, these costs may in fact be negligible since extra fat deposits do not appear to be otherwise wasteful ornaments, but can be directly used during periods of food shortage.

No matter what the exact logic of human mutual mate choice and both male–male and female–female competition may be, the consequences are rather obvious. Since both sexes are choosy, any mating decision is, at best, a compromise of conflicting interests which will be more difficult the more the sex-typical and individual preferences diverge. Sexual coercion and, in its extreme expression, rape, can be seen as a last resort for the male sex when no compromise can be achieved (compare Thornhill and Palmer 2000).

Individualistic Taste

The assessment of beauty is not always objective in the sense that it can be normed, but often simply intersubjective. The Fisherian objectivity of beauty is only an illusion which, like fashion everywhere, depends on the often arbitrary taste of the mass. The phenomenon of mate choice copying, which probably serves to reduce costs of mate choice (Westneat et al. 2000) is another example for the intersubjectivity of aesthetic taste: I simply choose what others find beautiful. Aesthetic preference can also be completely subjective: I am sensitive only to those who match my aesthetic sense irrespective of what others may find beautiful. Judging the attractiveness of a potential mate by his body odor is an example. At least in humans and mice, the odor is conveyed by the genes of the major histocompatibility complex (MHC), which is a highly diverse gene complex crucial to the immune response. Women prefer the body odor of MHC-dissimilar males (Wedekind et al. 1995), which ideally results in MHC-disassortative matings and a high degree of heterozygosity among offspring, which also decreases the risk of inbreeding depression (Wedekind and Füri 1997). Thus, at least in the context of MHC-based mate choice decisions, the heterozygosity hypothesis appears to be an excellent explanation for the biological functionality of disassortative mating (Penn and Potts 1999).

What an individual considers beautiful also depends on the context: 'good genes' are good only under certain conditions. Physical beauty, for example, is higher valued in environments with high levels of parasitic stress (Gangestad and Buss 1993), and the same appears to be true for low degrees of fluctuating asymmetry (Jones 1996). Moreover, in societies where females enjoy little education and freedom, they are more interested in the males' ability to provide resources (Cashdan 1996; Kasser and Sharma 1999). Finally, even individual standards are subject to change, suggesting that human mate choice is a conditional strategy (Buss 1998). Both sexes, for example, apply different criteria, depending on whether they are searching for a partner for a one-night stand or one for a long-term relationship. In the latter case, males prefer intelligent, friendly and somewhat attractive, but sexually inexperienced and relationship-oriented females, while in the former case females may be less intelligent, but should be sexually open and experienced as well as physically beautiful (Buss and Schmitt 1993). For females seeking short-term partners both physical attractiveness and immediate material benefits are important (Kenrick et al. 1990; Penton-Voak et al. 1999), while women seeking long-term partners prefer ambitious, older males with good financial prospects (Buss 1998). Female preferences are also cycle-dependent; dominance-signaling faces are favored in the fertile follicular phase (Penton-Voak et al. 1999), and preferences for the odor of males with a low degree of fluctuating asymmetry increase with increasing probability of conception (Gangestad and Thornhill 1998). These findings may support the view of a mixed female mating strategy, which generally values paternal investment, but during extra-pair matings also 'good genes' as important (e.g., Grammer 1993).

Even the fact that the 'best' choice often cannot be accomplished, influences human mate choice. The fact that highly attractive individuals usually do not only seek, but probably also get attractive partners (Waynforth and Dunbar 1995; Pawlowski and Dunbar 1999a), is not surprising. However, this rule even appears to affect our internal evaluation system; whom one finds attractive depends on how attractive one is (Little et al. 2001). Mate choice, therefore, always has several components of individuality, in which different cost-benefit ratios result in a rather high variability of partner preferences (Jennions and Petrie 1997).

The Dark Sides of Beauty

The life of the beauty is often expensive and dangerous. The most adorned individuals usually also suffer from the greatest rates of predation and the largest energetic and genetic costs (Cronin 1992; Brooks 2000). While the Fisherian runaway model suggests that beauty can evolve to a certain point despite its costs, the handicap hypothesis adopts the opposite viewpoint; beauty must be costly to function as a reliable indicator of quality for the choosy individual (Zahavi and Zahavi 1997). However, are aesthetic features always honest, as this model supposes? And, if so, are they automatically also costly?

Proponents of signal theory answer the first question clearly: The beauty signal can deceive (e.g., Krebs and Dawkins 1984; Dawkins and Guilford 1991; Johnstone and Grafen 1993; Hasson 1994; Hauser 1997; Møller and Swaddle 1997). Conflicts of interest in communication are common; what the sender of a signal wants is not always what the receiver wants. The sender, therefore, may benefit by manipulating the receiver. In an evolutionary stabile system, signals only have to be honest on average (Johnstone and Grafen 1993). If deception is rare, if the costs of ignoring potentially important signals such as warning calls are high, or if the costs of being deceived are low, dishonest signals will endure. From the deceiver's perspective, the deceiving signal should be cheaper to produce than the honest signal (Dawkins and Guilford 1991), or its benefit must be higher than its cost ('bluff'). There are various forms of deceptive signals among animals (Hasson 1997): mimicry disguises the identity of the sender, camouflage hides its presence, 'attenuators' mask particular features of the sender, and 'bluffs' signal a higher quality than actually present. Bluffs can persist alongside real handicaps because the cost-benefit ratio of the signal production is not identical for all members of a sender population (Hasson 1994). An example is the red mating color of male sticklebacks (*Gasterosteus aculeatus*), which shows a curvilinear distribution. Males in good condition, but also males in poor condition have larger red abdominal areas than males in ordinary condition. As a result, female sticklebacks are not able to distinguish 'fit' and 'unfit' males by their red coloring. Apparently, males in poor condition can invest more in the signal than males of medium fitness, probably because they no longer require survival reserves (Candolin 1999). In species living in complex social systems, concealing information (Cheney and Seyfarth 1990), 'tactical deception' (Byrne and Whiten 1985) and language-based lies are common and effective forms of deceptive signaling.

That cheating signals among species living in stable social systems occur frequently has been questioned for mainly two reasons. Firstly, among individuals familiar with each other cheating should not only be rare, but also much more subtle than in less social species (Cheney and Seyfarth 1990). Secondly, living in complex social groups not only promotes competition, but also cooperation which might reduce the frequency of deception up to the point that honesty prevails as the evolutionary more stable strategy (Markl 1985). However, it has been shown that deception is especially common and elaborated among primates (see Byrne and Whiten 1985, 1992; Savage-Rumbaugh and McDonald 1988; Cheney and Seyfarth 1990; Hauser 1997). Moreover, since the context of mate choice is predisposed for deception, it is not surprising to see that human females tend to present themselves as younger and more attractive, and males as more successful and relationship-oriented than they really are (Tooke and Camire 1991; Pawlowski and Dunbar 1999b). Moreover, deception is not limited to human language, which is composed of easily deceivable conventional signals (see Guilford and Dawkins 1995), and artificial alteration of natural beauty. Even 'natural' beauty may be deceptive; both the permanence of female breasts and the typical gynoid distribution of body fat (Low et al. 1987) and human facial attractiveness (Kalick et al. 1998) have been interpreted as potential deceptions (see below).

Must honest beauty always be expensive? According to Maynard Smith and Harper (Maynard Smith 1994; Maynard Smith and Harper 1995; Hasson 1997), there are several forms of a signal which guarantee its reliability without incurring any costs other than the actual information transmission. Reliability is principally possible through three evolutionary mechanisms (Hasson 1997): the design of the signal, the use of additional costs when forming the signal, or through convention. A 'cost-added signal' (Maynard Smith and Harper 1995) is a 'handicap' (Zahavi 1975), whose costs exceed the costs of signal transmission because high investment makes the signal much more a reliable indicator of the genetic quality of the sender. Thus, the reliability of the signal follows the principle: "The larger the investment, the more reliable the signal" (Zahavi 1993, 227). However, in signals which are reliable by design the costs do not exceed those for signal transmission. 'Indices', for example, are perfectly reliable because they are physically (and therefore inevitably) connected with the sender's quality (Maynard Smith and Harper 1995; see also Hasson 1997). To distinguish between indices and extra-cost signals may often be difficult, but the tail of male barn swallows (*Hirundo rustica*) provides an example (Maynard Smith and Harper 1995): females prefer males with long, symmetrical tails (Møller 1995). A long tail is a cost-added signal because it hinders while flying, whereas a symmetrical tail is an index since it aids flying and may signal genetic stability. If female swallows paid attention only to extra-cost signals, they would choose males with long, but asymmetrical tails. Traits produced by a Fisherian runaway process are a special case within signal theory because the question of reliability is relevant for condition-dependent traits only, but not for Fisherian traits which are preferred for other reasons. Fisherian traits do not advertise quality, but only beauty and this is the reason why they cannot be used for deception (but see Hasson 1994); they only promise what is already visible.

The Evolution of Human Beauty and Partner Preferences

While studies on nonhuman species greatly enhanced our knowledge about the evolution of sexually selected traits and preferences during the last decades (Andersson 1994), little is known about the exact evolutionary paths of such traits in the human species. Moreover, some authors even doubt that one could reliably assign a certain trait to one or another of the models discussed above (e.g., Cronin 1992; Ridley 1994), which are not necessarily mutually exclusive hypotheses, but might rather represent different points at a continuum (Kokko 2001) or explain different stages in the evolution of sexually selected traits (Pomiankowski and Isawa 1993; Miller 2000).

The Beard

Ever since Darwin (1871), the human beard is considered to be a result of female mate choice and, indeed, the degree of male facial hairiness appears to influence male attractiveness (Barber 1995; Etcoff 1999). Beards grow under the influence of testosterone, which also affect male height and muscularity. Thus, a beard could be an indicator of a male's overall condition (Thornhill and Gangestad 1993). But is it a handicap? It has been suggested that a beard "can make a man vulnerable in fight", that it is "a way of showing off male confidence", and that it reduces "the face's apparent size" and "lateral vision" (Zahavi and Zahavi 1997, p. 213). The beard, therefore, appears to be a costly signal. A somewhat more elegant solution to the cost problem, however, states that because testosterone suppresses the immune system and therefore increases vulnerability to pathogens (Folstad and Skarstein 1997), a beard is a reliable signal. High testosterone levels would be costly, therefore, and only high-quality males could afford the suppression of their immune system and the growth of a large beard. However, testosterone not only confers costs to the individual, but also benefits. In fact, the 'immunocompetence handicap' hypothesis (Folstad and Skarstein 1997) suggests that some suppression of the immune system by high levels of testosterone might be beneficial because it is required for spermatogenesis. The full, unsuppressed immune response would regard haploid sperm as antigens which are to be destroyed, which would considerably affect male fitness. According to this hypothesis, high levels of testosterone are costly, but necessary, and by inciting sperm competition females could evaluate the hereditary male pathogen resistance, because resistant males would also produce high quality sperm (Folstad and Skarstein 1997). From this point of view, the beard does not appear to be a handicap, but rather, just as any other testosterone-dependent trait, a signal that is honest by necessity, or what

Maynard Smith and Harper (1995) called an 'index'; testosterone levels are not raised to produce the signal, but the signal is merely an expression of a required, and therefore tolerated, evil.

There are at least two aspects that cast doubt on the good-genes interpretation of the human beard. First, the beard can hardly be used as a reliable indicator of phenotypic quality, since beards grow with increasing age, while phenotypic quality usually decreases with increasing age (Barber 1995). Moreover, although it has been argued that reaching old age by itself may be an indicator of good genes (Trivers 1972), this is not necessarily the case (Brooks and Kemp 2001). In addition, although older males may be, on average, socially dominant over younger males, within age cohorts there is no obvious relationship between characteristics of the beard and social dominance (Barber 1995). Second, characteristics of human facial hairiness show a high geographic variation (Kingdon 1993; see also Darwin 1871), which is difficult to explain by the good-genes hypothesis, but entirely consistent with a Fisherian runaway process (Barber 1995).

Male Voice

Why do human males' voices break, but not females'? Did the male voice evolve to attract females? Recent experiments suggest that this is the case (Collins 2000): When visual contact is precluded, women tend to associate deeper voices and closely spaced harmonics with masculinity, attractiveness, and older age. Unfortunately, none of these assessments appears to be true. Men with a more 'male' voice were neither more masculine, nor more attractive or older than others. Moreover, there was no correlation between testosterone levels and characteristics of the males' voices. Obviously, therefore, these vocal parameters are not honest signals of male quality, even though they apparently have great influence on the attractiveness of the sender, so what is the reason for this gross misevaluation? One possibility is a cognitive problem, the so-called peak-shift effect (Weary et al. 1993; Collins 2000) which leads to a stronger response to stimuli when they are located at the respective ends of a continuum. However, this approach does not explain why the cognitive error was maintained despite its costs; if the perceived signal does not advertise male quality, it would decrease, rather than increase female fitness. One possible solution is 'sensory exploitation' (Endler and Basolo 1998), where males exploit a sensory bias. Unlike women, men are not as easily fooled by the peak-shift effect, probably because there was a strong selective pressure to recognize potentially dangerous rivals by their voices (Fitch and Hauser 1995). Obviously, however, there is little reason to assume that it would not have been beneficial for women to be able to recognize a male's ability to protect females and their offspring, or to provide them with resources or just good genes (see Qvarnström and Forsgren 1998). Rather than reflecting mere quantitative differences in selection pressures, sex differential abilities in recognizing male traits from their voices appear to reflect some differences in the signaler–receiver relationship. However, if a woman cannot perceive a male's quality by his vocal attributes, what does she perceive? From a Fisherian view, perhaps pure attractiveness. This interpretation is supported by the fact that the extreme and rare form of the trait, a very deep, masculine voice, is considered more attractive than the average male voice (see also Cronin 1992; Barber 1995). In this view, the male's voice would not be part of an attractive signal complex which advertises male quality, but an attractive Fisherian trait of its own that may complement, but also contradict other signals.

Facial Attractiveness

Why does facial attractiveness influence mate choice decisions of both sexes? Humans consider faces attractive that are symmetrical (e.g., Gangestad et al. 1994; Grammer and Thornhill 1994; Mealey et al. 1999) and average (e.g., Langlois and Roggman 1990; Langlois et al. 1994; see also Perrett et al. 1994). Female faces are considered beautiful if they appear young (Johnson and Franklin 1993; Jones 1995), or if they have very 'feminine' features (e.g., Perrett et al. 1994), while male faces are valued as attractive if they emanate 'masculinity' (Cunningham et al. 1990; Scheib et al. 1999) and high socioeconomic status (Mueller and Mazur 1997; Hume and Montgomerie 2001). Once again the good-genes hypothesis, or the handicap model, may provide an explanation and, in fact, it has been suggested that an attractive female face – together with a beautiful body – is a reliable indicator for good health, fecundity and immunocompetence (Thornhill and Grammer 1999). Just as the male beard, female facial traits are developed under the influence of a hormone, in this case estrogen, and as with testosterone, estrogen production appears to have its costs; estrogen seems to promote certain hereditary diseases, such as cancer (Service 1998). Thus, it might be argued that only women with good resistance to pathogens can afford the negative influence of estrogen. However, once again, this argument is only valid as long as a high estrogen level is really 'costly' in the Zahavian sense. In fact, estrogen appears to be essential for the normal functioning of large parts of female reproductive physiology (e.g., Dixson 1998). Moreover, there are several indications that high levels of estrogen increase female fecundity by stocking up certain fat reserves (see below). If an attractive face merely signals a high estrogen level, which is required anyway, it is not a handicap.

As noted above, symmetrical faces are rated as more attractive than asymmetrical ones, and since fluctuating asymmetries are supposed to be the result of developmental stress (e.g., Parsons 1990), it has been assumed that facial symmetry is an indicator of good genes. In fact, the degree of facial fluctuating asymmetry has been found to correlate with individual health (Shackelford and Larsen 1997). However, the relationship between low degrees of fluctuating asymmetry and good genes is less clear than it appears. First, in a recent study Hume and Montgomerie (2001) also found a correlation between facial fluctuating asymmetry and health, but only for female, and not for male faces. Second, a comprehensive analysis of recent studies could find no consistent correlation between stress and fluctuating asymmetry (Bjorksten et al. 2000, cf. Jones 1996). Third, there seems to be a considerable publication bias in studies reporting fluctuating asymmetries which makes meta-analyses extremely difficult (Palmer 1999). Fourth, the female preference for symmetrical male traits may merely be an artifact of female neuronal structures (Enquist and Arak 1994; Johnstone 1994), as well as children's preferences for attractive faces may result from general information-processing mechanisms (Rubenstein et al. 1999; see Jansson et al. 2002), both without any link to sender quality. Thus, whether facial symmetry is an important parameter for evaluating facial attractiveness and, thereby, phenotypic quality, remains unclear (Scheib et al. 1999; Thornhill and Gangestad 1999; cf. Hume and Montgomerie 2001).

What is clear, however, is that facial symmetry, if it is a quality advertising signal, does not require additional costs to be honest. Rather than being a handicap, facial symmetry would be an index. Moreover, the hypothesis that facial attractiveness might be explained by the handicap hypothesis has been challenged by at least one other study which found no indication that facial attractiveness was related to phenotypic quality (Kalick et al. 1998; see also Hume and Montgomerie 2001, for male faces). Although attractive individuals in this study were rated also as the more healthy individuals, it turned out that this was not the case. It might be, the authors speculated, that during the 'environment of evolutionary adaptedness' attractive traits were reliable indicators of phenotypic quality, but that these traits later became subject to a Fisherian runaway selection process (Kalick et al. 1998).

Jones (1995) proposed that the male preference for attractive faces is merely a result of the preference for young females with a high reproductive value. A general preference for 'neotene' traits could then have led to a situation where faces accentuating apparent youth became supra-normal releasers for males, even though no female quality or true youth is indicated (Jones 1995). Although not stated explicitly, Jones' approach implies that females exploit male preferences, and that they deceive them about their actual age by developing traits whose attractiveness males cannot evade. Jones' hypothesis of facial neoteny as a supra-normal releaser might also easily be integrated into a Fisherian perspective, however.

Female Hips and Breasts

The female 'waist-to-hip ratio' (WHR) represents another case whose significance for (male) mate choice remains unclear. In many cultures, males prefer females with a low WHR, i.e., females with a small waist and large hips (Singh and Luis 1995; but see Yu and Shepard 1998; Wetsman and Marlowe 1999). Unlike children and post-reproductive females, females in their reproductive life phase store fat deposits unevenly, with most going to breasts, hips and buttocks, possibly signaling the existence of important energy reserves (Singh 1993; Low 1999). At the same time, the WHR signals female reproductive state; older and pregnant females have broader waists than younger and fecund females. Thus, "A relatively narrow waist means I'm female, I'm young, and I'm not pregnant" (Low 1999, 80). Moreover, a low WHR appears to be a reliable predictor of high fertility (e.g., Kaye et al. 1990; Zaadstra et al. 1993), which may be mediated by the morphological and physiological variability of fat cells in different regions of the body, as well as their regulation by sex hormones (Björntorp 1991; Rebuffé-Scrive 1991). While estrogen promotes the deposition of fat in the gluteo-femoral region, testosterone has an inhibitory effect, and it is this fat deposit that will be later used for reproduction, especially during late pregnancy and lactation (Björntorp 1987). It has also been argued that the distribution of female fat deposits advertises general condition and health status (reviewed by Singh 1993). In the 'environment of evolutionary adaptedness' the nutritional status of a female was probably of great importance for her value as a mate, although recent studies cast doubt on the view that it is the WHR that makes a woman attractive. Instead, it appears that body mass and the absolute width of the waist (Tassinary and Hansen 1998; Wetsman and Marlowe 1999), or some visual cues that are used to assess the body-mass index ('BMI'; Tovée et al. 1999) are the main factors determining attractivity of the female body.

In any case, as long as it is the stored fat that makes female attractive for a male, the argument appears valid; the gynoid fat distribution might be seen as an honest (but again not necessarily costly) signal of fecundity and health. Nevertheless, some caveats remain. Firstly, it is not yet clear why estrogen influences the fat distribution this way, since there is no physiological reason to expect a concentration of readily available energy in a specific region. Could it be possible that estrogen controls the distribution of fat in such a way that its signal function is increased? Secondly, is it only body fat that determines the attractiveness of the female body? Even if a low WHR is the result of fat storage, this is barely independent of the skeletal system, which is also an important trait with respect to female reproduction. During birth, a broad pelvis is of considerable advantage because, compared to other primates, the size of the neonate's head relative to the pelvic outlet makes human birth extremely difficult. Thus, early human males unquestionably would have gained a considerable advantage by choosing mates with a broad pelvis which allowed relatively easy births. These consideration gave rise to the view that the typical gynoid fat distribution may be a deceptive rather than an honest signal (Low et al. 1987). A narrow waist emphasizes broad hips, and the development of fat deposits in the hip and buttock region can emphasize this even more. According to this argument, women exploit the male preference for a broad pelvis by distributing fat deposits in such a way as to make themselves more attractive. Depending on whether males were more interested in a broad pelvis for an easier birth, or in well stocked energy reserves, or in the female's current reproductive state, broad hips might, therefore, be interpreted as a manipulative or an honest signal.

Empirical data do not appear to solve the issue. A study on Peruvian Matsigenka Indians revealed that males who were still rather isolated from Western influences prefer females with a high WHR, while those living more exposed to Western media also prefer the Western ideal, namely, females with a low WHR (Yu and Shepard 1998). The authors suggested that in traditional societies, where marriage is mainly determined by kinship structures and people have direct access to information about the age and health status of their prospective mates, physical attributes may be less important than in modern societies, where such information is more difficult to obtain. A completely different approach suggests that males in some cultures may prefer females with a high WHR because these females appear to produce relatively more male than female offspring (Manning et al. 1996; Singh and Zambarano 1997). Whether a high WHR is the cause or an effect of an overproduction of sons, is not clear, however. In a recent study by Tovée et al. (2001), no relationship between preconceptional WHR and offspring sex ratio has been found. These authors suggest that the female body mass index (BMI) is the main factor determining female attractiveness, but that male preferences for a given BMI vary according to ecological conditions.

The biological function of the human female breasts poses even more problems. Human females differ from other primates in developing large prominent breasts even before the first pregnancy (e.g., Dixson 1998). Some authors suggest that human breasts evolved by natural selection, because under the rough conditions of a savanna habitat they might have served as an important fat reserve (e.g., Reynolds 1991). As a consequence, breasts could easily have become subject to sexual selection as a signal advertising parasite resistance and fecundity, as some other authors believe (Gallup 1982). However, "breasts do not accurately reflect the reproductive capability of women" (Singh 1993, p. 315), since their development is independent of reproductive functions. Thus, Zahavi and Zahavi (1997) consider breasts to be a handicap signaling a good nutritional status, whereby the effectiveness of the signal is enhanced by the concentration of the fat. In their view, breasts are costly because mobility is impeded and heat-loss increased. However, whether these 'costs' are really costs is not so clear. In a savanna environment greater heat loss might have been beneficial. Hence, are breasts truly a handicap? The opposite seems more likely, namely, that breasts are a cheap signal that benefits its carrier beyond the signal function during recurring periods of droughts.

Due to the production of milk-producing tissue, breasts enlarge prior to and during lactation. Large breasts, therefore, indicate rather infertility than fecundity. Thus, large breasts may have been favored because they advertise maternal qualities such as the ability to nourish offspring. If so, sexual selection might have favored females who deceive males about their maternal qualities by storing extra fat into the milk-producing tissue (Low et al. 1987; Low 1999). Moreover, it has been argued that permanent breasts confuse males about the females' actual cycle state, thereby forcing them into permanent investment (Smith 1984). However, since most other female primates have no externally visible signs of ovulation (Pawlowski 1999), and male care is nevertheless rare (Paul 1998), this argument seems rather unlikely. Human females might also have evolved early appearing, prominent breasts to help hominid adolescents compete with other females for a 'good husband' (Hrdy 1999; see also Bielert and Anderson 1985). Whatever the best explanation may be, at least two findings support the notion of a deceptive signal: first, there is no evidence that breast size influences milk production (Barber 1995), and second, during periods when no more milk is required, the apparently costly milk-producing tissue is reduced and replaced by fat.

Another line of reasoning suggests that the degree of breast symmetry honestly signals female fecundity. Indeed, males prefer females with more symmetrical breasts (Singh 1995) and these females have also been found to be on average more fecund (Møller et al. 1995; Manning et al. 1997). Moreover, good-genes models predict that larger structures should be more susceptible to developmental stressors and should thus show a higher degree of fluctuating asymmetry (Watson and Thornhill 1994), which has been found in female breasts (Møller et al. 1995).

Despite these findings, however, the signal function of the human breasts remains obscure. Is it thus likely that they evolved as a result of a Fisherian runaway process? The high morphological variability between populations appears to support the argument (Barber 1995; see Pomiankowski and Isawa 1998), but the similar high variability of male preferences within populations, as well as the general preference for breasts of medium size, renders such an explanation unlikely.

Conclusion

Aesthetic sense, it has been argued, most probably evolved as an adaptation of choosiness in the context of sexual selection that allows individuals to assess the quality of potential mates by external signals (Thornhill 1998). The handicap principle (e.g., Zahavi 1975; Zahavi and Zahavi 1997) has been considered to be one potentially crucial mechanism for both the evolution of artificial aesthetics by social selection (see Voland, this Vol.) and socially beneficial provisioning (Hawkes and Bliege Bird 2002) in humans. However, at least as far as 'natural' aesthetic traits are concerned, the situation seems much more complex. As sexually selected aesthetes, humans do not appear to be only 'honest' signalers, but at least potentially, cheaters as well, and they appear to use both natural and artificial indicators as well as Fisherian traits for making their reproductive decisions. While most recent studies seem to take good-genes concepts (e.g., Gangestad and Buss 1993; Møller et al. 1995; Manning et al 1996; Scheib et al. 1999; Thornhill and Gangestad 1999; but see Barber 1995; Jones 1995) – or especially the handicap model (e.g., Zahavi and Zahavi 1997; Sluming and Manning 2000) – as the main functional background for interpreting human mate choice decisions, the alternatives must not be ruled out when trying to find a comprehensive perspective on the nature of human physical beauty.

Unfortunately, distinguishing between Fisherian traits and indicator traits in humans remains notoriously difficult, not least because of problems in measuring the reproductive consequences of a given sexual signal in modern societies. Moreover, up to now, there is no convincing evidence supporting the applicability of the handicap-model to the evolution of physical attractiveness in humans, largely because the interpretation of signal costs as mere reliability costs remains highly speculative for any given trait.

The fact that our own beauty ensures the representation of our genes in the next generation, and that we, for the same reason, perceive certain traits shown by members of the other sex as beautiful, does not entail that beauty is necessarily honest or that it advertises genetic quality. 'Adaptedness', used in the context of sexual selection or any other signaler-receiver relationship, is not the same as 'utility' or 'honesty'. The fitness of a sexual selected trait depends on its frequency within the population (Maynard Smith 1982); it depends on the behavior of other members of the population, and also whether these offer the same or completely different goods. This explains why sexual selection can produce not only beautiful, but also useless and completely arbitrary traits, which only enhance the fitness of the individual, sometimes even at the expense of its group (Jones 1995). We, and with us most other animals, can do nothing but pursue the highest amount of beauty and in turn be impressed by the beauty of others. We might be consolidated by the fact that beauty always swings between the two extremes of adaptedness and nonadaptedness (Ridley 1990), no matter whether it expresses hereditary quality, or it is an arbitrary, extravagant Fisherian or even a completely deceptive trait.

Summary

Ever since Darwin, evolutionary biologists have discussed how apparently useless physical beauty – and partner preferences for this beauty – can be functionally interpreted as products of selective processes. One possible answer is that human aesthetic sense in relation to physical attractiveness evolved as an adaptation in the context of sexual selection that allows individuals to assess the quality of potential mates by sexual signals (Thornhill 1998). This "good-genes" hypothesis is not the only convincing explanation for the evolution of traits by mate choice, however. Several current choice models have received empirical and theoretical support, but discrimination between single models often remains difficult when applying them to certain phenomena. While good-genes models and especially the "handicap"-model (Zahavi and Zahavi 1997) start from the principle of honestly advertising pheno- and genotypic "quality", the Fisherian runaway model refers to mere attractiveness, and the concepts of chase-away selection, sensory exploitation and signal theory stress the possibility of signal choice that lowers the fitness of the receiver. In contrast to the "handicap"-view, but consistent with signal theory and empirical data the sexual aesthetics of men and women is seen here as a part of a signaler-receiver complex, where sexual attractiveness can be both honest advertising and cheating. The evaluation of five human traits potentially relevant to partner choice shows to what extent different models are able to explain the evolution of human sexual aesthetics. Up to now, convincing evidence for the "handicap"-model is lacking in this context.

Acknowledgements. I thank Jan Beise, Athanasios Chasiotis, Eckart Voland, and an anonymous reviewer for critical comments. Numerous discussions with Eckart Voland sharpened my understanding of many aspects of the theory of sexual selection. I am most indebted to Andreas Paul for particularly helpful criticism of the manuscript. During preparation of the chapter, I was supported by the Deutsche Forschungsgemeinschaft (grants Vo 310/9-1 and Vo 310/9-2).

References

Alatalo RV, Kotiaho J, Mappes J, Parri S (1998) Mate choice for offspring performance: major benefits or minor costs? Proc R Soc Lond Ser B 265:2297–2301
Andersson MB (1994) Sexual selection. Princeton University Press, Princeton
Bakker TCM (1993) Positive genetic correlation between female preference and preferred male ornament in sticklebacks. Nature 363:255–257
Barber N (1995) The evolutionary psychology of physical attractiveness: Sexual selection and human morphology. Ethol Sociobiol 16:395–424
Barrett C, Warner RR (1997) Female influences on male reproductive success. In: Gowaty PG (ed) Feminism and evolutionary biology: boundaries, intersections, and frontiers. Chapman and Hall, New York, pp 334–350
Bateson P (1983) Optimal outbreeding. In: Bateson P (ed) Mate choice. Cambridge University Press, Cambridge, pp 257–278
Betzig L, Turke P (1992) Fatherhood by rank on Ifaluk. In: Hewlett BS (ed) Father-child relations. Cultural and biosocial contexts. de Gruyter, New York, pp 111–129
Bielert C, Anderson CM (1985) Baboon sexual swellings and male response: a possible operational mammalian supernormal stimulus and response interaction. Int J Primatol 6:377–393
Birkhead T (2000) Promiscuity. An evolutionary history of sperm competition. Harvard University Press, Cambridge, MA
Björntorp P (1987) Fat cell distribution and metabolism. In: Wurtman RJ, Wurtman JJ (eds) Human obesity. New York Academy of Sciences, New York, pp 66–72
Björntorp P (1991) Adipose tissue distribution and function. Int J Obesity 15:67–81
Bjorksten TA, Fowler K, Pomiankowsi A (2000) What does sexual trait FA tell us about stress? Trends Ecol Evol 15:163–166
Bowlby J (1969) Attachment and loss, vol 1. Attachment. The Hogarth Press, London
Bradbury JW, Vehrenkamp SL (2000) Economic models of animal communication. Anim Behav 59:259–268
Brooks R (2000) Negative genetic correlation between male sexual attractiveness and survival. Nature 406:67–70
Brooks R, Kemp DJ (2001) Can older males deliver the good genes? Trends Ecol Evol 16:308–313
Brown JL (1997) A theory of mate choice based on heterozygosity. Behav Ecol 8:60–65
Buss DM (1989) Sex differences in human mate preferences: evolutionary hypotheses tested in 37 cultures. Behav Brain Sci 12:1–14
Buss DM (1994) The evolution of desire: strategies of human mating. Basic Books, New York
Buss DM (1998) The psychology of human mate selection: exploring the complexity of the strategic repertoire. In: Crawford C, Krebs DL (eds) Handbook of evolutionary psychology. Ideas, issues, and applications. Lawrence Erlbaum, Mahwah, NJ, pp 405–429
Buss DM, Schmitt DP (1993) Sexual strategies theory: an evolutionary perspective on human mating. Psychol Rev 100:204–232
Buss DM, Shackelford T (1997) From vigilance to violence: mate retention tactics in married couples. J Personality Social Psychol 72:346–361
Byrne RW, Whiten A (1985) Tactical deception of familiar individuals in baboons. Anim Behav 33:669–673
Byrne RW, Whiten A (1992) Cognitive evolution in primates: evidence from tactical deception. Man 27:609–627

Candolin U (1999) The relationship between signal quality and physical condition: is sexual signalling honest in the three-spine stickleback? Anim Behav 58:1261–1267
Cashdan E (1995) Hormones, sex, and status in women. Hormones Behav 29:354–366
Cashdan E (1996) Women's mating strategies. Evol Anthropol 5:134–143
Cheney DL, Seyfarth RM (1990) How monkeys see the world. Inside the mind of another species. University of Chicago Press, Chicago
Clutton-Brock TH, Vincent ACJ (1991) Sexual selection and the potential reproductive rates of males and females. Nature 351:58–60
Collins S (2000) Men's voices and women's choices. Anim Behav 60:773–780
Crnokrak P, Roff DA (1999) Inbreeding depression in the wild. Heredity 83:260–270
Cronin H (1992) The ant and the peacock. Cambridge University Press, Cambridge
Cunningham EJA, Birkhead TR (1998) Sex roles and sexual selection. Anim Behav 56:1311–1321
Cunningham MR, Barbee AP, Pike CL (1990) What do women want? Facial metric assessment of multiple motives in the perception of male facial physical attractiveness. J Personality Social Psychol 59:61–72
Daly M, Wilson M (1983) Sex, evolution, and behavior, 2nd edn. Willard Grant, Boston, MA
Darwin C (1871/1998) The descent of man and selection in relation to sex. Prometheus Books, Amherst, NY
Davis GWF, O'Donald P (1976) Sexual selection for a handicap. A critical analysis of Zahavi's model. J Theor Biol 57:345–354
Dawkins MS, Guilford T (1991) The corruption of honest signalling. Anim Behav 41:865–873
Dixson AF (1998) Primate sexuality. Comparative studies of the prosimians, monkeys, apes, and human beings. Oxford University Press, Oxford
Eberhard WG (1996) Female control: sexual selection by cryptic female choice. Princeton University Press, Princeton, NJ
Ebert D, Hamilton WD (1996) Sex against virulence: the coevolution of parasitic diseases. Trends Ecol Evol 11:79–81
Endler JA, Basolo A (1998) Sensory ecology, receiver biases and sexual selection. Trends Ecol Evol 13:415–420
Enquist M, Arak A (1994) Symmetry, beauty, and evolution. Nature 372:169–172
Etcoff N (1999) Survival of the prettiest – the science of beauty. Doubleday, New York
Etges WJ (1996) Sexual selection operating in a wild population of *Drosophila robusta*. Evolution 50:2095–2100
Fisher RA (1958) The genetical theory of natural selection, 2nd edn. Dover, New York
Fitch WT, Hauser MD (1995) Vocal production in nonhuman primates: acoustics, physiology and functional constraints on honest advertisement. Am J Primatol 37:191–219
Folstad I, Skarstein F (1997) Is male germ line control creating avenues for female choice? Behav Ecol 8:109–112
Futuyama DJ (1990) Evolutionsbiologie. Birkhäuser, Basel
Gallup GG (1982) Permanent breast enlargement in human females: a sociobiological analysis. J Human Evol 11:597–601
Gangestad SW, Buss DM (1993) Pathogen prevalence and human mate preferences. Ethol Sociobiol 14:89–96
Gangestad SW, Thornhill R (1998) Menstrual cycle variation in women's preferences for the scent of symmetrical men. Proc R Soc Lond Ser B 265:927–933
Gangestad SW, Thornhill R, Yeo RA (1994) Facial attractiveness, developmental stability, and fluctuating asymmetry. Ethol Sociobiol 15:73–85

Gaulin SJC, Boster JS (1990) Dowry as female competition. Am Anthropol 93:994–1005
Geary DC (1998) Male, female. The evolution of human sex differences. American Psychological Association, Washington, DC
Getty T (1999) Chase away sexual selection as noisy reliable signalling. Evolution 53:299–302
Gilburn AS, Day TH (1994) Evolution of female choice in seaweed flies: fisherian and good genes mechanisms operate in different populations. Proc R Soc Lond Ser B 255:159–165
Grafen A (1990a) Sexual selection unhandicapped by the Fisher process. J Theor Biol 144:473–516
Grafen A (1990b) Biological signals as handicaps. J Theor Biol 144:517–546
Grammer K (1993) 5-α-androst-16en-3α-on: a male pheromone? A brief report. Ethol Sociobiol 14:201–208
Grammer K, Thornhill R (1994) Human (*Homo sapiens*) facial attractiveness and sexual selection: the role of symmetry and averageness. J Comp Psychol 108:233–242
Guilford T, Dawkins MS (1991) Receiver psychology and the evolution of animal signals. Anim Behav 42:1–14
Guilford T, Dawkins MS (1995) What are conventional signals? Anim Behav 49:1689–1695
Hames R (1996) Costs and benefits of monogamy and polygyny for Yanomamö women. Ethol Sociobiol 17:181–199
Hamilton WD, Zuk M (1982) Heritable true fitness and bright birds: a role for parasites? Science 218:384–387
Harcourt AH, Harvey PH, Larson SG, Short RV (1981) Testis weight, body weight and breeding systems in primates. Nature 293:55–57
Hasson O (1994) Cheating signals. J Theor Biol 167:223–238
Hasson O (1997) Towards a general theory of biological signaling. J Theor Biol 185:139–156
Hauser MD (1997) Minding the behaviour of deception. In: Whiten A, Byrne RW (eds) Machiavellian intelligence II. Extensions and evaluations. Cambridge University Press, Cambridge, pp 112–143
Hawkes K, Bliege Bird R (2002) Showing off, handicap signaling, and the evolution of men's work. Evol Anthropol 11:58–67
Hawkes K, O'Connell JF, Blurton Jones NG (2001) Hunting and nuclear families. Some lessons from the Hadza about men's work. Curr Anthropol 42:681–709
Holland B, Rice WR (1998) Chase-away selection: antagonistic seduction versus resistance. Evolution 52:1–7
Hrdy SB (1997) Raising Darwin's consciousness. Female sexuality and the prehominid origins of patriarchy. Human Nat 8:1–49
Hrdy SB (1999) Mother nature. A history of mothers, infants, and natural selection. Pantheon, New York
Hume DK, Montgomerie R (2001) Facial attractiveness signals different aspects of "quality" in women and men. Evol Human Behav 22:93–112
Huxley JS (1938) Darwin's theory of sexual selection and the data subsumed by it, in the light of recent research. Am Nat 72:416–433
Jackson LA (1992) Physical appearance and gender: sociobiological and sociocultural perspectives. State University of New York Press, Albany, NY
Jansson L, Forkman B, Enquist M (2002) Experimental evidence of receiver bias for symmetry. Anim Behav 63:617–621
Jennions MD, Petrie M (1997) Variation in mate choice and mating preferences: a review of causes and consequences. Biol Rev 72:283–327

Johnson VS, Franklin M (1993) Is beauty in the eye of the beholder? Ethol Sociobiol 14:183–199

Johnstone RA (1994) Female preference for symmetrical males as a by-product of selection for mate recognition. Nature 372:172–175

Johnstone RA, Grafen A (1993) Dishonesty and the handicap principle. Anim Behav 46:759–764

Jones D (1995) Sexual selection, physical attractiveness, and facial neotony. Curr Anthropol 36:723–748

Jones D (1996) Physical attractiveness and the theory of sexual selection: results from five populations. Anthropological Papers 90, Museum of Anthropology, University of Michigan, Ann Arbor

Jones TM, Quinnell RJ, Balmford A (1998) Fisherian flies: the benefits of female choice in a lekking sandfly. Proc R Soc Lond Ser B 265:1651–1657

Kalick SM, Zebrowitz LA, Langlois JH, Johnson RM (1998) Does human facial attractiveness honestly advertise health? Psychol Sci 9:8–13

Kasser T, Sharma YS (1999) Reproductive freedom, educational equality, and females' preference for resource-acquisition characteristics in mates. Psychol Sci 10:374–377

Kaye SA, Folsom AR, Prineas RJ, Potter JD, Gapstur SM (1990) The association of body fat distribution with lifestyle and reproductive factors in a population study of postmenopausal women. Int J Obesity 14:583–591

Kenrick DT, Sadalla EK, Groth G, Trost MR (1990) Evolution, traits and the stages of human courtship: qualifying the parental investment model. In: Betzig L (ed) Human nature: a critical reader. Oxford University Press, Oxford, pp 213–222

Kingdon J (1993) Self-made man: human evolution from Eden to extinction. Wiley, New York

Kirkpatrick M (1982) Sexual selection and the evolution of female choice. Evolution 36:1–12

Kirkpatrick M (1986) The handicap mechanism of sexual selection does not work. Am Nat 127:222–240

Kirkpatrick M, Ryan MJ (1991) The evolution of mating preferences and the paradox of the lek. Nature 350:33–38

Kokko H (2001) Fisherian and 'good genes' benefits of mate choice: how (not) to distinguish between them. Ecol Lett 4:322–326

Kose M, Møller AP (1999) Sexual selection, feather breakage and parasites: the importance of white spots in the tail of the barn swallow (*Hirundo rustica*). Behav Ecol Sociobiol 45:430–436

Krebs JR, Dawkins R (1984) Animal signals: mindreading and manipulation. In: Krebs JR, Davies NB (eds) Behavioural ecology: an evolutionary approach., 2nd edn. Blackwell Scientific Publications, Oxford, pp 380–402

Lande R (1980) Sexual dimorphism, sexual selection and adaptation in polygenic characters. Evolution 34:292–305

Lande R (1981) Models of speciation by sexual selection on polygenic traits. Proc Natl Acad Sci USA 78:3721–3725

Langlois JR, Roggman LA (1990) Attractive faces are only average. Psychol Sci 1:115–121

Langlois JR, Roggman LA, Musselman L (1994) What is average and what is not average about attractive faces? Psychol Sci 5:214–220

Little AC, Burt DM, Penton-Voak IS, Perrett DI (2001) Self-perceived attractiveness influences human female preferences for sexual dimorphism and symmetry in male faces. Proc R Soc Lond Ser B 268:39–44

Low BS (1999) Why sex matters. A Darwinian look at human behavior. Princeton University Press, Princeton, NJ
Low BS, Alexander RD, Noonan KM (1987) Human hips, breasts and buttocks: is fat deceptive? Ethol Sociobiol 8:249–257
Manning JT, Anderton R, Washington SM (1996) Women's waists and the sex ratio of their progeny: evolutionary aspects of the ideal female body shape. J Human Evol 31:41–47
Manning JT, Koukourakis K, Brodie DA (1997) Fluctuating asymmetry, metabolic rate and sexual selection in human males. Evol Human Behav 18:15–21
Markl H (1985) Manipulation, modulation, information, cognition: some of the riddles of communication. In: Hölldobler B, Lindauer M (eds) Experimental behavioral ecology and sociobiology. Sinauer, Sunderland, MA
Maynard Smith J (1976) Sexual selection and the handicap principle. J Theor Biol 57:239–242
Maynard Smith J (1982) Evolution and the theory of games. Cambridge University Press, Cambridge
Maynard Smith J (1987) Sexual selection – a classification of models. In: Bradbury JW, Andersson MB (eds) Sexual selection: testing the alternatives. John Wiley, Chichester, pp 9–20
Maynard Smith J (1991) Honest signaling: the Philip Sidney game. Anim Behav 42:1034–1035
Maynard Smith J (1994) Must reliable signals always be costly? Anim Behav 47:1115–1120
Maynard Smith J, Harper DGC (1995) Animal signals: models and terminology. J Theor Biol 177:305–311
McLain DK (1998) Non-genetic benefits of mate choice: fecundity enhancement and sexy sons. Anim Behav 55:1191–1201
Mealey L, Bridgestock R, Townsend GC (1999) Symmetry and perceived facial attractiveness: a monozygotic cotwin comparison. J Pers Soc Psychol 76:151–158
Miller GF (2000) The mating mind. How sexual choice shaped the evolution of human nature. Doubleday, New York
Møller AP (1995) Sexual selection in the Barn Swallow. Oxford University Press, Oxford
Møller AP, Alatalo RV (1999) Good-genes effects in sexual selection. Proc R Soc Lond Ser B 266:85–91
Møller AP, Soler M, Thornhill R (1995) Breast asymmetry, sexual selection and human reproductive success. Ethol Sociobiol 16:207–219
Møller AP, Swaddle JP (1997) Asymmetry, developmental stability, and evolution. Oxford University Press, Oxford
Mueller U, Mazur A (1997) Facial dominance in *Homo sapiens* as honest signalling of male quality. Behav Ecol 8:569–579
Murdock GP (1981) Atlas of world cultures. University of Pittsburgh Press, Pittsburgh
O'Donald P (1962) The theory of sexual and natural selection. Heredity 22:499–518
O'Donald P (1980) Genetic models of sexual selection. Cambridge University Press, Cambridge
Palmer AR (1999) Detecting publication bias in meta-analyses: a case study of fluctuating asymmetry and sexual selection. Am Nat 154:220–233
Parker GA (1983) Mate quality and mating decisions. In: Bateson PPG (ed) Mate choice. Cambridge University Press, Cambridge, pp 141–166
Parsons PA (1990) Fluctuating asymmetry: an epigenetic measure of stress. Biol Rev 65:131–145

Paul A (1998) Von Affen und Menschen: Verhaltensbiologie der Primaten. Wissenschaftliche Buchgesellschaft, Darmstadt
Pawlowski B (1999) Loss of oestrus and concealed ovulation in human evolution. The case against the sexual selection hypothesis. Curr Anthropol 40:257–275
Pawlowski B, Dunbar R (1999a) Impact of market value on human mate choice decisions. Proc R Soc Lond Ser B 266:281–285
Pawlowski B, Dunbar R (1999b) Withholding age as putative deception in mate search tactics. Evol Human Behav 20:53–69
Pawlowski B, Dunbar RIM, Lipowicz A (2000) Tall men have more reproductive success. Nature 403:156
Penn DJ, Potts WK (1999) The evolution of mating preferences and major histocompatibility complex genes. Am Nat 153:145–164
Penton-Voak IS, Perrett DI, Castles DL, Kobayashi T, Burt DM, Murray LK, Minamisawa R (1999) Menstrual cycle alters face preference. Nature 399:741–742
Perrett DI, May KA, Yoshikawa S (1994) Facial shape and judgements of female attractiveness: Preferences for non-average characteristics. Nature 368:239–242
Pomiankowski A, Isawa Y (1993) Evolution of multiple sexual ornaments by Fisher's runaway process of sexual selection. Proc R Soc Lond Ser B 253:173–181
Pomiankowski A, Isawa Y (1998) Runaway ornament diversity caused by Fisherian sexual selection. Proc Natl Acad Sci USA 95:5106–5111
Pomiankowski A, Isawa Y, Nee S (1991) The evolution of costly mate preferences. I. Fisher and biased mutation. Evolution 45:1422–1430
Presgraves DC, Baker RH, Wilkinson GS (1999) Coevolution of sperm and female reproductive tract morphology in stalk-eyed flies. Proc R Soc Lond Ser B 266:1041–1047
Qvarnström A, Forsgren E (1998) Should females prefer dominant males? Trends Ecol Evol 13:498–501
Rebuffé-Scrive M (1991) Neuroregulation of adipose tissue: molecular and hormonal mechanisms. Int J Obesity 15:83–86
Reynolds V (1991) The biological basis of human patterns of mating and marriage. In: Reynolds V, Kellett J (eds) Mating and marriage. Oxford University Press, New York, pp 46–90
Rice WR, Holland B (1999) Reply to comments on the chase-away model of sexual selection. Evolution 53:302–306
Ridley M (1990) Evolution. Probleme – Themen – Fragen. Birkhäuser, Basel
Ridley M (1994) The red queen. Sex and the evolution of human nature. Penguin Books, London
Rubenstein AJ, Kalakanis L, Langlois JH (1999) Infant preferences for attractive faces: a cognitive explanation. Dev Psychol 35:848–855
Ryan MJ, Keddy-Hector A (1992) Directional patterns of female mate choice and the role of sensory biases. Am Nat 139:S4-S35
Savage-Rumbaugh S, McDonald K (1988) Deception and social manipulation in symbol-using apes. In: Byrne RW, Whiten A (eds) Machiavellian intelligence. Social expertise and the evolution of intellect in monkeys, apes, and humans. Oxford Science Publications, Oxford, pp 224–2370
Scheib JE, Gangestad SW, Thornhill R (1999) Facial attractiveness, symmetry and cues of good genes. Proc R Soc Lond Ser B 266:1913–1917
Schmitt DP, Buss DM (1996) Strategic self-promotion and competitor derogation: sex and context effects on the perceived effectiveness of mate attraction tactics. J Pers Soc Psychol 70:1185–1204

Searcy WA, Andersson M (1986) Sexual selection and the evolution of male song. Annu Rev Ecol Syst 17:507–533

Service R (1998) New role of estrogen in cancer. Science 279:1631–1632

Shackelford TK, Larsen RJ (1997) Facial asymmetry as an indicator of psychological, emotional, and physiological distress. J Pers Soc Psychol 72:456–466

Singh D (1993) Body shape and women's attractiveness: the critical role of waist-to-hip ratio. Human Nat 4:297–321

Singh D (1995) Female health, attractiveness and desirability for relationships: role of breast asymmetry and waist-to-hip ratio. Ethol Sociobiol 16:465–481

Singh D, Luis S (1995) Ethnic and gender consensus for the effect of waist-to-hip-ratio on judgement of womens' attractiveness. Human Nat 6:51–65

Singh D, Zambarano RJ (1997) Offspring sex ratio in women with android body fat distribution. Human Biol 69:545–556

Sluming VA, Manning JT (2000) Second to fourth digit ratio in elite musicians: evidence for musical ability as an honest signal of male fitness. Evol Human Behav 21:1–9

Smith RL (1984) Human sperm competition. In: Smith RL (ed) Sperm competition and the evolution of animal mating systems. Academic Press, London, pp 601–659

Symons D (1979) The evolution of human sexuality. Oxford University Press, New York

Symons D (1995) Beauty is in the adaptations of the beholder: the evolutionary psychology of human female sexual attractiveness. In: Abramson PR, Symons D (eds) Sexual nature, sexual culture. University of Chicago Press, Chicago, pp 80–118

Tassinary LG, Hansen KA (1998) A critical test of the waist-to-hip ratio hypothesis of female physical attractiveness. Psychol Sci 9:150–155

Thiessen D, Young RK, Burroughs R (1992) Lonely hearts advertisements reflect sexually dimorphic mating strategies. Ethol Sociobiol 14:209–229

Thornhill R (1998) Darwinian aesthetics. In: Crawford C, Krebs DL (eds) Handbook of evolutionary psychology. Ideas, issues, and applications. Lawrence Erlbaum, Mahwah, NJ, pp 543–572

Thornhill R, Gangestad SW (1993) Human facial beauty: averageness, symmetry and parasite resistance. Human Nat 4:237–269

Thornhill R, Gangestad SW (1999) Facial attractiveness. Trends Cognitive Sci 3:452–460

Thornhill R, Grammer K (1999) The body and face of woman: one ornament that signals quality? Evol Human Behav 20:105–120

Thornhill R, Palmer C (2000) A natural history of rape: biological bases of sexual coercion. MIT Press, Cambridge, MA

Tooke W, Camire L (1991) Patterns of deception in intersexual and intrasexual mating strategies. Ethol Sociobiol 12:345–364

Tovée MJ, Maisey DS, Emery JL, Cornelissen PL (1999) Visual cues to female physical attractiveness. Proc R Soc Lond Ser B 266:211–218

Tovée MJ, Brown JE, Jacobs D (2001) Maternal waist-to-hip ratio does not predict child gender. Proc R Soc Lond Ser B 268:1007–1010

Trivers RL (1972) Parental investment and sexual selection. In: Campbell B (ed) Sexual selection and the descent of man, 1871–1971. Heinemann, London, pp 136–179

Voland E, Engel C (1990) Female choice in humans. A conditional mate selection strategy of the Krummhörn women. (Germany, 1720–1874). Ethology 84:144–154

Wallace AR (1889) Darwinism. Macmillan, London

Watson PJ, Thornhill R (1994) Fluctuating asymmetry and sexual selection. Trends Ecol Evol 9:21–25

Waynforth D, Dunbar RIM (1995) Conditional mate choice strategies in humans: evidence from lonely hearts advertisements. Behaviour 132:755–779

Weary DM, Guilford TC, Weisman RG (1993) Peak shifts produced by correlated response to selection. Evolution 47:280–290

Wedekind C, Füri S (1997) Body odour preferences in men and women: do they aim for specific MHC combinations or simply heterozygosity? Proc R Soc Lond Ser B 264:1471–1479

Wedekind C, Seebeck T, Bettens F, Paepke A (1995) MHC-dependent mate preferences in humans. Proc R Soc Lond Ser B 260:245–249

Wedell N, Tregenza T (1999) Successful fathers sire successful sons. Evolution 53:620–625

Westneat DF, Walters A, McCarthy TM, Hatch MI, Hein WK (2000) Alternative mechanisms of nonindependent mate choice. Anim Behav 59:467–476

Wetsman A, Marlowe F (1999) How universal are preferences for female waist-to-hip ratios? Evidence from the Hadza of Tanzania. Evol Human Behav 20:219–228

Whittier TS, Kaneshiro KY (1995) Intersexual selection in the Mediterranean fruit fly: does female choice enhance fitness? Evolution 49:990–996

Wiederman MW (1993) Evolved gender differences in mate preferences: evidence from personal advertisements. Ethol Sociobiol 14:331–352

Wiley RH, Poston J (1996) Indirect mate choice, competition for mates, and coevolution of the sexes. Evolution 50:1371–1381

Wilkinson GS, Reillo PR (1994) Female choice response to artificial selection of an exaggerated male trait in a stalk-eyed fly. Proc R Soc Lond Ser B 255:1–6

Williams GC (1966) Adaptation and natural selection: a critique of some current evolutionary thought. Princeton University Press, Princeton

Willson MF (1990) Sexual selection in plants and animals. Trends Ecol Evol 5:210–214

Yu DW, Shepard GH Jr (1998) Is beauty in the eye of the beholder? Nature 396:321–322

Zaadstra BM, Seidell JC, van Noord PAH, te Velde ER, Habbema JDF, Vrieswijk B, Karbaat J (1993) Fat and female fecundity: prospective study of effect of body fat distribution on conception rates. Br Med J 306:484–487

Zahavi A (1975) Mate selection – a selection for a handicap. J Theor Biol 53:205–214

Zahavi A (1977) The cost of honesty: further remarks on the handicap principle. J Theor Biol 67:603–605

Zahavi A (1991) On the definition of sexual selection. Fisher's model and the evolution of waste and of signals in general. Anim Behav 42:501–503

Zahavi A (1993) The fallacy of conventional signalling. Philos Trans R Soc Lond Ser B 338:227–230

Zahavi A, Zahavi A (1997) The handicap principle. A missing piece of Darwin's puzzle. Oxford University Press, New York

Beyond Nature Versus Culture: A Multiple Fitness Analysis of Variations in Grooming

MICHAEL R. CUNNINGHAM and STEPHEN R. SHAMBLEN

Beyond Nature Versus Culture: A Multiple Fitness Analysis of Variations in Grooming

Discussion of the nature of personal aesthetics, or physical attractiveness, seems to alternate between two positions, which we term Natural Classicism and Cultural Constructivism. These two positions are illustrated in fashion historian Julian Robinson's (1998) account of a conversation that he had with art historian Sir Kenneth Clark:

> "Sir Kenneth Clark said that...his own interests lay in classical styles of beauty as seen from a purely Mediterranean viewpoint, which had reached its zenith in the marble sculptures of ancient Greece. He explained that to a great extent the aesthetic appeal of such beauty depended on perfect symmetry, regular features, and an unvarying adherence to the prevailing classical ideals of shape, form and measurable proportions. In turn, I explained that my inclinations and convictions had become firmly rooted in the notion that human beauty is a reflection of cultural perceptions and inherited ideas of aesthetics, and that such aesthetics were not immutable...I went on to say that all human ideals and notions of beauty appeared to be inextricably linked to the varying forms of symbolism to which cultural groups appear to become "addicted" and which by ritual becomes an important aspect of their lives, and that each new generation learns these notions and addictions in the same way as it learns all other cultural matters – thus human beauty exists only in the eyes of those with the specific knowledge and cultural heritage to perceive it." (pp. 13–14)

Robinson continued:

> "Today, in this same quest for beauty and an admired social identity, men and women of the Western world...diet to the point of anorexia, have parts of their intestines surgically removed, so as to reduce the waistline and lower their caloric intake, or have their buttocks, breasts, thighs or nose reshaped...In other parts of the world...where traditional

ideals and symbols of beauty differ quite markedly from those of Western society, cultural groups call on the aid of other techniques...Some mark their bodies with welts or branding irons, or distort parts of them with tight bindings or compression boards. And, although such notions of beauty are at variance with our own, the intention is the same, to obtain the approval of their peers, the respect of their elders, and the attention of members of the opposite sex. In some countries, such as with some tribal groups in Cameroon, West Africa, the Sudan, Mongolia, Nigeria, the Amazon, and on some Melanesian islands in the Bismarck Sea fatness is equated with beauty, so at puberty young women are placed in a specially constructed communal "fattening house", where they are fed an enriched diet. Later, they will proudly display their corpulence...Other cultures believe that the body without its traditional marks of civilization and symbols of beautification – such as tattooing and decorative scars...is not worthy of attention. The natural, unadorned human body is merely animal. It is the tattooing or scars that make a body human, beautiful and desirable." (pp. 20–21)

The Natural-Classical position, represented by Clark, suggests that physical attractiveness is a manifestation of one, or a small number, of ideal properties. That ideal property may be, for example, the golden proportion, symmetry, or averageness. Clark's view is inherently Platonic, in that the desirability of that property is believed to transcend historical time period and specific culture. Supporting the Natural-Classical view is the fact that the Grecian sculptures that he admired remain pleasing to the eye today, 4000 years after their creation, and nearly half a century after Clark's (1958) commentary. In its most recent manifestation, the Natural-Classical position has been embodied in evolutionary analyses of physical attractiveness (Cunningham 1981; Zebrowitz 1997; Buss 1999; Etcoff 1999; Langlois et al. 2000), which we will describe in more detail below.

The Cultural-Constructivist position, represented by Robinson, suggests that physical attractiveness is an arbitrary social creation. A quick review of Robinson's (1998) volume, National Geographic, or any anthropology text, will quickly convince the reader that people have presented themselves in a dizzying array of costumes and decorations over the millennia. The diversity in human physical self-presentation rivals the diversity seen around the world in food, language, music, and money.

Grooming and self-adornment behaviors are similar to language and music in that they convey meaningful self-expression, but are influenced by a symbolic tradition, and some knowledge of the symbolism is necessary for full understanding and appreciation. Grooming and self-adornment behaviors are similar to preparing, serving and consuming food, in that they have their origins in meeting basic biological needs, but also serve as the opportunity to demonstrate creativity, skill, individuality, and elicit pleasure from the recipient. Finally, grooming and self-adornment behaviors are similar to money, in that an authority that is distant in time or space may seem to dictate what shall, and what shall not, be seen as desirable or valuable. Indeed, Naomi Wolf (1991) saw a conspiracy behind current Western standards of beauty:

> "The more legal and material hindrances women have broken through, the more strictly and heavily and cruelly images of female beauty have come to weigh upon us....We are in a violent backlash against feminism that uses images of female beauty as a weapon against women's advancement: the beauty myth." (p. 10)
> The beauty myth tells a story: the quality called beauty objectively and universally exists. Women must want to embody it and men must want to possess women who embody it. This embodiment is an imperative for women but not men. Which situation is necessary and natural because it is biological, sexual and evolutionary. Strong men battle for beautiful women, and beautiful women are more reproductively successful. Women's beauty must correlate with their fertility, and since this system is based on sexual selection, it is inevitable and changeless. None of this is true. "'Beauty' is a currency system like the gold standard. Like any economy, it is determined by politics, and in the West, it is the last best belief system that keeps male dominance intact. In assigning value to women in a vertical hierarchy, according to a culturally imposed physical standard, it is an expression of power relations in which women must unnaturally compete for resources that men have appropriated to themselves.
> Beauty is not universal or changeless, though the West pretends that all ideals of female beauty stem from one Platonic ideal" (p. 12)

Despite Wolf's assertion of conclusiveness, this chapter will address the issue of whether beauty is an arbitrary, perhaps malevolent, social construction, or a universal ideal, by suggesting that the wrong question is being asked. We suggest that dividing the question, and focusing on the different needs served by beauty, will lead to a clearer understanding of human aesthetic judgment.

Early Scientific Formulations of Physical Attractiveness

Before advancing our differentiated reframing of the Natural-Classical vs. Cultural-Constructivist question, it may be helpful to briefly recapitulate the scientific progress on the perception of physical attractiveness. In the pioneering Minnesota studies of physical attractiveness (Walster et al. 1966; Dion et al. 1972; Dion and Berscheid 1974), physical attractiveness was implicitly treated as an arbitrary biosocial stereotype, akin to race or ethnic group. That is, facial appearance elicited acquired perceptual schemas or stereotypes that stimulated differential treatment of attractive and unattractive people. Such differential social treatment may cause attractive and unattractive targets to internalize the social responses that they receive, develop differential self-perceptions and behaviors, and provide a "kernel of truth" to perpetuate the stereotype (Snyder et al. 1977).

The power of physical attractiveness was confirmed in subsequent studies and literature reviews (Feingold 1990, 1992; Jackson 1992). The recent Langlois et al. (2000) meta-analytic review confirmed that attractive children and adults were judged more positively, and treated more positively, than unattractive children and adults, even by those who know them, and that attractive children and adults exhibited more positive behaviors and traits than unattractive children and adults.

A strength, and a weakness, of early research is that it treated physical attractiveness as a unitary dimension on which a given target might be rated low, medium or high. That was a strength, in that judges displayed substantial consensus in their attractiveness ratings, and rapid research progress was made on the social perception and consequences of physical attractiveness. That approach was a weakness because the unidimensional perspective was unable to account for the complexity of the features and attributes that caused a person to be seen as attractive, which lead to a certain agnosticism about the nature of beauty. Berscheid and Walster (1974), for example, cited Darwin's informal cross-cultural survey of beauty standards, which noted variation in the appreciation of facial hair, tattoos, tooth filing, and scarification, and quoted Darwin's conclusion: "It is certainly not true that there is any universal standard with respect to the human body" (p. 752).

Yet, Darwin was willing to consider the possibility of universal standards for judgment of the face, as in his statement:

> "Mr. Winwood Reade... who has had ample opportunities for observation... with the Negroes of...Africa...who have never associated with Europeans is convinced that their ideas of beauty are, on the whole, the same as ours; and Dr. Rohlfs writes to me the same effect with respect to Bornu and the countries inhabited by the Pullo tribes.... Capt. Burton, believes that a woman whom we consider beautiful is admired throughout the world" (pp. 748–749)

Just as Darwin was equivocal about cross-cultural standards of beauty, Berscheid and Walster (1974) were ambivalent about the contribution of specific physical characteristics to the judgment of physical attractiveness. They suggested that:

"Identification of the physical characteristics considered attractive in Western culture, or any other, seems a hopeless task...It is the total Gestalt which is important...Despite the probability that people respond to characteristics in combination, to the complete configuration, one individual characteristic which may account for a fair portion of attractiveness variance, particularly in men, is height." (p. 178)

Recent research indicates that features, as well as configural information, is involved in face processing (Cabeza and Kato 2000). Indeed, Berscheid and her associates demonstrated that height affected judgments of male physical attractiveness (Graziano et al. 1978). It follows that if one physical attribute affected the perception of attractiveness, then it would be reasonable to examine other such attributes. However, if one were to contemplate embarking on a program of research to explore the physical bases of physical attractiveness, the Natural-Classical vs. Cultural-Constructivist debate becomes germane. One would not want to spend a year studying the impact of red lips on judgments of physical attractiveness, only to have cosmetic fashions change, such that red lips are seen as unattractively out of style and pale pink are seen as glamorous. Therefore, an appropriate reframing of the Natural-Classical vs. Cultural-Constructivist question may be:

1. Which components of physical appearance are consistently favored cross-culturally, and what human needs are served by differential responses to such consistently favored attributes?
2. Which components of physical appearance show substantial variation across different times and cultures, and do specific ecological or social factors covary with such social constructions?

Although much of our own prior work focused on the first of these questions, this presentation will focus primarily on the latter issue, by examining the determinants of grooming and self-presentation. However, some attention to the first question is necessary to place matters in an appropriate evolutionary context.

Variability in the Perception of Physical Attractiveness

Langlois et al. (2000) noted that the social interpretation of physical attractiveness leads to several differential predictions. If attractiveness standards are arbitrary constructs, then there should be a great deal of variability in physical attractiveness judgments across cultures. In addition, because adults have been socialized to a greater extent than children, adults should make more consistent attractiveness judgments than children. That prediction was not supported in the Langlois et al. meta-analyses. Indeed, we previously noted (Cunningham et al. 2002) that consistencies in judgments of attractiveness were shown by Chinese, Indian, and English females judging Greek males (Thakerar and Iwawaki 1979); Cruzans and Americans rating White males and females (Maret and Harling 1985); Whites, Blacks and Chinese evaluating White and Chinese males and females (Bernstein et al. 1982); White South Africans and Americans judging White males and females (Morse et al. 1976); Whites, Asians and Hispanics judging Whites, Blacks, Asians and Hispanics (Cunningham et al. 1995); and Whites, Blacks, and Koreans judging Whites, Blacks, and Koreans (Zebrowitz et al. 1993). The high judges' consensus obtained in some studies was criticized because extremely attractive or unattractive stimuli were carefully pre-selected to elicit consistent judgments (Langlois et al. 2000), but stimulus pre-selection artifacts were not evident in all attractiveness perception studies (Cunningham et al. 1995).

Despite high levels of cross-cultural agreement, it should not be concluded from the foregoing that culture has no impact on responses to physical attractiveness. Dion et al. (1990) reported that, compared to individualistic research participants, collectivist research participants living in Toronto were less likely to display the physical attractiveness stereotype, or the assumption that a highly physically attractive target is more kind, considerate, sincere, and easy to communicate with than a less physically attractive target. However, while their social stereotypes diverged, it is interesting to note that collectivists and individualists made comparable judgments of who was, and who was not, physically attractive.

It might be argued that collectivist and individualist adults living in North America make comparable physical attractiveness judgments because they have been immersed in Western aesthetics. Cunningham et al. (1995), however, compared Taiwanese students who had varying degrees of exposure to Western media, and found that these judges made comparable evaluations of physical attractiveness regardless of their exposure to Western films, television and magazines.

Because adults have been socialized and young children have not, adults should make consistent attractiveness judgments, whereas children's responses should be inconsistent. However, contrary to the notion that the perception of attractiveness is solely attributable to cultural learning are the findings that 2–3-month-olds and 6–8-month-olds looked longer at attractive than unattractive faces (Langlois et al. 1987, 1991), that 12-month-old infants responded more positively to adults wearing attractive than unattractive masks, and played longer with attractive than unattractive dolls (Langlois et al. 1990).

Although the foregoing studies supported the Natural-Classical over the Cultural-Constructivist position, most of the foregoing studies did not specify which aspects of the target photos were responsible for eliciting the judgments of attractiveness. As a consequence, the existence of an underlying Natural-Classical ideal was primarily established by default.

Multiple Fitness Model

Cunningham and his associates (Cunningham 1981, 1985; Cunningham et al. 1990, 1995, 1997) drew upon social psychological (Berscheid and Walster 1974), ethological (Guthrie 1976) and evolutionary (Symons 1979) analyses to offer an integrative theoretical account of the perception of physical appearance, called the multiple fitness model. The multiple fitness model was developed to account for the perception of physical attractiveness ranging from the romantically unappealing to the stunningly beautiful world-class model. Although youthfulness and health (including freedom from disfiguring genetic disorders) have long been assumed to be desirable qualities afforded by physical attractiveness (Symons 1979), the multiple fitness model went beyond those expectations. The model has several tenets, supporting evidence for which is reviewed in detail in Cunningham et al. (2002).

1. Different categories of features suggest different forms of fitness in the perceptual target, which are desirable underlying personal qualities that aid adaptation to the physical and social environment. The perception of fitness as an ideal romantic partner involves a specific combination of features, including:
   a) Neonate features, such as large eyes and a small nose that conveys an exaggerated appearance of youthfulness, freshness, naïveté, and openness, and suggests fitness as a beneficiary of resources.
   b) Sexual maturity features, such as thick eyebrows and a large chin and shoulders in males, and high cheekbones, visible breasts and hips in females, which convey reproductive health, strength and ascendancy, and suggest fitness to fulfill sex role responsibilities.
   c) Senescence features, such as male pattern baldness and gray hair, which convey increased social maturity, reduced competitiveness, and fitness as a mentor. The absence of senescence features increases romantic attractiveness.
   d) Expressive features, such as a large smile and dilated pupils, which convey agreeableness, supportiveness, low likelihood of antisocial behavior (Dabbs 1997) and suggest fitness as a friend (Harker and Keltner 2001).
   e) Grooming features convey a variety of biological and symbolic meanings, including cultural standing and appropriateness and fitness as a compatriot, which will be discussed in more detail below.

2. Desirable features are most attractive at an optimal deviation from the average in the general population. Males may be more attractive when they are above average in height, rather than nine feet tall (Shepard and Strathman 1989).
3. Alternate combinations of facial features may stimulate different forms of attraction in the perceiver. For example, romantic attractiveness is what first comes to mind when physical attractiveness is evaluated, but that is not the only form of attraction. When one is traveling alone overseas, another individual's clothing and hairstyle may be sufficient to elicit the recognition of a compatriot, and attraction to a potential traveling partner.
4. The relative importance that an individual places on maturity and expressive features may be influenced by personal mood (Wong and Cunningham 1990; Cunningham et al. 1997) and hormone level (Penton-Voak et al. 1999). However, the variations that are preferred under such circumstances remain relatively close to the multiple fitness model's ideal.
5. The value of some facial and bodily qualities is universal, whereas the values of other qualities are influenced by the local culture and ecology, in a manner that will be described below.
6. Human faces may be primarily selected to enhance interactions with other people. This is supported by the finding that human faces have a greater impact on attractiveness judgments than bodies (Alicke et al. 1986). By contrast, human bodies may be primarily selected for adaptation and survival in the ecologies in which they exist. Humans live in a wide variety of ecologies, which may lead to more variation in preferred bodies across time and groups than variation in preferred faces (cf. Ford and Beach 1951). Thus, humans may prefer bodies that are best suited to the particular demands and opportunities afforded by their ecology.

The multiple fitness model focused on combinations of exceptional features and other characteristics that create desirable impressions such as youthfulness, sexiness and friendliness. Alternate models of facial attractiveness emphasize other qualities that may be related to attractiveness ratings, including symmetry and averageness. Our review of the literature on exceptional, symmetrical, and average features suggests that each quality may be effective in accounting for a segment of the continuum of physical attractiveness (Perrett et al. 1994; Cunningham et al. 2002). Extreme asymmetry is unattractive, because it conveys developmental abnormalities such as a weak immune system or genetic abnormalities. Attractiveness increases with increasing symmetry, but perfect symmetry does not appear to produce the pinnacle of attractiveness. Computer averaging of a normal distribution of faces produces a composite that is more attractive than many individuals, both because asymmetries and facial irregularities are deleted, and because the mean of the population is familiar and unoffending. We suggest that a symmetrical and average face is more attractive than a face with small eyes, large nose and low cheekbones. However, an averaged and symmetrical face is not, in our judgment, as attractive as a face with exceptional features, including large eyes, small nose, high cheekbones, small chin on females, and large chin on males, and a big smile.

Although we emphasize universal qualities in the attractive face, we acknowledge subtle cross-historical and cross-cultural variations in the ideal face. These seem to constitute variations on a theme, however, rather than major deviations. Pettijohn and Tesser (1999) hypothesized that female facial maturity would be in greater demand during economic downturns than during times of prosperity. The investigators gathered social and economic indicators, such as changes in consumer prices and disposable income. They also measured the faces of the top 81 American actresses for each year between 1932 and 1995. They found that popular actresses had more neoteny and less facial maturity during good times, including larger eyes, rounder cheeks, and smaller chins. The converse was evident during economic downturns, in which stronger and more rectangular female faces were popular. It is important to note that female chin size, for example, was only relatively larger during hard times; it probably did not fall into a range that would be considered unattractively masculine during times of prosperity. Because we reviewed additional information on the face in Cunningham et al. (2002), we will concentrate here on clothing, adornment, and the body.

Grooming Variables

The multiple fitness model uses the term "grooming variables" to refer to all controllable modifications of appearance, especially those encouraged by the culture, such as clothing, cosmetics, orthodontics, hair styling and coloring, scarification, tattooing, body piercing, jewelry, clothing, head and feet shaping during infancy, cosmetic surgery and fattening or dieting. Given the prospect of eliciting differentially favorable treatment by potential mates, by potential employers, and even by potential juries (Jackson 1992), it is not surprising that individuals would seek to modify their physical appearance. Osborne (1996) demonstrated, for example, that judgments of physical attractiveness could be altered simply by cosmetic use. However, it should be noted that self-modifications may be employed for motives other than beauty enhancement. Darwin (1871) observed:

> "Hardly any part of the body which can be unnaturally modified has escaped. The amount of suffering thus caused must have been extreme,

for many of the operations require several years for their completion, so that the idea of their necessity must be imperative. The motives are various; the men paint their bodies to make themselves appear terrible in battle: certain mutilations are connected with religious rites, or they mark the age of puberty, or the rank of the man, or they serve to distinguish the tribes. Among savages the same fashion prevails for long periods, and thus mutilations, from whatever cause first made, soon come to be valued as distinctive marks. But, self-adornment, vanity, and the admiration of others seem to be the commonest motives." (pp. 741–742)

Darwin correctly recognized that a variety of unusual self-adornments originated to signify tribal membership, or the attainment of status, rather than to enhance physical attractiveness, per se. Further, an individual can seek the admiration of others through self-adornment while recognizing that such self-adornments are bizarre rather than beautiful. An individual who paints his face and torso in his football team's colors probably thinks of the action as demonstrating his loyalty and increasing his acceptance among fellow sports fans, rather than as increasing his handsomeness. Similarly, an observer who lacked a native informant might mistakenly presume that the circumcision rituals performed on Jewish and many Christian babies were designed to enhance their sexual attractiveness.

The foregoing suggests the importance of establishing the cultural context of a grooming practice before drawing any conclusions. When we are puzzled by the grooming practices of other cultures, it is useful to recall that they may not have been striving to enhance their romantic attractiveness at the moment that they were being photographed.

Novelty and Stimulation

Repeated exposures to a stimulus, such as a line drawing or piece of music, may cause attraction (Zajonc 1968), but too much exposure can cause boredom and disinterest (Brickman and D'Amato 1975). The need for stimulation and variety may motivate the development and adoption of new grooming practices, at least within limits, as Darwin (1871) recognized:

"The men of each race prefer what they are accustomed to; they cannot endure any great change; but they like variety, and admire each characteristic carried to a moderate extreme. If all our women were to become as beautiful as the Venus di Medici, we should for a time be charmed, but we should soon wish for variety; and as soon as we had obtained variety, we should wish to see certain characteristics a little exaggerated beyond the then existing common standard." (p. 752)

We previously noted that some exaggerated facial characteristics, such as larger eyes or chin, may function both as attractive supernormal stimuli, and draw disproportionate attributions of the positive qualities associated with those characteristics. Beyond those effects, attraction to novelty may have adaptive value by contributing to increased immunocompetence in the offspring. The major histocompatibility complex (MHC) is a group of genes which are responsible for immune system functioning and influence body odor (Wedekind et al. 1995). Wedekind and Furi (1997) had men and women rate the pleasantness of the odor of T-shirts worn by men and women. Male and female participants rated T-shirts to smell more pleasant when the person who had worn the T-shirt had an MHC combination opposite from their own. This is beneficial, as it likely leads to the production of offspring with a heterozygous, and more efficient, combination of MHC genes.

Similarly, novelty in grooming practices may advertise to potential partners that one is bold and adventurous, and perhaps likely to produce robust and adventurous offspring (cf. Godin and Dugatkin 1996). Many aspects of fashion change simply to generate increased attention and a refreshed aesthetic response (and increase sales). However, social and ecological variables also seem to play a role in changes in fashion, as will be described below.

Fingernails and Neckties

A wide range of grooming practices may be employed because they suggest that the wearer has privileged access to valuable objects and costly services, and does not need to be concerned with durability or efficiency. Veblen's (1912) theory of the leisure class suggested that women from a number of cultures wear long fingernails as a status symbol to indicate "industrial exemption", or financial independence from the need to work in a farm field or factory. The practice of fingernail grooming was carried to extremes at the turn of the twenty-first century, with multicolored enamels and faux jewels. Decorative nails advertise that the wearer has the time and money to commission such miniature art. It remains unclear whether women attract the opposite gender as much as they enhance their status with their own through such displays.

Males generally devote less attention to their hair and clothing than females, but that does not mean that male grooming is inconsequential. Western men who wear costly clothing, such as a dark suit with a necktie, were generally rated more positively than those wearing less expensive apparel, such as a T-shirt (Hill et al. 1987; Cunningham et al. 1990). It is unlikely, however, that a suit and necktie are intrinsically attractive. The suit and necktie may have value in signifying industrial exemption by being expensive, uncomfortable, and awkward to wear when performing manual chores. Such social and economic cues may account for most of the increase in rated attractiveness of the individual who appears in such clothing, especially in women's judgments of men (Townsend and Levy 1990).

Female Hairstyle

Women's hairstyle also has been the subject of extensive elaboration, including the presentation of extensive height and decoration in eighteenth-century France. Vestiges of that style are evident in formal occasions in the West, where women often have their hair styled, sprayed and pinned to be displayed above the head. In less formal circumstances, the hair is let down, but is still subjected to substantial investment. Some of that investment pays off in the form of increased social attention. Cunningham et al. (1995) found that women with greater hair width, due to hair styling, received higher attractiveness ratings.

A portion of the positive ratings associated with a pleasing hairstyle may be due to assumptions about the personality of the individual who devotes the effort to good grooming. Women who wore their hair stylishly combed were seen as more reliable than women who wore their hair more casually arranged (Graham and Jouhar 1981). Perceivers also make judgments about personality based on the color of a person's hair. Blondes were seen as warmer, but also as weaker-willed, and less intelligent and complex, than brunettes (Lawson 1971). Blond hair is more common in children than adults, so perceivers may have generalized their stereotypes about naïve and compliant young children to their perception of blondes (Cunningham et al. 1989). An unexplored question is whether women who engage in the grooming behavior of changing their hair color to blond consciously wish to be seen as younger, warmer or more helpless, and women who change their hair color to brunette or black wish to be seen as more intelligent or mature.

There is little doubt that women who change their hair color to conceal gray do so in order to appear younger, which carries with it a variety of desirable attributes including higher mate value. Hinsz et al. (2001) argued that two dimensions of women's hair could serve as a reliable clue to mate value. The investigators collected samples of hair from more than two hundred women age 13–73, as well as self-reports. Younger women, with greater reproductive potential, tended to wear their hair longer than older women, with lower reproductive potential. In addition, women whose hair was rated as possessing higher quality by beauticians reported themselves to be in better overall health than women whose hair was rated lower. Although it was not clear how much of the relation between hair quality and health was due to age, and how many of the women had changed their hair color in this study, it seems reasonable that young healthy women would advertise their high reproductive potential by displaying long and lustrous hair. A variety of other features, such as the brightness of teeth, may be similar to hair in conveying an underlying biological fitness that is enhanced by careful grooming.

Male Hairstyle and Baldness

Synott (1987) noted the symbolism of hair, suggesting that opposite sexes, and opposite ideologies, often have opposite hair styles. If long hair was seen as attractively feminine, short hair was seen as masculine, at least in the West (Pancer and Meindl 1978). The Western association of short hair with masculinity, however, was not evident among Native Americans, such as the Lakota Sioux, and probably had its origins in Western military discipline, which expected uniformity in the soldier's hair and his clothing. Western men who reject hierarchy and discipline may display the opposite hairstyle, and wear their hair long. As a consequence, men of similar tastes, temperaments and political persuasions may recognize kindred spirits and compatriots by their hairstyles and other grooming features.

Yet, while short scalp hair may be seen as highly masculine in the West, the senescence feature of male pattern baldness may have the opposite effect. Muscarella and Cunningham (1996) suggested that male pattern baldness may have evolved as an appeasement signal, to convey that the male possesses valuable experience and wisdom and should not be the target of intrasexual competitiveness. Consistent with this reasoning, male baldness reduced perceived masculinity, potency, assertiveness, and romantic attractiveness, but increased the appearance of age, appeasement, and social maturity (Roll and Verinis 1971; Cash 1990; Muscarella and Cunningham 1996; Henss 2001). It might be noted that these effects were obtained for the appearance of androgenetic alopecia, which entails baldness on the top and a band of darker, thicker hair on the sides. Males who shave their heads, including the hair on the sides may be perceived in a manner similar to males who wear a very short crewcut, as highly masculine and not very concerned about their appearance.

The multiple fitness model suggested that grooming features can convey biological and cultural appropriateness, and contribute to the perception of fitness as a compatriot. Long fingernails, neckties, elaborate hairstyles, and other grooming features may convey socially desirable status and wealth by being costly and difficult to maintain. Once a grooming practice has been established as a status symbol, however, some individuals may seek to increase their attractiveness by cloning the symbol without possessing the requisite status. Distinguishing oneself from the copy-cats requires that the elite periodically change their fashions, which provides a second impetus, besides novelty, for random changes in grooming behavior. However, fashion behavior is not always random, as we will describe below.

Ecological Variation and Female Dress Styles

The multiple fitness model echoed prior analyses that suggested that grooming features that convey sexual availability, such as a higher skirt hem, plunging neckline, and a voluptuous figure, are a reflection of ecological and social dynamics. Richardson and Kroeber (1940) assessed the length and width of skirts and other aspects of the dress each year across 3 centuries. Recurring cycles of fashion for such variables as widest possible skirt were evident, with a periodicity of approximately a century. No inferential tests were presented demonstrating periodicity, but moving averages were used to smooth the time series function, which would reduce the impact of random fluctuations in dress styles. Periodicity was demonstrated by examining the trough to trough or peak to peak in skirt width as one periodic function. The mean number of years between periodic functions was 100.6 years (50, 95, 132, 111, and 115 years). Mabry (1971) extended that observation by correlating the length of dresses shown in women's magazines over a 50-year-period with the value of the Dow Jones Industrial average, finding that as stock prices rose and fell, so did hemlines. Morris (1977) offered two possible explanations for that relationship:

> "Exactly why females should want to expose more of their legs when the economy is healthier, it is hard to understand, unless a sense of financial security makes them feel more brazenly invitatory towards the males. Perhaps the general atmosphere of financial activity makes them feel more physically active – a condition favored by shorter, less hampering skirt-lengths." (p. 221)

Barber (1999) offered a third explanation, based on Guttentag and Secord's (1983) account of the impact of the sex ratio on relations between the genders. The operational sex ratio is typically expressed in terms of the number of men per 100 women in a population. If the sex ratio is high, say 110 men for every 100 women, then many men will have difficulty in obtaining mates. By contrast, if the sex ratio is low, say 90 men for every 100 women, then many women will have difficulty obtaining men, at least in monogamous cultures. Guttentag and Secord suggested that when the sex ratio is high and men outnumber women, then women will be more selective, enforce standards of refinement and decorum in men, and men will enforce norms against premarital and extramarital sexual relationships. By contrast, when the sex ratio is low and there is an undersupply of men, some women will compete more strongly to attract men, advertising their reproductive value and engaging in more permissive sexual behavior. Some men will respond to such opportunities by reluctance to make a commitment, multiple liaisons, family abandonment, or divorce. Some women may disdain men in favor of fulfilling alternatives such as professional careers, and advocating structural change in the balance of political and economic power.

Barber (1999) extended the Guttentag and Secord analysis to dress styles, suggesting:

> "Assuming that long dresses convey signals of sexual restrictedness, these should covary with a good marriage market for women. Similarly, a narrow waist, which is indicative of high reproductive value, and low necklines, which call attention to the protrusion of the breasts, should be emphasized in dress styles at the same periods. Conversely, when women are opting to enter professions in large numbers, dresses should be shorter, necklines higher, and waists wider." (p. 451)

Barber (1999) used data from Richardson and Kroeber (1940), Mabry (1971) and Weeden (1977) to conduct a series of intriguing analyses. He found that a high population sex ratio (greater number of males relative to females) was positively correlated with longer and wider décolletage, narrower waists, greater skirt length, and inversely related to the number of BA degrees awarded (indicating lower female deployment of effort in careers), and lower divorce rates. Conversely, a lower population sex ratio (fewer males) was associated with less exposure of the bust, looser waists, shorter skirts, more female BA degrees and a higher divorce rate.

It seem reasonable that favorable marriage prospects for women would encourage them to advertise their reproductive value from the hips upward, while conveying restricted sexual access below, with long skirts. It is less clear why a poor marriage market for women would induce them to conceal their assets above the waist while showing them below. Barber (1999) suggested that women became more demonstrative to attract the limited number of available men that occurs when the sex ratio is low. However, it is unclear why women in such circumstances only show off below the waist and not above. It is possible that an asexual top combined with an inviting skirt conveys the message of availability for a short-term relationship, instead of marriage. Alternately, perhaps an asexual top transforms the meaning of a short skirt from one of sexual invitation to dynamic athleticism. Athleticism may be appropriate for a single career woman (although it is not clear that a short skirt is more athletic than trousers).

Barber's (1999) study also did not include such ecological variables as stock market value or infant mortality rates, so it was not possible to evaluate the causal effect of the material ecology on the marital economy. Poor economic circumstances tend to have more adverse impact on male than on female fetuses and newborns (Cunningham 1981), which skews the sex ratio and makes the causal variables difficult to untangle. Nonetheless, the possible relationships of sex ratio to education, divorce and dress styles are intriguing.

The foregoing focused on dress cut, but did not address the issue of dress color. Although clothing color influences social perception (Damhorst and Reed 1986; Frank and Gilovich 1988), no systematic research was located identifying the impact of ecological dynamics on clothing colors. Nonetheless, McClelland (1961) reported that preferences for blues and greens were associated with a higher need for achievement at the individual level, and greater economic progress at the state level, whereas preferences for reds and yellows were associated with a lower need for achievement and less economic development. It follows that sedate, dark clothing colors might be expected when times are tough and hard work is demanded, and vivid, brighter colors might be expected during times of prosperity. Informal observation, however, indicates that women's clothing tends to be more colorful then men's, and athletic, tropical and night wear are all more colorful than business attire (Lurie 1981). Individuals who are particularly dynamic, carefree or passionate, or wish to appear so, may attract people who desire such qualities through the display of such colors.

Female Body Weight and Curvaceousness

Body weight was classified as a grooming variable, because it is relatively controllable through voluntary exercise and dietary choice, at least when supplies of food and transportation are relatively secure. When the ecology offers few such resources, a body weight that displays stored capital in the form of pulchritude may be seen as attractive. Consistent with that logic, Sobal and Stunkard (1989) reported that thinness was associated with higher status in industrialized countries, whereas obesity was associated with higher status in developing countries.

Anderson et al. (1992) extended that analysis by systematically analyzing the correlates of desired female body shape in a cross-cultural survey. Of the 62 cultures for which data were available, 19.4% preferred slim women, 37.1% preferred women of moderate weight and 43.5% preferred plump women. Cultural preferences for slenderness were associated with the physical ecology variables of a reliable and effectively distributed food supply (although the relation of unreliability with obesity was stronger than the converse) and a warmer climate (based on greater proximity to the equator). A preference for slenderness was also evident in four social structural indicators of female power: anti-machismo attitudes; wives tending to dominate husbands; a lack of menstrual taboos, and a trend for higher value of female labor. Preference for slenderness was not an unmixed indicator of a desirable female social role, however, in that it was also correlated with adolescent female stress, including contradictory messages about female sexuality. Unfortunately, the small number of cultures with complete data prevented effective multiple regression analyses and determination of the most powerful predictors of slenderness.

Silverstein et al. (1986) examined slenderness in terms of bust to waist measures in Vogue and Ladies Home Journal from 1901 to 1981. Other researchers had observed that women were perceived as more feminine and attractive, but less competent, when they displayed a sexually mature full figure (Kleinke and Staneski 1980). Silverstein et al. found that when large numbers of women obtained college degrees and sought professional success, the slender and active female body was emphasized by the media, in preference to a more curvaceous body that appeared ready for marriage and child bearing. The emphasis on a slender and active female body may reflect sex-ratios that entail a poor marriage market for women (Barber 1998a, b), as will be described below.

Barber (1999) replicated Silverstein's findings that a slender, linear figure was seen as higher in career potential, but lower in reproductive potential, than a curvaceous, hourglass figure with high bust-to-waist and hip-to-waist ratios. Independent of curvaceousness, overweight figures also were seen as less attractive, but not less professionally competent, than slender figures. In addition, Barber (1998a,b) reanalyzed bust-to-hip ratios in Vogue from 1901 to 1993, and reported a general decline in curvaceousness over that time period. The decline in curvaceousness was correlated with a falling birth rate, a rise in the Standard and Poor stock market index, as well as increases in per capita gross national product, the proportion of women enrolled in higher education, and the percentage of women in the work force. Media de-emphasis of curvaceousness was not related to the population sex ratio, or the proportion of single women, but it was related to the ratio of single women to men in the prime marriageable age bracket of 20–24 years old. Given reduced prospects for marriage, young adult women may have been implicitly encouraged by the media to enhance their career prospects through slenderness, rather than display their fertility through curvaceousness.

As long as the job market is good, focusing on controllable aspects of one's life, such as grooming oneself to be slender for one's career, seems more adaptive than dressing for romantic partners who may never materialize. Unfortunately, some women emphasize their self-control, and deemphasize their sexuality, to the point of anorexia. It would be interesting to know if female anorexic tendencies could be reduced by exposure to an environment with a favorable ratio of men to women.

The Barber (1998a, b) studies did not determine the strongest predictor of the preference for slenderness, or exclude alternate interpretations based on other social dynamics. Increased medical knowledge of the dangers of obesity, for example, could induce an historical decline in curvaceousness that was independent of the sex ratio. If an emphasis on career over romance was operative, it might be evident in other aspects of female grooming, such as a de-emphasis of feminine long hair, or youthful blond hair, but such analyses have not yet been reported.

The strongest argument against a spurious relationship is the apparently cyclical pattern of both curvaceousness, and the ratio of men to women in college. Both variables fell from the early decades of the twentieth century to the 1920s, then rose from the 1920s to the late 1940s, only to decline again from the 1950s to the 1990s. Barber (1998b) suggested that: "The relationship depicted is not explainable in terms of any simple secular trend in a third variable because the graphs are curvilinear rather than linear" (p. 298). Barber's graphs were visually compelling, but the magnitude of the curvilinear relations, after controlling for linear effects, was not reported.

Personal Ecology and Within-Culture Variation in Romantic Grooming

Individuals in the same society may grow up and live in very different personal ecologies, with different degrees of access to resources. Such personal ecologies may influence an individual's approach to romance and self-presentation. Impoverished inner city neighborhoods have a constellation of variables that may destabilize long-term relationships (Geary 2000). These variables include: (1) males with low access to material resources, (2) high rates of male imprisonment due to both crime and selective law enforcement, (3) a low sex ratio due to poor health care and high rates of male infant mortality, which reduces females' marital opportunities, and (4) remaining males have increased sexual opportunities due to the excess of women.

Barber (2000) suggested that when the sex ratio is low, rates of adolescent pregnancy tend to be high, because "If a woman has no point in marrying a high investing spouse, there is little point in delaying reproduction while awaiting a favorable marriage" (p. 26). Barber (2000) conducted a survey of 185 countries, examining the impact of sex ratio, gross national product (GNP), percent urbanization, and population density and distance from the equator. He found that sex ratio and proximity to the equator were associated with an increased likelihood of teen pregnancy, whereas GNP, urbanization and population density were associated with decreased likelihood of teen pregnancy. Such results suggest that a general perception of a poor ecology, including decreased marriage prospects, few material resources, and a short life span due to equatorial pathogens, all may contribute to the decision to reproduce early rather than later in life. Such reproductive strategies, in turn, may influence the individual's emotional dynamics, self-perception, and the self-presentation of appearance, as the following will suggest.

Draper, Belsky and their associates (Draper and Harpending 1982, 1987, 1988; Draper and Belsky 1990; Belsky et al. 1991) described the impact of ecological variables on psychological dynamics. They suggested that when resources are poor or unpredictable, then parents are often stressed and in conflict with each other, which may cause the father to abandon the family. Impoverishment or father absence may cause the mother to be rejecting or coercive toward her children, as well. Father abandonment and poor treatment by mothers may cause females to develop an anxious or avoidant attachment style, making it difficult for them to form and sustain loving relationships (Kirkpatrick 1998). In such an ecologically challenging and emotionally loveless environment, young women may seek solace both through adolescent sexual liaisons, and through seeking impregnation, to produce a little person who will provide continuing affection. It may seem paradoxical that females with an avoidant and cynical attitude would engage in sexual activity at an early age, but these investigators suggest that females are pursuing a short-term reproductive strategy. That is, if the environment is harsh and people are untrustworthy, it may be more adaptive for a female to reproduce early, while she can, rather than take the unpredictable risk that she will be alive and attractive at a later date.

The relative experience of nurturance and stress during childhood may contribute to biological changes that encourage, or deter, the teens' display of sexual maturity (Belsky et al. 1991). In an 8-year prospective study, Ellis et al. (1999) found that "fathers' presence in the home, more time spent by fathers in child care, greater supportiveness in the parental dyad, more father-daughter affection, and more mother-daughter affection, as assessed prior to kindergarten, each predicted later pubertal timing by daughters in 7th grade" (p. 387).

The foregoing suggests a number of predictions for female grooming behavior. Females who experienced a harsh childhood and developed an avoidant attachment style may be likely to have more curvaceous figures, which advertise a readiness for child-bearing (Silverstein et al. 1986) due to early pubertal development (Ellis et al. 1999). Such women may engage in assertive presentations of their physical attractiveness, such as by wearing skirts that are shorter than other women, clothing that is cut to display more curvaceousness, and hair that is styled to be more eye-catching, compared to females with a secure attachment style. Such women also may be more attracted to men with transferable assets, such as money, rather than resources that may leave with the person, such as empathy or a good sense of humor. That tendency may be particularly extreme if the female's immediate personal ecology is acutely unsupportive, with a local shortage of males, or a rising level of unemployment levels. By contrast, females who were raised by supportive parents, and who developed a secure attachment style, may respond to a low sex ratio by displaying slenderness and focusing on their careers (Silverstein et al. 1986), rather than on early reproductive opportunities or a wealthy mate.

African-Americans live in a harsher ecology, with lower economic opportunities, and a substantially higher infant mortality rate than White Americans (Cunningham and Barbee 1991). Given such differences in ecology, it might be expected that what African-Americans perceive to be physically attractive differs from that of Whites. In fact, African-American men specified a body weight for the ideal date that was 7 pounds heavier than that preferred by European-Americans (Cunningham et al. 1995).

African-American's rejection of White American's asceticism may reflect group differences in implicit perceptions of access to nutrition (Furnham and Alibhai 1983), or it may be a function of the operational sex ratio. Guttentag and Secord (1983) noted that "there is an acute shortage of black men" (p. 199). Following the logic developed above, African-American women may have developed the grooming style of voluptuousness, rather than slenderness. Not only does the voluptuous aesthetic protect black women from anorexia and bulimia, but black women appear to maintain a generally more positive body image relative to their white counterparts (Thomas and James 1988; Rucker and Cash 1992; Shorter-Gooden and Washington 1996). It is possible that black women's recognition that they are in a disadvantaged ecology relieves them of some of the self-imposed stress and intrasexual competition experienced by white women. Such an external locus of control expectancy, however, may lead to feelings of helplessness, and put some women at greater risk for adolescent pregnancy, obesity, and inattention to personal health maintenance.

It might be noted that acceptance of pulchritude in the face of a harsh ecology may become a cultural value, and may not always be affected by personal prosperity. African-American college students from relatively advantaged circumstances were similar to others from their culture in being tolerant of higher levels of body weight, and protected from eating disorders (Wildes and Emery 2001). The reduced impact of socioeconomic status of body aesthetics in the African-American population should not, however, be seen as evidence of that group's insensitivity to ecological dynamics. Quite the contrary, data examined over a 25-year period indicated that as black infant mortality rates declined, black birth rates also declined, at rates parallel to that of whites (Cunningham and Barbee 1991). Data on mean body weight were not examined in that study, although similar declines might be expected.

Yet, while African-American and European-Americans showed some divergence in ideal female body weight, it should not be concluded that the two groups possess very different aesthetics. Cunningham et al. (1995) found that African-American men gave highly similar attractiveness ratings to female faces compared to European-American men, $r = 0.94$. This outcome is consistent with the multiple fitness model's suggestion that body aesthetics reflect adaptation to the physical ecology, whereas face aesthetics reflect general social tendencies.

It should be noted that the multiple fitness analysis of personal ecology is not limited to the plight of women in inner city neighborhoods who face a shortage of men. Women in forward-based military units, engineering schools, and some corporations may find themselves in the minority, and have the luxury of being highly selective about their romantic partners. Men who serve on airline flight crews, on cheerleading squads or who attend historically women's colleges may have many potential partners from whom to choose, regardless of the population sex ratio.

Social-Structural and Personality Variations

The foregoing suggested that individual differences, such as ethnic group membership, attachment style and locus of control may moderate the relationship between an abundant or harsh ecology, sexual behavior, and personal display. Social-structural variables, such as religion and marriage norms, also may play a substantial role in determining whether a voluptuous or slender body style is best suited to attracting a mate in ecologies that favor short- or long-term mating strategies, respectively. In the Irish island community of Inis Baeg (Messinger 1971), for example, money is scarce and marriage prospects are poor, but that did not produce an inclination toward premarital sexual activity. Conservative Christian religious traditions, enforced through neighborhood observation of young adults, deterred such liaisons. Instead, sexual activity and the age of marriage was delayed until men accumulated sufficient assets to support a family.

When males are scarce, females may have a greater likelihood of marriage if they can mate with a male who already has a wife. Low (1990) found that high levels of life-threatening parasites and diseases were strongly correlated with the likelihood of polygyny, perhaps because such pathogens are differentially lethal to males. Polygyny is more common in societies with a low sex ratio (Ember 1974). However, there is some evidence that polygyny contributes to a skewed sex ratio, by increasing competition and insuring that assets and reproductive success is concentrated in a smaller number of families (Low 1996).

Islamic and Mormon religious traditions support polygyny, but Judeo-Christian religious traditions, which are embedded in Western marriage laws, prevent such arrangements. Western females who have little chance of a husband, therefore, may be forced to choose between nonmarital pregnancy or celibacy. Western moral and political sentiment advocates the latter, but it is interesting to note that Christian ministers scored the lowest on testosterone levels (Dabbs et al. 1990).

Low testosterone males were found to be more likely to marry, less likely to have extramarital affairs or divorce, and experienced higher quality marital interaction than other men (Dabbs 1993). Such males, including ministers, may have some difficulty understanding females who have limited access to responsible males such as themselves, and who may settle for short-term male attention, rather than endure none at all.

Male Facial Hair

Male facial hair indicates post-pubescent sexual maturity, and may have evolved to facilitate intrasexual competition, by concealing the expressive mouth and enhancing the apparent size of the masculine chin and jaw. Bearded males have consistently been seen as more dominant and more masculine than nonbearded males (Kenny and Fletcher 1973; Pellegrini 1973; Wood 1986). Some studies have found faces with beards to be desirable (Pancer and Meindl 1978), whereas others found bearded faces to be unattractive (Feinman and Gill 1978). Cunningham et al. (1990) reported that males displaying moustaches were rated as less attractive than males without moustaches. This observation was extended by Muscarella and Cunningham (1996) who used computer imagery to create the same faces both with a moustache and beard, and clean shaven. In that study, the male with a moustache and beard was rated as more aggressive, older, less appeasing and less attractive than the same male displayed with a clean-shaven face.

Robinson (1976) extended Richardson and Kroeber's (1940) analysis of female fashion by analyzing men's hair fashions in the Illustrated London News from 1842 to 1971. An intriguing association between female and male fashions was reported, such that 21 years after women's skirts were at their fullest, men's beards also reached their zenith. Robinson suggested that the men in the London News were prominent and older, and presumably developed their tastes two decades earlier, whereas female fashion models were predominately younger and likely reflected the style of the times. This suggests that males formed their preference for full beards during the time when women's skirts were at their fullest, but the men were only features in the London News after they attained prominence. Thus, the finding of a 21-year time lag is likely a methodological artifact, and actually represents direct covariation between skirt width and beardedness.

Barber (2001) extended the analysis performed by Robinson (1976) to examine the relations between facial hair and indicators of the marriage market for women. Barber argued that moustaches and beards are attractive to women, and that men are more likely to display them during times of greater competition. Men have greater competition during decades in which the sex ratio is high, and women have many men from which to choose. A low sex ratio, creating an unfavorable marriage market for women, combined with high illegitimacy rates, predicted an absence of moustache and beard fashions. Barber argued that facial hair, particularly a mustache, serves to conceal the nonverbal signs of deception. Women may prefer clean-shaven men when men may have greater opportunity for philandering, because clean-shaven men are easier to read, and are less attractive to other women.

It is not clear from Barber's (2001) analysis why men would voluntarily surrender their physical attractive beards in circumstances in which there are low sex ratios, given that they have many women from which to choose. However, Barber's (2001) suggestion that a moustache and beard are inherently attractive to women may overstate the case, given the findings of Feinman and Gill (1978) and Cunningham et al. (1990). Instead, females may possess an ambivalent attraction to aggressive symbols of masculinity, such as facial hair or physical power. When the local ecology is threatening and male aggressiveness may be particularly helpful, then beards may be desirable. When males are untrustworthy, or the local ecology is benign or prosperous, then a beard may cause a face that already appears mature to look too competitive and threatening to the perceiver. It is interesting to note that a hint of beard stubble, which conveys masculinity without being overpowering, was associated with higher attractiveness ratings in Cunningham et al. (1990), a study which was conducted during relatively prosperous times. When a substantial economic downturn occurs, males may need to let the stubble sprout into a full beard to be attractive.

Bodybuilding and Athleticism

In a landmark series of studies, Sales (1972, 1973) demonstrated that ecological challenges, in the form of economic recession, stimulated the expression of authoritarian and power-assertive responses, including increased sales of attack dogs, increased funding for police, and the debut of hypermasculine comic strip figures such as Superman and Batman.

Some individuals may not be satisfied to simply view comic strip superheroes, but instead may strive to look like one, by investing their time and energy in bodybuilding. Bodybuilding activity may produce some social benefits. Male physiques with broader shoulders and chests were rated as more attractive than average physiques (Lavrakas 1975; Horvath 1979), Furthermore, a well-developed abdomen may be seen as particularly desirable in other cultures, as the ancient Chinese and Japanese art of Judo teaches that the life force/strength, or *ki*, emerges from the abdomen (Tohei 1976).

It should be noted that a pumped-up bodybuilder physique was not seen as highly attractive by most females (Beck et al. 1976). Gillett and White (1992) suggested that the hypermasculine body cultivated by male bodybuilders symbolizes an attempt by men to restore feelings of masculine self-control and self-worth in the face of feminist threats to male hegemony. Such speculation received some support in Maier and Lavrakas (1984), who found that men with negative attitudes towards women endorsed the attractiveness of the idealized, tapering V-shape male physique.

Maier and Lavrakas (1984) also found that men with authoritarian personalities also endorsed the hypermasculine body ideal. Such results suggest the possibility of moving the analysis from the individual psychological to the ecological. If authoritarianism is stimulated by ecological stress, it may be that rates of bodybuilding may increase, and a hypermasculine body may be seen as more attractive in periods and places with high levels of stress. Thus, male bodybuilding also may increase when the economy is contracting, causing males to feel under threat (cf. Sales 1972, 1973). Bodybuilding may be particularly likely in unemployed or underemployed males, given that the time and effort demanded by weight lifting is difficult to reconcile with a professional career. Men also may be more inclined towards bodybuilding where the threat of physical violence is high, such as in the inner city, or in prison.

When there is a high level of mate competition between males, such as when the sex ratio is high, or in the gay male subculture (Canarelli et al. 1999), men may engage in bodybuilding to enhance their chances. It is interesting to note that male bodybuilders often remove their chest and shoulder hair. This may be done to enhance the visibility of their muscles, but it also may increase the appearance of youthfulness, and decrease the appearance of mature masculinity.

When the sex ratio is low, bodybuilding may increase due to the increased possibility of short-term relationships based on physical attractiveness (Guttentag and Secord 1983). More selective women, with an internal locus of control and the capacity for delay of gratification, however, may reject buff men. If such men appear to be narcissistic, and potentially unfaithful, then women may exert a contrary selection pressure toward less muscular male bodies. Although this is similar to the dynamics that Barber (2001) suggested concerning mustaches, the interaction of sex ratio and female dynamics on the popularity of a hypermuscular male body remains to be empirically investigated.

With respect to female bodies, Banner (1983) traced seven distinct physical ideals in white female beauty in the period between 1800 and 1960:

"In the Antebellum years, the frail, pale, willowy woman... predominated. In the decades after the Civil War she was challenged by a buxom, hearty and heavy model who made her first appearance on the British music hall stage...the "voluptuous" woman. In turn, she was challenged by the tall, athletic, patrician Gibson girl of the 1890s, whose vogue was superseded in the 1910s by a small, model of beauty...This "flapper" model of beauty was predominant throughout the 1920s. By 1930 a new, less youthful and frivolous beauty ideal came into being and remained popular throughout the 1960s, culminating in a renewed vogue of voluptuousness that bore resemblance to nineteenth-century types. That same decade, however, a youthful adolescent model reappeared and continued in popularity through the 1960s." (p. 5)

Informal observation suggests an emphasis on an athletic female body in the 1890s, 1920s, 1940s, 1970s and 1990s in the US, but work is needed to document if changes in female bodybuilding correspond to changes in ecological and economic variables.

Females may respond to a low sex ratio and a challenging ecology by accentuating their slenderness, even to the point of anorexia (Barber 1998a, b; Voland and Voland 1989). However, female interest in sports (Daniels 1992) and bodybuilding (Guthrie and Castelnuovo 1992) has increased in the West in recent years. This change may be an alternate form of self-empowerment in the self-controlled female's response to a favorable career, but unfavorable marriage, market. If both the career and the marriage market is unfavorable, however, or the female lacks self-restraint, she may emphasize voluptuousness and early pregnancy over either slenderness or muscularity. Again, research is needed to document these speculations.

Scarification and Tatooing

Individuals may paint, tattoo or scarify their bodies to draw attention to their fitness, or perhaps conceals signs of unfitness, such as disease-pocked skin. Singh and Bronstad (1997) used the Standard Cross-Cultural Sample of 186 societies to examine the relation between pathogen presence and the grooming behavior of facial and body scarification and tattooing. Such practices were expected to be more likely in areas of the world where disease was rampant. The investigators reported that females tended to scarify their faces more than men overall, perhaps due to a greater emphasis placed on female than male physical attractiveness. In addition, females were more likely to scarify their stomachs and breasts in areas of the world where pathogens were high compared to areas where pathogens were low, perhaps to draw attention to their fitness, potential fertility, and capacity to endure the pain of childbirth.

Gangestad and Buss (1993) suggested that physical attractiveness is regarded as more important in a potential mate in areas in which pathogens are rampant, because it may indicate a healthy immune system. They reanalyzed data from 29 cultures in which 7139 individuals were asked to indicate the relative importance that they placed on physical attractiveness when selecting a mate. They reported that people in parts of the world with a relatively greater prevalence of pathogens tended to value a mate's physical attractiveness more than people in geographical areas with lower incidence of pathogens. Gangestad and Buss further found that the relationship between pathogen prevalence and the value of physical attractiveness remained evident even after distance from the equator and average income were controlled.

Recent research has raised questions about whether attractive features reliably signify better than average medical status. Perception of a person as looking older (Borkan et al. 1982) and measurements of facial asymmetry (Shackleford and Larson 1997, 1999) have been shown to be accurate predictors of health. There was a moderate relationship between physical attractiveness and health in the five studies conducted on the topic ($d = 0.38$, $p < 0.05$; Langlois et al. 2000). Ratings of perceived health and actual health are related, but this relationship was attenuated when ratings of physical attractiveness were statistically controlled (Kalick et al. 1998). Thus, it appears that humans are better and more direct judges of physical attractiveness than they are of health. Nevertheless, physical attractiveness may be a moderately reliable indicator of fitness (Thornhill and Grammer 1999), especially in extreme cases, such as with genetic anomalies that produce mental retardation and other disabilities. Modern advances in the quality of healthcare may have reduced more subtle relations between appearance and health that may have existed earlier in human evolution.

Declining Importance of Physical Attractiveness?

Interpretations of cross-cultural variation in the importance placed on various material qualities, such as physical attractiveness and earning capacity, have been challenged. Eagly and Wood (1999) reanalyzed Buss' 37 culture data using United Nations indices of the status of women in each culture. They reported that increased gender equality was associated with men decreasing the importance that they placed on women's skill as a domestic worker, with women's decreased rating of the importance of men's earning potential, and with women's decreased preference for older men. Eagly and Wood interpreted that outcome as indicating that social structural differences, such as gender roles and the distribution of material wealth, were responsible for many gender differences.

Eagly and Wood (1999) reported that in cultures in which females had more empowerment, both males and females rated physical attractiveness as less important in the choice of the mate than in cultures with less female development. If female empowerment is associated with greater industrialization and control of pathogens, however, then physical attractiveness may decline in importance to the extent that it provides less information about health and fertility. The relation between pathogen prevalence, female empowerment, and the importance of physical attractiveness has not yet been reported.

Mate-Copying

There is evidence that social variables may partially override the importance of physical attractiveness in other species. Female Trinidadian guppies prefer to mate with bright orange-colored males, who often behave more boldly in the presence of predators than their muted brethren. Female guppies may partially ignore coloration, however, when other variables are salient. Female guppies who witnessed a muted male behave boldly toward a potential fish predator demonstrated a clear preference for that male despite his being less colorful or physically attractive than other males (Godin and Dugatkin 1996).

Bold behavior is context-specific, and may not always be demonstrated (Coleman and Wilson 1998), so females may rely on other types of information about the desirability of a potential mate than their own observation. Female guppies tend to engage in "mate-copying" or displaying increased mating interest toward a male who receives disproportionate attention from other females. In an experimental test, a less colorful male appeared to be "popular" with other females, whereas a more colorful male appeared to receive no female attention. In most cases, the female guppies preferred the more popular but less physically attractive male. When the two males differed by 40% in orange body color, however, observed females preferred the more colorful and physically attractive male and did not copy the mate choice of other females. Thus, imitation can override physical attractiveness preferences when the loss of physical attractiveness is small to moderate, but preferences for ideal appearance may override imitation effects when the difference in physical attractiveness is large (Dugatkin 1996).

The evaluations of a potential date's social desirability, as assessed by other people, may be as influential as objective physical attractiveness qualities in human mate choice. Graziano et al. (1993) reported that social influence caused attractiveness ratings to shift a few points, especially among females. Females may be more influenced by social information than males, because a portion of the criteria that females emphasize when selecting mates are intangible. The male's ambitiousness, reliability, and acceptance by the larger community are not overtly apparent, which may increase the value of social evaluation information.

Cunningham et al. (Cunningham et al., in prep.) examined mate-copying in humans. In study 1, males and females reported their interest in short- and long-term relationships, and their social perceptions of six targets. The targets were described as high or low in physical attractiveness, and as receiving high, medium, or low levels of attention from peers. Although physical attractiveness increased mating interest, subjects also reported greater mating interest in popular members of the opposite sex, consistent with the behavior of other species. Peer attention was also associated with the perception of greater social skills, sense of humor, and wealth to the socially popular mate. The mate-copying tendency was particularly evident in females. Study 2 manipulated peer attention, physical attractiveness, and wealth that stemmed from the parent's luck. Peer attention again increased mating interest, especially for females. Physical attractiveness and wealth also influenced mating responses, but peer attention and perceived personality were more reliable predictors of mating interest.

Mate-copying presumably evolved as a way to make use of what others have learned about mating in the local ecology. If all of her female acquaintances are prosperously married to men who are somewhat thin and babyfaced, rather than exceptionally rugged or mature in body or visage, then an unmarried woman might reasonably conclude that the local ecology does not require hypermasculinity for mating success. However, that does not mean that the woman would seek out the scrawniest of men, or snub a well-muscled man who displayed interest. It seems likely that it would be adaptive for human mating criteria to be responsive to social influence within a moderate range of physical attractiveness values.

Trading Off Physical Attractiveness for What It Represents

The partial plasticity of attraction responses is not in conflict with evolutionary interpretations of the perceptions of physical attractiveness. Evolved dispositions orient the individual to some basic dimensions of a fit mate, but awareness of the ecology, the behavior of other people, and one's own goals, all may influence responses to physical features. Quite simply, attention to physical attractiveness is in the service of adaptation, and if adaptation can be achieved through attention to more reliable information, then physical attractiveness cues may be disregarded. It is conceivable that the perceiver might be capable of ignoring neoteny features if the perceiver is certain about a target's youth and openness, or partially disregard sexual maturity cues if there is other evidence of strength, fertility or virility, and extroversion. Direct experience with a target's friendliness and emotional stability might suffice instead of expressive features, and reliable information about a target's social status, cultural adaptation and conscientiousness might substitute for grooming cues. Nevertheless, because physical attractiveness is highly vivid, the alternative information would have to be equally salient to override the impact of overt appearance.

Sprecher (1989) experimentally manipulated written descriptions of an opposite sex target's physical attractiveness, earning potential, and expressiveness. She found that women and men were both most affected by physical appearance descriptions when judging the attractiveness of the person, but females incorrectly believed that they were more affected by earning potential and expressiveness than by attractiveness. Using a comparable, but forced-choice, approach, Cunningham et al. (2000) manipulated physical attractiveness using photographs and written descriptions of personality and wealth. For a date and for marriage, both males and females were most likely to choose a person who was physically attractive and possessed a good personality, but was financially disadvantaged, over one who was physically attractive and wealthy but had an undesirable personality, or had wealth and a good personality, but was physically unattractive.

Cunningham et al. (Cunningham et al. submitted) surveyed 157 men and 230 women on the importance of 195 mate selection criteria in 25 categories. They found that genders were much more similar than different in the value that they place on various qualities in a romantic partner. Across the 25 mate preference categories collected in this study, the correlation between male and female mean importance ratings for mate selection criteria was $r = 0.98$, $p < 0.0001$. Men were found to be more interested in care variables, such as social support (ranked second of 25) than in physical appearance (ranked fifth). They valued a woman who 'accepts you for who you are' more than one who is physically attractive. Similarly, they preferred a woman who 'makes you feel unique and special' more than one with 'physical attractiveness'. Men also placed just as much importance on the characteristic of 'honest', and on someone who 'tells you her innermost thoughts' as on physical attractiveness. Women also valued social support (ranked second) as much as did men, and rated physical attractiveness (ranked seventh) nearly as importance as affluence (ranked fifth) in a prospective mate.

We conducted similar analyses using Buss et al.'s (1990) 37 culture data, and produced similarities between male and female mate selection criteria rankings that ranged from $r = 0.63$ in Nigeria to $r = 0.97$ in Brazil, with a mean of $r = 0.87$ across cultures. Such results suggest that physical attractiveness is used around the world as one piece of information in choosing a mate who will, ideally, provide care and comfort for oneself and one's children.

Conclusions

The multiple fitness model suggested that some elements of facial physical attractiveness are universal, supporting the Natural-Classical position. Other aspects of physical attractiveness, particularly grooming variables, show substantial cross-cultural variation, supporting the Cultural-Constructivist position. The model recognized that some grooming practices, such as the wearing of various clothing colors, could be done to communicate one's personality, or one's group membership. The use of other grooming practices, such as the wearing of long fingernails, elaborate hairstyles, silk neckties or wool suits in the summer could be attractive simply because they are expensive, and thereby provide a means of signifying that one has desired resources, such as wealth, knowledge, or time. Novel or extreme grooming practices may be used to signal potential partners that one is different from the masses, and likely to produce healthy children. The advertisement of reproductive fitness may be intensified when population sex ratios put individuals at a disadvantage for reproduction.

The multiple fitness model, which has consistently emphasized the importance of ecological variables for attraction, has endeavored to integrate research and extend speculation on the impact of poverty, low sex ratios and pathogens on physical and sexual self-presentation. The model echoed prior research, which noted that when a female's marriage and career prospects are both poor, she may emphasize her curvaceousness, and engage in less restrained eating and sexual behavior. The females may choose males who emphasize their masculine qualities, such as with a highly muscular body and facial hair, and who might produce children who are adaptive in that ecology. Females in such circumstances may be particularly attracted to men who display a willingness to engage in intense short-term investment, by spending money on dinner and jewelry for them.

The multiple fitness model extended prior discussions by suggesting an interaction effect between marriage prospects and career prospects. When a female's marriage prospects are poor, but her career prospects are good, she may wear her hair short, her collar buttoned, and emphasize her slenderness and athleticism. However, when a female's marriage prospects are better, she may delay both puberty and marriage, and send a mixed message by presenting voluptuousness above the waist, and demureness below the waist, in hopes of attracting an excellent husband. The male grooming that is most appealing to selective women may include a clean-shaven face that is intermediate in babyfacedness and sexual maturity, a body that is tall and fit, but not conspicuously muscular, and is encased in higher status clothing. Such men seem to appear with some regularity on the pages of elite women's magazines, but systematic analyses have yet to be performed.

Although we have endeavored to provide some insight into the determinants of attraction to physical appearance, we have not discussed such variables as eyeglasses, tinted contact lenses, decorative gold or other modifications of the teeth, body piercing, foot deformation, or a host of other grooming behaviors. Nonetheless, it should be clear that what makes people look good is not simply attributable conspiracies of the fashion industry. Such determinants of grooming behavior may not always be evident in focused experiments or surveys, but may be evident when investigators adopt a broader cross-cultural and trans-historical perspective.

Summary

Traditional approaches to aesthetics attribute beauty either to natural, universal qualities, or to socially constructed, culturally arbitrary qualities. As an alternative to both positions, the multiple fitness model specifies certain dimensions of physical attractiveness, including neonate, sexually mature, and some expressive features that appear to be universally desirable, and other dimensions, especially grooming and self-adornment behaviors, which appear to vary substantially from culture to culture, and across time within a given culture. The model further stipulates that a portion of the variation in grooming behaviors is attributable to status competition and the desire for novelty, whereas other aspects of grooming can be linked to ecological dynamics. The heuristic value of the multiple fitness model is illustrated in a review of current research on the impact of such ecological variables as sex ratios, economic prosperity, father absence, and attachment style on female dress length, female slenderness vs. curvaceousness, the display of male facial hair, bodybuilding, and tattooing.

References

Alicke MD, Smith RH, Klotz ML (1986) Judgments of physical attractiveness: the role of faces and bodies. Personality Social Psychol Bull 12:381–389

Anderson JL, Crawford CB, Nadeau J, Lindberg T (1992) Was the Duchess of Windsor right: a cross-cultural review of the socioecology of ideals of female body shape. Ethol Sociobiol 13:197–227

Banner LW (1983) American beauty. University of Chicago Press, Chicago

Barber N (1998a) Secular changes in standards of bodily attractiveness in American women: different masculine and feminine ideals. J Psychol 132:87–94

Barber N (1998b) Secular changes in standards of bodily attractiveness in women: tests of a reproductive model. Int J Eating Disorders 23:449–453

Barber N (1999) Women's dress fashions as a function of reproductive strategy. Sex Roles 40:459–471

Barber N (2000) On the relationship between country sex ratios and teen pregnancy rates: a replication. Cross Cult Res J Comp Social Sci 34:26–37

Barber N (2001) Moustache fashion covaries with a good marriage market for women. J Nonverbal Behav 25:261–272

Beck S, Ward-Hull C, McClear P (1976) Variables related to women's somatic preferences of the male and female body. J Personality Social Psychol 34:1200–1210

Belsky J, Steinberg L, Draper P (1991) Childhood experience, interpersonal development, and reproductive strategy: an evolutionary theory of socialization. Child Dev 62:647–670

Bernstein IH, Tsai-Ding L, McClellan P (1982) Cross- vs. within-racial judgments of attractiveness. Perception Psychophys 32:495–503

Berscheid E, Walster E (1974) Physical attractiveness. In: Berkowitz L (ed) Advances in experimental social psychology, vol 7. Academic Press, New York, pp 158–216

Borkan GA, Bachman SS, Norris AH (1982) Comparison of visually estimated age with physiologically predicted age as indicators of rates of aging. Social Sci Med 16:197–204

Brickman P, D'Amato B (1975) Exposure effects in a free-choice situation. J Personality Social Psychol 32:415–420

Buss DM (1999) Evolutionary psychology: the new science of the mind. Allyn and Bacon, Needham Heights, MA

Buss DM, Abbott M, Angleitner A, Asherian A, Biaggio A (1990). International preferences in selecting mates: a study of 37 cultures. J Cross-Cult Psychol 21:5–47

Cabeza R, Kato T (2000) Features are also important: contributions of featural and configural processing to face recognition. Psychol Sci 11:429–433

Canarelli J, Cole G, Rizzuto C (1999) Attention vs. acceptance: some dynamic issues in gay male development. Gender Psychoanal 4:47–70

Cash TF (1990) Losing hair, losing points? The effects of male pattern baldness on social impression formation. J Appl Social Psychol 20:154–167

Clark K (1958) The nude: a study in ideal form. Doubleday, Garden City, NY

Coleman K, Wilson DS (1998) Shyness and boldness in pumpkinseed sunfish: individual differences are context-specific. Anim Behav 56:927–936

Cunningham MR (1981) Sociobiology as a supplementary paradigm for social psychological research. In: Wheeler L (ed) Review of personality and social psychology, vol 2. Sage, Beverly Hills, pp 69–106

Cunningham MR (1985) Levites and brother's keepers: a sociobiological perspective on prosocial behavior. Humboldt J Social Relations 13:35–67

Cunningham MR, Barbee AP (1991) Differential K- selection versus ecological determinants of race differences in sexual behavior. J Res Personality 25:205–217

Cunningham MR, Wong DT, Rodenhiser JM, Roberts RA, Richardson T (1989) Facialmetric analyses of physical attractiveness. Paper presented at the Convention of the Kentucky Psychological Association, Louisville, KY

Cunningham MR, Barbee AP, Pike CL (1990) What do women want? Facialmetric assessment of multiple motives in the perception of male facial physical attractiveness. J Personality Social Psychol 59:61–72

Cunningham MR, Roberts AR, Barbee AP, Druen PB, Wu C (1995) "Their ideas of beauty are, on the whole, the same as ours": consistency and variability in the cross-cultural perception of female physical attractiveness. J Personality Social Psychol 68:261–279

Cunningham MR, Druen PB, Barbee AP (1997) Angels, mentors, and friends: trade-offs among evolutionary, social, and individual variables in physical appearance. In: Simpson JA, Kenrick DT (eds) Evolutionary social psychology. Lawrence Erlbaum, Mahwah, NJ, pp 109–140

Cunningham MR, Barbee AP, Graves CR, Lundy DE, Lister SC, Rowatt W (2000) Can't buy me love: the effects of male wealth and personal qualities on female attraction. University of Louisville, Louisville, KY

Cunningham MR, Barbee AP, Philhower CL (2002) Dimensions of facial physical attractiveness: the intersection of biology and culture. In: Rhodes G, Zebrowitz LA (eds) Facial attractiveness: evolutionary, cognitive, and social perspectives. Advances in visual cognition, vol 1. Ablex Publishing, Westport, CT, pp 193–238

Dabbs JM (1993) Salivary testosterone measurements in behavioral studies. In: Malamud D, Tabak LA (eds) Saliva as a diagnostic fluid. New York Academy of Sciences, New York, NY, pp 177–183

Dabbs JM (1997) Testosterone, smiling, and facial appearance. J Nonverbal Behav 21:45–55

Dabbs JM, de la Rue D, Williams PM (1990) Testosterone and occupational choice: actors, ministers, and other men. J Personality Social Psychol 59:1261–1265

Damhorst ML, Reed AP (1986) Clothing color value and facial expression: effects on evaluations of female job applicants. Social Behav Personality 14:89–98

Daniels DB (1992) Gender (body) verification (building). Play Cult 5:370–377

Darwin C (1871) The descent of man and selection in relation to sex. Murray, London

Dion KK, Berscheid E (1974) Physical attractiveness and peer perception among children. Sociometry 37:1–12

Dion KK, Berscheid E, Walster E (1972) What is beautiful is good. J Personality Social Psychol 14:97–108

Dion KK, Pak AW, Dion KL (1990) Stereotyping physical attractiveness: a sociocultural perspective. J Cross-Cult Psychol 21:158–179

Draper P, Belsky J (1990) Personality development in evolutionary perspective. J Personality 58:141–161

Draper P, Harpending H (1982) Father absence and reproductive strategy: an evolutionary perspective. J Anthropol Res 38:255–279

Draper P, Harpending H (1987) Parent investment and the child's environment. In: Lancaster JB, Altmann J (eds) Parenting across the life span: biosocial dimensions. De Gruyter, Hawthorne, NY, pp 207–235

Draper P, Harpending H (1988) A sociobiological perspective on the development of human reproductive strategies. In: McDonald K (ed) Sociobiological perspectives on human development. Springer, Berlin Heidelberg New York, pp 340–372

Dugatkin LA (1996) Interface between culturally based preferences and genetic preferences: female mate choice in *Poecilia reticulata*. Proc Natl Acad Sci USA 93:2770–2773

Eagly AH, Wood W (1999) The origins of sex differences in human behavior: evolved dispositions versus social roles. Am Psychol 54:408–423

Ellis BJ, McFadyen-Ketchum S, Dodge KA, Pettit GS, Bates JE (1999) Quality of early family relationships and individual differences in the timing of pubertal maturation in girls: a longitudinal test of an evolutionary model. J Personality Social Psychol 77:387–401

Ember CR (1974) An evaluation of alternative theories of matrilocal versus patrilocal residence. Behav Sci Res 9:135–149

Etcoff N (1999) Survival of the prettiest. Random House, New York
Feingold A (1990) Gender differences in effects of physical attractiveness on romantic attraction: a comparison across five research paradigms. J Personality Social Psychol 59:981–993
Feingold A (1992) Good looking people are not what we think. Psychol Bull 111:304–341
Feinman S, Gill GW (1978) Sex differences in physical attractiveness preferences. J Social Psychol 105:43–52
Ford CS, Beach FA (1951) Patterns of sexual behavior. Harper, New York
Frank MG, Gilovich T (1988) The dark side of self- and social perception: black uniforms and aggression in professional sports. J Personality Social Psychol 54:74–85
Furnham A, Alibhai N (1983) Cross-cultural differences in the perception of female body shape. Psychol Med 13:829–837
Gangestad SW, Buss DM (1993) Pathogen prevalence and human mate preferences. Ethol Sociobiol 14:89–96
Geary DC (2000) Evolution and proximate expression of human paternal investment. Psychol Bull 126:55–77
Gillett J, White PG (1992) Male bodybuilding and the reassertion of hegemonic masculinity: a critical feminist perspective. Play Cult 5:358–369
Godin JJ, Dugatkin LA (1996) Female mating preference for bold males in the guppy, *Poecilia reticulata*. Proc Natl Acad Sci USA 93:10262–10267
Graham JA, Jouhar AJ (1981) The effects of cosmetics on person perception. Int J Cosmetic Sci 3:199–210
Graziano W, Brothen T, Berscheid E (1978) Height and attraction: do men and women see eye-to-eye? J Personality 46:128–145
Graziano WG, Jensen-Campbell L, Shebilski L, Lundgren S (1993) Social influence, sex differences and judgments of beauty. Putting the "interpersonal" back in interpersonal attraction. J Personality Social Psychol 65:522–531
Guthrie RD (1976) Body hot spots: the anatomy of human social organs and behavior. Van Nostrand Reinhold, New York, NY
Guthrie SR, Castelnuovo S (1992) Elite women bodybuilders: models of resistance or compliance? Play Cult 5:401–408
Guttentag M, Secord PF (1983) Too many women: the sex ratio question. Sage, Beverly Hills, CA
Harker L, Keltner D (2001) Expressions of positive emotion in women's college yearbook pictures and their relationship to personality and life outcomes across adulthood. J Personality Social Psychol 80:112–124
Henss R (2001) Social perceptions of male pattern baldness: a review. Dermatol Psychosomatics 2:63–71
Hill EM, Nocks ES, Gardner L (1987) Physical attractiveness: manipulation by physique and status displays. Ethol Sociobiol 8:143–154
Hinsz VB, Matz DC, Patience RA (2001) Does women's hair signal reproductive potential? J Exp Soc Psychol 37:166–172
Horvath T (1979) Correlates of physical beauty in men and women. Soc Behav Personality 7:145–151
Jackson LA (1992) Physical appearance and gender: sociobiological and sociocultural perspectives. SUNY Press, Albany, NY
Kalick SM, Zebrowitz LA, Langlois JH, Johnson RM (1998) Does human facial attractiveness honestly advertise health? Psychol Sci 9:8–13

Kenny CT, Fletcher D (1973) Effects of beardedness on person perception. Perceptual Motor Skills 37:413–414

Kirkpatrick LA (1998) Evolution, pair-bonding, and reproductive strategies: a reconceptualization of adult attachment. In: Simpson JA, Rholes WS (eds) Attachment theory and close relationships. Guilford Press, New York, pp 353–393

Kleinke CL, Staneski RA (1980) First impressions of female bust size. J Social Psychol 110:123–134

Langlois JH, Roggman LA, Casey RJ, Ritter JM, Rieser-Danner LA, Jenkins VY (1987) Infant preferences for attractive faces: rudiments of a stereotype? Dev Psychol 23:363–369

Langlois JH, Roggman LA, Rieser-Danner LA (1990) Infants' differential social responses to attractive and unattractive faces. Dev Psychol 26:153–159

Langlois JH, Ritter JM, Roggman LA, Vaughn LS (1991) Facial diversity and infant preferences for attractive faces. Dev Psychol 27:79–84

Langlois JH, Kalakanis L, Rubenstein AJ, Larson A, Hallam M, Smoot M (2000) Maxims or myths of beauty? A meta-analytic and theoretical review. Psychol Bull 126:390–423

Lavrakas PJ (1975) Female preferences for male physiques. J Res Personality 9:324–334

Lawson ED (1971) Hair color, personality and the observer. Psychol Rep 28:311–322

Low BS (1990) Marriage systems and pathogen stress in human societies. Am Zool 30:325–339

Low BS (1996) Behavioral ecology of conservation in traditional societies. Human Nat 7:353–379

Lurie A (1981) The language of clothes. Random House, New York

Mabry MA (1971) The relationship between fluctuations in hemlines and stock market averages from 1921–1971. Master Thesis, University of Tennessee, Knoxville

Maier RA, Lavrakas PJ (1984) Attitudes toward women, personality rigidity, and idealized physique preferences in males. Sex Roles 11:425–433

Maret SM, Harling CA (1985) Cross-cultural perceptions of physical attractiveness: ratings of photographs of Whites by Cruzans and Americans. Perceptual Motor Skills 60:163–166

McClelland D (1961) The achieving society. Van Nostrand, Princeton, NJ

Messinger JC (1971) Sex and repression in an Irish folk community. In: Marshall DS, Suggs RC (eds) Human sexual behavior. Prentice Hall, London, pp 3–37

Morris D (1977) Manwatching: a field guide to human behavior. Harry N Abrams, New York

Morse SJ, Gruzen J, Reis HT (1976) The nature of equity-restoration: some approval-seeking considerations. J Exp Social Psychol 12:1–8

Muscarella F, Cunningham MR (1996) The evolutionary significance and social perception of male pattern baldness and facial hair. Ethol Sociobiol 17:99–117

Osborne DR (1996) Beauty is as beauty does? Makeup and posture effects on physical attractiveness judgments. J Appl Social Psychol 26:31–40

Pancer SM, Meindl JR (1978) Length of hair and beardedness as determinants of personality impressions. Perceptual Motor Skills 46:1328–1330

Pellegrini RJ (1973) Impressions of the male personality as a function of beardedness. Psychology 10:29–33

Penton-Voak IS, Perrett DI, Castles DL, Kobayashi T, Burt DM, Murray LK, Minamisawa R (1999) Menstrual cycle alters face preference. Nature 399:741–742

Perrett DI, May KA, Yoshikawa S (1994) Facial shape and judgments of female attractiveness: preferences for non-average characteristics. Nature 386:239–242

Pettijohn TF, Tesser A (1999) Popularity in environmental context: facial feature assessment of American movie actresses. Media Psychol 1:229–247
Richardson J, Kroeber AL (1940) Three centuries of women's dress fashions: a quantitative analysis. Anthropol Rec 5:111–153
Robinson DE (1976) Fashion in shaving and trimming of the beard: the men of the Illustrated London News, 1842–1972. Am J Sociol 81:1133–1141
Robinson J (1998) The quest for human beauty: an illustrated history. WW Norton, New York, NY
Roll S, Verinis JS (1971) Stereotypes of scalp and facial hair as measured by the semantic differential. Psychol Rep 28:975–980
Rucker CE, Cash TF (1992) Body images, body-size perceptions, and eating behaviors among African-American and White college women. Int J Eating Disorders 12:291–299
Sales SM (1972) Economic threat as a determinant of conversion rates in authoritarian and nonauthoritarian churches. J Personality Social Psychol 23:420–428
Sales SM (1973) Threat as a factor in authoritarianism: an analysis of archival data. J Personality Social Psychol 28:44–57
Shackleford TK, Larsen RJ (1997) Facial asymmetry as an indicator of psychological, emotional and physiological distress. J Personality Social Psychol 72:456–466
Shackleford TK, Larsen RJ (1999) Facial attractiveness and physical health. Evol Human Behav 20:71–76
Shepard JA, Strathman AJ (1989) Attractiveness and height: the role of stature in dating preference, frequency of dating and perceptions of attractiveness. Personality Social Psychol Bull 15:617–627
Shorter-Gooden K, Washington NC (1996) Young, black, and female: the challenge of weaving an identity. J Adolescence 19:465–475
Silverstein B, Peterson B, Perdue L, Vogel L, Fantini S (1986) Some correlates of the thin standard of bodily attractiveness for women. Int J Eating Disorders 5:895–905
Singh D, Bronstad PM (1997) Sex differences in the anatomical locations of human body scarification and tattooing as a function of pathogen prevalence. Evol Human Behav 18:403–416
Snyder M, Tanke ED, Berscheid E (1977) Social perception and interpersonal behavior: on the self-fulfilling nature of social stereotypes. J Personality Social Psychol 35:656–666
Sobal J, Stunkard AJ (1989) Socioeconomic status and obesity: a review of the literature. Psychol Bull 105:260–275
Sprecher S (1989) The importance to males and females of physical attractiveness, earning potential, and expressiveness in initial attraction. Sex Roles 21:591–607
Symons D (1979) The evolution of human sexuality. Oxford University Press, New York
Synott A (1987) Shame and glory: a sociology of hair. Br J Sociol 38:381–413
Thakerar JN, Iwawaki S (1979) Cross-cultural comparisons in interpersonal attraction of females toward males. J Social Psychol 108:121–122
Thomas VG, James MD (1988) Body image, dieting tendencies, and sex role traits in urban Black women. Sex Roles 18:523–529
Thornhill R, Grammer K (1999) The body and face of woman: one ornament that signals quality? Evol Human Behav 20:105–120
Tohai K (1976) Book of Ki: coordinating mind and body in daily life. Japan Publications, New York
Townsend JM, Levy GD (1990) Effects of potential partners' physical attractiveness and socioeconomic status on sexuality and partner selection. Arch Sex Behav 19:149–164
Veblen T (1912) The theory of the leisure class. Macmillan, New York

Voland E, Voland R (1989) Evolutionary biology and psychiatry: the case of anorexia nervosa. Ethol Sociobiol 10:223–240

Walster E, Aronson V, Abrahams D, Rottmann L (1966) Importance of physical attractiveness in dating behavior. J Personality Social Psychol 4:508–516

Wedekind C, Furi S (1997) Body odour preferences in men and women: do they aim for specific MHC combinations or simply heterozygosity? Proc R Soc Lond Ser B 264:1471–1479

Wedekind C, Seebeck T, Bettens F, Paepke AJ (1995) MHC-dependent mate preferences in humans. Proc R Soc Lond Ser B 260:245–249

Weeden P (1977) Study patterned on Kroeber's investigation of style. Dress 3:9–19

Wildes J, Emery R (2001) The roles of ethnicity and culture in the development of eating disturbance and body dissatisfaction: a meta-analytic review. Clin Psychol Rev 21(4):521–551

Wolf N (1991) The beauty myth: how images of beauty are used against women. Morrow, New York

Wong DT, Cunningham MR (1990) Interior versus exterior beauty: the effects of mood on dating preferences for different types of physically attractive women. Paper presented at the Southeastern Psychological Association, Atlanta. Data reported in: Cunningham M, Druen P, Barbee A (1997) Evolutionary, social and personality variables in the evaluation of physical attractiveness. In: Simpson J, Kenrick F (eds) Evolutionary social psychology. Lawrence Erlbaum, Mahwah, NJ, pp 109–140

Wood DR (1986) Self perceived masculinity between bearded and nonbearded males. Perceptual Motor Skills 62:769–770

Zajonc RB (1968) Attitudinal effects of mere exposure. J Personality Social Psychol 9:1–27

Zebrowitz LA (1997) Reading faces: window to the soul? Westview, Boulder, CO

Zebrowitz LA, Montepare JM, Lee HK (1993) They don't all look alike: individual impressions of other racial groups. J Personality Social Psychol 65:85–101

Aesthetic Preferences in the World of Artifacts – Adaptations for the Evaluation of 'Honest Signals'?

Eckart Voland

Introduction

However the faculty of aesthetic judgment may be defined in detail, on the psychological level it is founded on the evaluation of things and scenarios as "attractive" or "repulsive". Beauty attracts, ugliness repels, and thus the aesthetic judgment appears to be the reflection of a psychological preference. Psychological preferences, however, are evolved mental mechanisms, whose adaptive function consists in the making of decisions in the face of the varying problems of life. By processing relevant information, preferences help to recognize adaptive problems and to motivate the finding of advantageous solutions (Buss 1999). The evolutionary logic of some preferences is understood fairly well. The tendency to prefer fresh, aromatic foods, safe and ecologically productive habitats, a more pleasant climate and sexually attractive mates, for example, can easily be grasped as the evolutionary cumulative result of fitness-promoting. In these cases, aesthetic judgment functions as an aid to survival and reproduction (Orians 2001). Whatever attracts us, offers us opportunities to increase fitness; whatever repels us, holds fitness risks.

Aesthetics would surely not continue to form this huge problem of evolutionary biology that is still rather resistant to theorizing, if it were restricted to the fields mentioned above, since the analysis of diet, habitat and mating preferences provides insights of the type that evolutionary biologists are seeking. However – and this is where the actual problem arises – it is not only the "spontaneous" features of the living world in which we find ourselves, which attract or repel us, but also the intentional features of an artificial world of artifacts. Everywhere in the world, people are engaged in activities which Dissanayake (1992) referred to as "making special". They consume time, energy and resources solely in order to allow the things around them to appear different to the usual. The results of "making special" are verified with the attributes of beautiful or ugly, when they attract or repel us. But what biological sense does such a judgment make? At first glance, it does not appear to be very plausible that aesthetic preferences in artificial contexts could also serve as a guide to orientation in fitness-relevant life decisions. And yet if they do, in which? Or has the aesthetics of the artificial world freed itself from "tyranny of functionality" (Lumsden 1991) instead – and thus established creative freedom for the tyranny of unconventional memes?

Costly Signals

In the context of sexuality and eroticism, the evolution of aesthetic preferences is increasingly better understood. Those features which attract and arouse, which stimulate mating effort and trigger the willingness to take on the (opportunity) costs and risks of sex, contain a reliable signal of the gains to be expected. Sexual beauty signals – besides other information – what biologists have called "good genes", meaning the genotypically determined ability to cope with pathogen stress and other perils of life. "Good genes" reveal themselves phenotypically in those features which are of outstanding importance in sexual aesthetics precisely for this reason, e.g. in facial and body symmetry (Thornhill and Gangestad 1999). Following aesthetic standards in mate selection ultimately means searching for fit partners. Moreover, because fitness has a heritable component (Møller and Alatalo 1999), there is automatically an evolutionary coupling of indicator and preference. Whoever seeks beauty will be rewarded with reproductive fitness on average and at the same time, will provide for the genetic distribution of his/her preference. Beauty itself is not useful, it only signals usefulness. The train of a peacock is not useful – on the contrary – it hinders the peacock in flight – but the genotype is useful, which makes the train so very imposing: The most attractive peacocks produce the most vital chicks – because they have "good genes" (Petrie 1994; but see Brooks 2000).

Of course, the elaboration of male courtship features can only serve as a reliable indicator of "good genes", if deception is impossible. If it were possible, i.e., if males with inferior genetic quality were also able to form the respective traits to the same degree, they would definitely do so in their own mating interests. As a consequence, the correlation between trait and fitness would disappear and according to the logic of frequency-dependent selection, the traits would lose their value as a reliable signal of quality. The fact that this is not actually the case is attributable in many cases to the costs of these traits. An attractive peacock train, the powerful red color of the stickleback, the firmly erect cockscomb, the strength-sapping song of the nightingale and all of the other sexually effective indicators of "good genes" cannot simply be imitated. They are expensive to produce and/or maintain and, therefore "honest signals". Males, who elaborately display their expensive features must necessarily be able to afford to do so. This observation let Zahavi (1975) to formulate the "handicap principle" (Grafen 1990; Iwasa et al. 1991; Maynard Smith 1991; Johnstone

1995; Zahavi and Zahavi 1997) – which is frequently also known by the name of "costly signaling theory". The logic is based on the observation that an equal increase in the elaboration of a signal is more expensive for less fit individuals than for fitter individuals. Whoever can afford to produce costly signals, i.e., to accept additional expenses (thus handicapping themselves, so to speak) must logically be really fit, which is why females link their mate-selection preferences to these not infrequently bizarre and lavish traits, thus ensuring the evolutionary persistence of these signals.

In every single case it is necessary to ask why these traits cannot be imitated cheaply, i.e., why they are actually so costly (Olson and Owens 1998). It appears that testosterone frequently plays an important role for male sexual displays (Folstad and Karter 1992; Wedekind and Folstad 1994; Zuk 1994; Lindström et al. 2001), because this sexual hormone packs a physiological double effect: although it facilitates the production of some of the epigamic traits (Buchanan 2000; Owens and Short 1995), yet it simultaneously increases one's susceptibility to parasites, so that only healthy males, i.e., those with good immunocompetence will have enough testosterone "left over" to elaborate their sexual traits. Accordingly, a trade-off problem between investment in the immune system and investment in an extravagant "handicap" would ultimately be a crucial stage, on which the evolution of mate-choice preferences is played.

However, not only "good genes", but also qualities with solely phenotypical benefits can be publicly announced with costly and showy signals: political power, material wealth, fighting strength, hunting skills, etc., briefly, everything, which, depending on the predominant lifestyle, contributes to economic and social competitiveness and which brings prestige through this public display.

In order to understand the evolution of a biological trait, we first have to clarify whether it is a useful trait, i.e., a trait, with the aid of which self-preservation and reproduction are more likely to be successful, or whether it is instead a trait, which without being directly useful, exists only because it indicates hidden usefulness as an honest signal. This difference is grave. Whereas utility selection requires economic efficiency, signal selection maximizes reliability (Zahavi and Zahavi 1997). Efficiency is achieved if a maximum of utility is reached with a minimum of investment. On the other hand, selection for reliability leads to that which at first glance appears to be an uneconomic waste of precious resources, namely to presumably non-functional redundancy and extravagance of the signals – and this only because signals have to be costly in order to be convincing. Or more precisely, the signals have to be costly for potential cheaters, while actually for the high quality signaler they are relatively cheap.

For useful traits, their production costs are disadvantageous, but unavoidable. With signals, however, the additional costs are what count. Contrary to long-held economic rationality, demand increases together with its price. Useful traits do not lose utility if their price falls. Signals, on the other hand, lose their function if their production becomes cheap (Zahavi and Zahavi 1997).

Even if signals are shaped by signal evolution, the ability to understand these signals and to react appropriately, is the result of utility evolution (Zahavi and Zahavi 1997). Ignorant individuals will be less successful biologically on average than those who know how to read the signs and to make their social tendencies dependent on what their counterparts communicate via signals.

This paper explores the possibility of whether the functional logic of the handicap principle could perhaps also explain the evolution of aesthetics in the world of artifacts. It picks up on the idea which occurred to, but was not systematically pursued by Zahavi (1978). Perhaps adaptive interpretations of artificial aesthetics have previously been so dissatisfying, because they have sought utility where there is none, and thus failed to appropriately take the signal character of beauty into account. Could it be that artificial beauty – like sexual beauty – is not subject to utility evolution, but to signal evolution? Is something beautiful because it is costly and because in a social and competitive world, it is useful to develop preferences for persons who can afford lavish expenditure?

The Beautiful Handicap

Although a connection between art and communication is seen in many theories of aesthetics, including the naturalistic ones (e.g., Eibl-Eibesfeldt 1988; Coe 1992; Dissanayake 1992, 1995; Aiken 1998; Sütterlin 1998a; Cooke and Turner 1999), yet the issue of the adaptive logic of this relationship requires more analyses. What precisely is being communicated for what purpose and what are the selective advantages for signaler and signalee, who agree to this exchange of aesthetic communication? Why can it be fitness-promoting to communicate through extravagant, but useless signals? These are questions which a Zahavian perspective of artificial aesthetics attempts to answer. If that which we consider to be beautiful is the result of a signal evolution and our preferences for beauty are to be understood as the result of a utility evolution, three hypotheses have to be correct:

- Beauty must be costly to be an honest signal.
- As an honest signal for the quality of the signaler, beauty must woo the attention of certain target signalees, and
- it must be useful for these target signalees to be able to assess the quality of the signaler via beauty.

The following sections will deal with these three hypotheses.

Hypothesis 1: "Making Special" Means "Making Expensive"

The first hypothesis requires that the production of beauty, i.e., what Dissanayake (1992) has referred to as "making special", must be "making expensive". Only if the act of "making special" requires a costly effort, does it have a chance of being awarded the attribute of "beautiful". In order to be valid as costs in the logic of the handicap principle, they must be able to be balanced in the currency of natural selection and reproductive fitness. Therefore, life effort determines the price and life effort can be invested in basically three ways: either in the form of (1) precious resources, (2) as a risk to health and life, and (3) as time, i.e., as lost opportunities for alternative routes of fitness maximization. What precisely determines the price can be very different from culture to culture, from group to group, and from individual to individual, depending on the prevailing life situation.

Resources

Who is not familiar with childish disappointment when the beautiful, shiny jewel on the beach turns out to be a ground fragment of glass? The fact that we consider expensive jewelry to be more beautiful is more likely to be the rule than the exception. Think of "kitsch", which is stigmatized merely because it attempts to cheaply reproduce what is elegant and expensive, while, on the other hand, what is monetarily more expensive is only rarely devalued as being kitsch. The correlation between material preciousness and beauty can repeatedly be found in cultural history and could easily be illuminated through additional examples. For our topic, the causality of the correlation between preciousness and beauty is of special interest. Is something beautiful, because it is costly or is it costly, because it is beautiful?

In his "Theory of the Leisure Class", Thorstein Veblen (originally: 1899) thoroughly investigated this correlation and concluded: "The superior gratification derived from the use and contemplation of costly and supposedly beautiful products is, commonly, in great measure a gratification of our sense of costliness masquerading under the name of beauty" (Veblen 1953, p. 95). It is not enough, therefore, to appeal to our sensory apparatus in a special way – kitsch does this too – instead preciousness is required for something to be considered beautiful. It may be that gold is more sensually appealing for certain neurophysiological reasons than other metals, such as aluminum, nevertheless it is beautiful because of its value. Although the bluish-gray shading of basalt may be sensually appealing, who would consider loose chippings to be aesthetically appealing? In short: "The marks of expensiveness come to be accepted as beautiful features of the expensive articles" (Veblen 1953, p. 97).

A critical objection at this point could now be that Veblen's point of view, which matches the Zahavian handicap principle surprisingly well despite its academic and historically different origin, possibly applies to developed agrarian and industrial societies with their money economies and their pronounced social stratifications, but in the rather different socio-ecological milieus of hominization, Veblen's observation might be invalid. After all, the Pleistocene hunter-gatherer societies are considered to be more or less socially egalitarian, whose subsistence economies did not permit the accumulation of resources or even surplus production. Nevertheless, and despite this contrast, it has "always" seemed to be true – according to everything that we know – and that means since the beginning of the history of art approximately 50,000–60,000 years ago that it is especially the precious and not the usual which is considered to be beautiful. Body painting (Watts 1999) is one of the historically oldest forms of "making special", and for this purpose, precious soils, which could only be obtained with much effort, were used as a matter of fact. Coe (1992) reports of Australian aborigines who had to risk crossing enemy territory in order to obtain their ocher, and the ethnography of recent hunter-gatherer and planter societies has collected numerous pieces of evidence for the fact how preciousness constitutes what is beautiful (Eibl-Eibesfeldt 1988; Power 2000). In sum: one of the cost factors that constitutes beauty consists of the degree to which goods in short supply are used.

Risks to Life (Vitality)

However, beauty is not solely constituted by means of material splendor. Over and above this, signals can also become costly and thus honest, because risks to one's health and life have to be taken to produce them. In many historical and traditional societies, those whose bodies have been deformed in a special way are considered to be beautiful, be it by filing teeth, through scarification and tattoos, by piercing or by skull, genital or foot deformations. The well-known practice of lacing the female waist in Western fashions belongs to this category. All of these somatic manifestations of "making special" demand their more or less high price in health. Skull deformations reduce vitality, ablations cause pain and problems with chewing and speaking, scarifications can end fatally due to the risks of infection, and the reference that the corseted "wasp waist" can be unhealthy can be found in the first issues of progressive women's magazines.

In part, it is the deformed persons themselves who mutilate themselves, in part it is the parents who begin with the reshaping of their children soon after their birth. Even if the pain associated with the deformation is suffered by the children, it is the parents who bear the costs for this example of "making special" in accordance with the logic of the handicap principle. They are ultimately spending part of their reproductive fitness. The formula is: "Beauty is something that wastes vitality". The emphasis lies on the word "wastes", because this does not refer to all of those scars which involuntarily, but automatically occur when someone passes through the perils of life. A face that is deformed by an accident will only rarely be perceived as being beautiful, because it ultimately does not testify to waste, and therefore has nothing to do with the signals dealt with by the handicap principle. The situation is different with regard to the desired dueling scar – in the corresponding circles it was considered a prestigious and thus "beautiful handicap"!

Time (Opportunities)

Finally, it is time which determines the price – the time that must be devoted to artistically shape an object or which one needs to learn and perfect certain techniques. Only those who practice a lot can play music, sing, carve, sculpt, paint, embroider, dance, etc. beautifully. Because not everyone who practices for a long time eventually masters these techniques, the time/performance ratio is a useful measure for that which is generally referred to as "talent" and with which certain aspects of intelligence and creativity are expressed. While it is the practice that makes the signal costly, it is important that some are prodigies at music, carving, writing, painting etc. To these people a masterpiece can be fashioned in a remarkably short space of time and apparently without much effort. It is precisely that these things are not costly to the fortunate few, but are to the unfortunate many that they are worth signaling.

Nevertheless, time is so precious because its waste can cause opportunity costs. One could work productively, accumulate resources, and – as is to be expected – invest these in mating and reproduction – instead of constantly practicing or designing things in a time-consuming fashion. Whoever does not do this, i.e., a Stone Age hunter who does not hunt the gazelle, but plays the flute instead, pays a price, and because not everyone can pay this price purely for reasons of self-preservation, this waste of time becomes an honest signal. "Conspicuous leisure" is what Thorstein Veblen calls this form of investment in social prestige in his "Theory of the Leisure Class". Whoever wants to be somebody should not be productive, because otherwise he would not be distinguishable from the masses of poor suckers on the margins of their existence. Whoever wants to be somebody must instead waste things that are in short supply, and that includes time.

Something is beautiful if it consumes valuable time for its production, and therefore dance nets are considered beautiful among the Eipo from the Highlands of New Guinea – which have no practical use at all – but require much time and practiced manual dexterity for their production and for this reason alone decorate their wearers and give them prestige (Eibl-Eibesfeldt 1988). For the same reason, wares made by craftsmen are presumably considered by many to be more beautiful than the same wares made using industrial production methods – and this, even though industrial goods are mostly more perfect when they come from a modern production line than those made manually by craftsmen, which always have some slight irregularities. It is this investment through extra work which gives these things aesthetic value. Only what is expensive can be beautiful, and time (or more precisely, the opportunity costs) is one of the three components of price.

If, however, the price falls due to technical developments, i.e., if "making special" no longer requires any noteworthy investment from a limited personal budget of time, things lose their aesthetic appeal. Beauty is subject to an inflationary devaluation completely apart from its sensory appeal. The possibility for the automatic reproduction of lace has allowed the interest in lace to disappear almost completely. Velvet and silk, which in the past were certain guarantees for a beautiful outfit, have more or less fallen into the category of common textiles. Carvings from the Oberammergau region in Germany are today suspected of being made by machine and thus are regarded as being kitsch instead of as art in contrast to the past. There are computer programs which can produce music in the style of the old masters, namely in a quality which even makes it difficult for experts to recognize the new productions as such. Yet we are going to find Vivaldi in the original much more beautiful in a very irrational way, than the stylistically equivalent artificial products of computer technology, because our aesthetic algorithms whisper to us: Whatever can be produced cheaply cannot be beautiful.

Conclusion: Only What Is Costly Is Perceived as Being Beautiful

Only what is expensive is perceived as being beautiful. This applies, of course, only when taking the various forms of cost, including resource values and the time invested and health, into consideration. As different forms of life effort, they are convertible into a common key currency, namely "reproductive fitness". All organisms, including humans, are designed by nature to maximize this currency of the Darwinian principle. For this purpose, persons develop and accumulate resources, invest in their health and self-preservation and in that of their offspring and other kin and deal with time in a cost-benefit oriented way. Whenever resources, vitality and time are spent, lifetime and reproductive opportunities are being drawn upon, and therefore, this triad of biological costs pushes the ultimate price tag up on biological traits. In the final analysis, the price of signals is determined by the investment of reproductive fitness, which is necessary for their production – even if the costs can be expressed in various phenomenological ways.

Whoever is in a position to indicate his status in a forgery-proof way will acquire social prestige, that precious good for which one competes on the psychologically perceived level. Those wishing to increase their prestige must show that they have earned it, and this can be done unmistakably through waste – in Veblen's diction through "conspicuous consumption and leisure". From Zahavi's sociobiological perspective, waste as an investment in prestige (and this includes the investment in "making special") clearly is an honest signal. Hypothesis 1 appears to be documented through numerous examples: "Making special" is essentially "making expensive", and on the psychological level competition for prestige.

Hypothesis 2: Beauty Signals Quality

Whoever bears the costs of the signal is considered to be the signaler. First of all, this can be the producer of the signal himself or herself, i.e., that person who makes something special using his or her life effort. However, it can also be a kin group, who jointly invest in a signal (the totem poles of Northwest American Indians are an example here). It can also be those who as sponsors or clients pay for the working time and practiced artisan skills of clever and creative persons in order to have something beautiful produced. In such cases, one will not wish to deny that the artisans themselves are talented artists – however, they do not stand at the core of the explanation for the handicap principle, because they do not pay the price of the signal, and this all the less, the more economically successful they are in marketing their talents.

In a barter society, in which life effort becomes a commodity and is subject to the law of supply and demand, it, therefore, does not necessarily have to be the producer of something beautiful who bears the costs of the signal. What is crucial in the sense of Darwinian aesthetics is instead the issue of who ultimately pays for the investment in beauty. Therefore, the magnificent facade is primarily a signal sent by the prince and possibly secondarily – but in a much narrower context – also a signal sent by the stonemason. A park is primarily a signal sent by the lord of the manor and not by the gardener; the altar is not the signal of the woodcarver, but of the church. Of course, their qualities are expressed in the work of the stonemason, gardener and woodcarver, but in these cases, the quality of the work may not be understood as a signal within the meaning of the handicap principle, because good work is rewarded economically and is therefore directly useful. The work is not invested in an expensive "handicap" associated with personal extra-costs, but exchanged for economic benefits. The best economic utilization of skills and capabilities is nothing which costly signaling theory pretends to explain, because here the direct benefit is in the foreground, not the expensive and clearly visible investment in something that is obviously useless in order to gain mere prestige.

Honest signals have evolved in three contexts in the animal kingdom (Zahavi and Zahavi 1997), namely in the communication between prey and their predators; in the social competition for ranking positions, where they help to negotiate hierarchies without battles; and finally, in sexual competition, where they permit conclusions pertaining to mate quality. Apart from prey/predator communication, which no longer plays a significant role due to humans' ecological dominance, the other two arenas of signal evolution have seamlessly fit into human cultural history. Social ranking positions and mate quality are publicly announced, especially through artificial, but no less necessarily costly signals. Moreover, one signal system is a typical human accomplishment, namely, that one which indicates moral virtue.

If the hypothesis is correct that aesthetic preferences have evolved as useful decoders of "honest signals", we should expect that beautiful artifacts as a medium of forgery-proof information either play a role in sexual competition or on the various stages of social competition for power outwardly and solidarity inwardly. There is some evidence for all three surmises.

Beautiful and Sexy

In response to the question of why they actually and arduously decorated their clothing and articles of everyday use, the Kalahari San interviewed by Polly Wiessner (1983) said that they wanted "to present a positive image to partners in reciprocity and to members of the opposite sex" (p. 258). Here, a lot of everyday knowledge from presumably all the cultures of this world are reflected. When it comes to staging one's self as sexually and socially attractive, not only natural beauty, but also artificial beauty plays an important role.

The function of what has been made beautiful as an eye-catcher in the arena of sexual competition applies not only to one's personal outfit, but finds its parallels in other forms of "making special". Of poetry, music, the fine arts, Miller (1999, 2000) claims in a direct follow-up to Darwin (1871) that they have solely evolved as advertising mediums of male mating interests. With a beautiful song, a beautiful poem, a beautiful painting, (primarily) men create courtship displays, with which they woo for the favors of the (mostly) other sex. Cultural displays increase sexual access, to which sources, such as Jimi Hendrix, Pablo Picasso, Charlie Chaplin, Honoré de Balzac and others who have lived their artistic lives as insatiable lovers bear testimony. Even the upright Max Frisch is made to notice this – obviously unwillingly – because as he complains as a 64-year-old in Montauk: "Someone who has 'made it' could look like a walrus and women will not only associate with him, but also develop their charms unasked with almost no inhibitions. Only on the street, anonymous in a crowd, do I perceive myself utterly to be a walrus" (Frisch 1975, p. 20, my translation).

More impressive than such anecdotes, for which surely counter-examples can be found, the undoubtedly noticeable age and sex distribution of artists speak for Miller's (1999, 2000) interpretation of art as sexually evolved courtship displays. As is known, men produce much more art than women, which is to be expected in view of human sexual strategies with their higher share of male than female competition. Moreover, most works of art are produced by young adult males, i.e., precisely at a stage in life when their sexual competition is the largest. To the degree that artists age and their life effort increasingly reflects parental and nepotistic effort, cultural displays lose meaning for them. They pass the zenith of their creative powers, once this sexual competition subsides.

In many traditional and modern societies, initiation rites form the traditional platform for sexual self-portrayal. The personal costs of the initiation can be considerable, inter alia, because they coincide with physical, e.g. genital mutilations (Low 1988; Rowanchilde 1996; Power 2000). The level of the costs is co-determined by the local ecology. Physical mutilations are expensive especially in regions with pronounced pathogen stress, because here, they are especially risky. Moreover, in perfect agreement with costly signaling theory, they are traditionally widespread in tropical regions with an above-average risk of infection. Precisely where they are the most expensive, scars contribute to the beauty of (mostly) women (Singh and Bronstad 1997). The message is obvious: the healing process allows deductions pertaining to the immunocompetence of the women, and this is one of the decisive fitness factors in a pathogenically risky habitat. Mate quality rises with the body's defenses, and this is indicated in a honest way by artificial injuries.

In many societies, only those men who have properly completed their initiation can expect to find a wife or even social respect. Whoever was unable to bear the pain of the initiation, on the other hand, "never enjoyed the least respect, women despised them, chiefs refused to accept food from their hands which they called stinking ... (and) many a father refuses him his daughter's hand" (Marquard 1984 for Samoa, cited according to Coe 1992). A pierced penis signals courage and the ability to suffer on the part of the male and according to Rowanchilde (1996), it carries a second message as well. Because male genital modifications at the expense of the male intensify the sexual experience of the female; they also provide information about the male's commitment to cooperation in the relationship.

In these contexts, artificial beauty signals mate value, whereby it remains to be studied, which components of mate value are indicated by which signals. A couple of these messages appear to be decoded – others are not. Physical mutilations signal vitality (especially a strong immune system) and expensive outfits signal material wealth. Both are recurring correlates of reproductive fitness, about which naturally evolved signals already provide information in the animal kingdom, but for what fitness component does artistic creativity stand? One could surmise with Miller (1999, 2000) that it indicates cognitive eloquence, even if no close correlation between intelligence and creativity appears to exist. One only needs to think of the artistic masterpieces produced by emotionally unbalanced persons. Undoubtedly there is still an extensive need for more research here.

Beautiful and Strong

"In the Renaissance ... the best artists were to be found close to the most powerful rulers: Dürer was the court painter of Emperor Maximilian; the second-best painter, Cranach, was the court painter of the second most powerful ruler, namely the Prince-Elector of Saxony; the third-best painter Grünewald, was the court painter at the court of the third most powerful man, Albrecht, the Prince-Bishop of Mainz and Duke of Brandenburg; the fourth-best painter, Holbein, was unable to find an adequate place in Germany. He was the court painter of the English king". Even if art historians are likely to debate whether this interpretation of Baumann (1998, p. 200, my translation) actually applies in its rigid strictness, there is one thing that cannot be disputed, namely the regularly found correlation in history between power and the beauty of its insignia (Sütterlin 1998a). Powerful people will probably always and everywhere want to portray themselves, and this can best be done with forgery-proof signals. Only those who are able to afford the best artists are believed to be rich (or powerful, mostly both at the same time). Because demonstrative waste in these contexts announces competitive status in power systems (cf. Neiman 1997 for a case study on the Maya architecture), it automatically sets off a spiral of an aesthetic arms race. The most powerful man must always invest a bit more in his status signals, that is, he must always have a bit more beauty produced than the second most powerful man is capable of doing.

The message of this "showing-off" is obvious. Whoever has something expensive produced, must be able to control people and resources and is therefore influential and powerful. The logic of this signal system functions equally well, regardless of whether individuals, dynasties, clans, tribes, political parties, states, churches or conglomerates are competing with each other. Consequently, beauty is more likely to arise, the more prevalent social or economic competition is. This is why it is surely no accident when art historians see the numerous small principalities with the corresponding density in competing rulers as the reason for why music developed in more diversity at this time in history in Germany, than in neighboring centralist France, for example. In addition, the most beautiful bank skyscrapers are being built where their competitors already stand. Only an honest signal, i.e., the proof of power over people and resources gives a publicly recognized justification for a claim to power.

Beautiful and Morally Good

Early human history was characterized by constant competition among autonomous kin groups for ecological opportunities. The evolutionary outcome of this intergroup competition was a reinforced in-group/out-group moral, whose most essential function consisted of binding the members of one group in a social alliance and then committing them to a "we feeling" However, like all public goods, this group solidarity is also subject to the free-rider problem. In a conflict between one's own interests and the welfare of the group, there is a large probability that one's own interests are going to be the victor. Although one may be inclined to harvest the benefits of group membership for one's self to the best degree possible, on the other hand, there are strong incentives as a personal benefits maximizer to avoid the costs arising from such a social alliance, if possible. Thus, group solidarity always runs the risk of being exploited – unless its members and especially any new adherents announce their moral integrity with "honest signals". This function is assumed by rituals. Whoever is willing to take the high costs of an initiation rite upon himself or herself is acknowledging his or her in-group in a manner perceivable by all and demonstrates the required loyalty. The high costs of initiation prevent opportunistic persons free-riding, because the net benefits of group affiliation are only realized once the initial entry costs have been compensated for. Moreover, outer visible signs block their wearers permanently from the possibility of an opportunistic or tactical change in group and thus justify the high probability for a lifelong commitment on the part of the initiate. With this costly signal, he or she demonstrates a credible interest in affiliation with this group. One buys prestige and recommends oneself as a reliable and morally good partner (Knight 1998; Power 2000).

Thus there is an evolutionary coupling of "beautiful" and "morally good", which has ultimately left behind deep traces in Western cultural history. Only in the recent past has the enlightened insight begun to take hold that external beauty and "inner values" are to be observed uncoupled from each another. From ancient times through the beginning of the modern era, the linking of moral judgment with aesthetic judgment was "common sense". "The good is inherent in the beautiful, because beauty has the same substrate as the good" says Albertus Magnus (quoted according to Henss 1992, p. 86). Correspondingly, the good heroes of our histories are good-looking and the bad are ugly – both in cheap paperbacks and in Hollywood productions. The tendency to mingle aesthetic and moral judgments is obviously anchored very deeply in our emotional and cognitive mechanisms (Henss 1992), and it really looks like there is an evolutionary logical background for this link.

Conclusion: "Making Special" Arises from the Competition for Sexual, Political or Moral Respect

Humans are able to compete with almost all of their activities, which is why Dissanayake (1995) doubts that there is an evolutionary link between "making special" and any one form of competition. She went on to indicate that if men grow beards, drink beer or engage in shot-putting for a bet, this surely does not mean that hair growth, drinking or shot-putting have evolved to become competitive purposes. Correspondingly, even if people compete with artistic displays, this did not necessarily mean that art has evolved for this reason. One has, however, to take possible function transformations in evolutionary processes into account. It is not ruled out, of course that signals contain elements which originally arose from utility evolution, but were then co-opted by signal evolution. Even if tail feathers are likely to have originally evolved because of their utility, later on – as in the case of the peacock – they could be monopolized by signal evolution. Even if the temporal arts (music, poetry and dance) possibly have their origin in the context of the mother–child interactions of early hominids (Dissanayake 2000), yet they have also been co-opted by signal evolution, and now fulfill – after an evolutionary process of functional transformation – the role of an "honest signal". Of course, it would also be conceivable that the earliest mother–child communication also consisted of an exchange of honest signals, and thus that the temporal arts were already signals in their origins. As the case may be, one does not understand the train of a peacock when one investigates the mechanics of flying. One does not understand the function of a track and field competition, if one studies the natural history of running, jumping, or shot-putting and one does not understand the erotic impact of Carlos Santana's guitar playing, if one studies the ontogenesis of the mother and child bond in early hominids. On the other hand, ritualization, i.e., the transformation of original utilities to signals, creates phylogenetic continuities. Therefore, it is surely no accident that the temporal arts not only play such as significant role in the emotional communion of mother and child, but also in the emotional communion of adults.

Beauty has a signal function. It communicates the sociobiological qualities of those who have invested in the production of beauty. Here, we are dealing with a triad of messages, which at first glance appear to have little to do with each other. The "I am fit" comes from sexuality, the "I am strong" from the competition for power, and the "I am good" from morality. What these messages have in common is that their core information is not so obviously recognisable. "Good genes" cannot be seen; the power and strength of political protagonists can only occasionally be perceived – and then possibly at high personal risk. Moreover, moral integrity cannot only not be seen, but it is even unlikely a priori, in a world of personal benefit maximizers. Therefore, all three messages require proof of their substance and provide costly signals – which alone due to their mere existence document hidden qualities of the signaler.

Signalers have no other choice, but to enter into a competition that is so expensive for them, because they are wooing for partners, and these want to be convinced. Wasteful "making special" documents one's own market value as a sexual and social partner. Superficially, this has to do with gaining attention, but ultimately with gaining respect in the critical eyes of potential partners – which brings us to the third hypothesis.

Hypothesis 3: Aesthetic Preferences Are Useful

Signalers have a personal interest in reaching an audience to whom they can show their hidden qualities. Simple, economically rare emblems with messages, such as "I am fit, strong and morally good" were unable to evolve because a naive, untested acceptance of socially motivated information would be highly risky for signalees, since the interests of signaler and signalee are not identical. After all, signalees would have to expect to be exploited. On the other hand, they also have a vital interest in learning about the hidden qualities of their social partners, because they are seeking the best possible sexual mates, highly potent "Machiavellists" as patrons and solidarity partners who are reliable. Therefore, a detector which checks the signaler's personal statements for "substance" and assesses the signalers according to their usefulness as partners in sexual or social contexts is advantageous for signalees and precisely this function is fulfilled by aesthetic judgment. It assesses "honest signals" about hidden qualities. In the world of artifacts, aesthetic preferences have the same function that they have always had in a world of natural features, namely as aids to orientation for decisions in sexual and social affairs.

In the selection processes during hominization, undoubtedly those individuals are rewarded who have made the "right" partner decisions in a world of sexual and machiavellian competition. They are those individuals who did not indiscriminately mate, but learned to estimate the mate value of possible partners, those who did not join just any coalitions, but were able to recognize the power and influence of the political figures, and those who did not commit indiscriminate acts of solidarity, but understood how to verify the moral reliability of possible alliance partners. This was only possible for them, because the suppliers of expensive and thus honest signals informed others about their respective qualities as possible sexual and social partners. The psychobiological control mechanisms of social affects process this information and ensure important decisions on this basis in a competitive world: To whom should one feel attracted – sexually, politically and morally – and with what intensity? Aesthetic perception is essentially an evaluation process and aesthetic judgment in its core is a judgment about the sociobiological quality of those who produce or sponsor costly signals and thus advertise for partners for sexual, political, or moral forms of cooperation.

Aesthetic Preferences in Human Culture: Adaptation or By-Product?

The evolutionary history of biological traits can have taken basically two different routes. Either a trait owes its evolutionary persistence to a specific selective advantage – which would be an adaptation in this case – or the trait is carried along through evolutionary periods as a non-functional by-product of another functional adaptation. The fact that bones are white is not an adaptation, because the color of bones was never selected. Instead, the calcium deposits in bone tissue are biologically functional and adapted, and calcium just happens to be white. The color of bones is, therefore, a "coincidental" by-product of a trait that is adaptive for other reasons. With respect to aesthetic preferences, a highly significant question arises, namely of whether in the world of artifacts they are non-functional by-products or biologically functional adaptations.

There are theoretically two possibilities in which aesthetic judgment could occur as a non-selected and thus non-functional by-product. In the "hard version", a by-product is useless in every respect. Its existence is so superfluous, but at the same time as automatic as that of the naval. One could live without it, but that is simply not possible. According to this point of view, it could be hypothesized, for example that our brain physiology performs so much vital operational work that aesthetic cognitions accidentally occur as unavoidable by-products, so to speak, which cannot be used for anything, but are also not harmful, or at least not so harmful that it would be worth it to the organism to take the time and effort to prevent such cognitions. This image is comparable with an engine, which also provides useless warmth, even though its actual function lies in the production of power.

In the "soft version", by-products occur through the transfer of functional traits to contexts in which they did not evolve. We prefer mating partners in accordance with certain criteria, for example, according to their facial symmetries. Symmetries can also be found in the world of artifacts, of course, and because we are unable to avoid going through life with open eyes and a willingness to make aesthetic judgments, we also perceive things as being beautiful, which had nothing to do with the evolution of aesthetic preferences. Accordingly, the beauty of a symmetrically designed hand-axe could perhaps be understood as the non-functional by-product of an evolved and biologically functional preference for symmetrical faces, for example, and the artistic precision of a pas de deux possibly activates adaptive preferences for youth and vitality outside the context of their evolution. Furthermore, because a powerful motivation grows out of the pleasure in fulfilling these preferences, we learn to deal creatively with the mechanisms of aesthetics which have been handed down to us. We make things beautiful, solely because beauty is beautiful – without being rewarded for this biologically. In this view, art would be so successful, because it leads us insatiable beauty addicts to believe promising illusions concerning a presumably attractive world. Artistic aesthetic judgment would thus be a transfer of an adaptive, naturally aesthetic judgment – without itself being of any relevance to fitness. From this point of view, evolved preferences overshoot their actual mark. We would not only be like a moth circling a lantern at night; but like a moth which has succeeded in inventing a lantern in order to have fun circling it at night.

"Soft" views of aesthetics as a biological by-product have played, either explicitly or implicitly, the dominating role in evolutionary aesthetics to date. One of its basic assumptions has been formulated and investigated repeatedly: Aesthetic perception is linked to the perception of especially configured patterns of stimuli (Rensch 1984; Eibl-Eibesfeldt 1988; Green 1995; Baukus 1996; Grinde 1996; Aiken 1998; Sütterlin 1998b; Pinker 1999; Richter 1999; Sitte 1999; Ralevski 2000), whereby the "triggering mechanisms" originate from the dealing of humans with their social and ecological challenges. The integrated ensemble of these biological triggering mechanisms forms a universal grammar of aesthetics (Richter 1999), which is comparable to the linguistic deep structure found by Chomsky. As a genetically fixed universal, it is typical of all humans regardless of culture and fashion, and because of its "innate" a priori character, it ensures "that we humans are capable of doing more than we learned with our aesthetic judgment" (Richter 1999, p. 50, my translation).

A comparison with another by-product, namely the ability to be aroused by pornography, allows doubts to arise as to whether artificial aesthetics actually is due to nothing more than a simple transfer. Pornography was able to enter the world – even though its consumption does not lead to generative consequences – because it is based on both sensory and emotional aspects of an evolved sexuality. Depictions of sexual stimuli thus find the same sensory appeal as do the stimuli themselves, and in both cases, the reaction is sexual arousal. Sensory appeal and emotional arousal are identical in both cases (the by-product and the adaptation). The situation is clearly different with aesthetics, however. Even if it were true that things that have been made beautiful consist of evolved stimuli configurations and, therefore possess an a priori sensory attractiveness, the question remains of whether the emotional reactivity is comparable. Do we get a sexual appetite on seeing a symmetrical artifact as we perhaps do when seeing a symmetrical face? Do we find the still-life works of the Flemish master painters more beautiful when we are hungry? Even if this actually cannot be excluded, yet on the whole, it seems likely to be improbable that the perception of beautiful art is going to trigger the same emotional reactions as upon the perception of naturally beautiful items. In the world made by humans, aesthetic preferences are obviously more than pure transfers of archaic competencies and hence, more than pure by-products.

If aesthetic preferences in artificial contexts were merely non-functional by-products of adaptive aesthetic preferences in natural contexts, they would have remained by-products which, remarkably, had not been co-opted by any other mechanism and to which, therefore, no selective processes could attach. However, how probable would such a case be? How probable is it that by-products have remained by-products "forever" i.e., that they could have been able to be "overlooked" by natural selection, so to speak? With such complex (and costly, because they consume lots of operational energy) phenomena, such as aesthetic cognitions, it appears to be extremely doubtful at any rate that they could have been transferred from their original domains, e.g., eroticism, to another, e.g. the fine arts without coinciding with a gain in function. It appears to be more likely that with their transfer into another context (into the world of artifacts here), evolved biological preferences acquire the character of predispositions for the evolution of new traits and thus very quickly grow out of their original nature as a by-product. To the degree to which the naturally selected aesthetic preferences have penetrated the world of artifacts, they have been exploited by numerous social processes and undergone signal evolution. They began to play a new and significant role.

Finally, why should humans produce beauty requiring an effort that is occasionally immense? On the assumption of the by-product hypothesis, it would merely be expected that we spontaneously judge what we find in our living world – including artifacts – aesthetically, but not that we lose ourselves in non-functional activities. Why should we accept costs by investing time, resources and risks in "making special", if this investment ultimately does not pay – "pay" in the currency in which the success of every biological trait is measured, namely, in reproductive fitness? On balance, it seems safer to regard aesthetic preferences even with respect to artifacts as adaptations than as non-functional by-products of evolved aesthetic preferences with respect to natural features, with their typical playing fields of eroticism and biophilia.

The sociobiological perspective of aesthetics take a stand accidentally and wholly unintentionally on the ancient philosophical debate on the origins of beauty. Is beauty a category of the objects or the recognizing subjects? In the almost two-and-one-half thousand-year-old philosophical debate, realistic positions which view beauty as objectively existing in reality clash irreconcilably with idealistic positions, which – in their harshest versions – interpret aesthetic perception as solely a subjective act, which is unjustifiable, and cannot be objectified or even communicated. It should have become clear that neither of these two positions finds support from the sociobiological perspective. Of course, beauty adheres to things to a certain degree. Signals are real, objective and perceptible facts, and that which constitutes their beauty, is determinable (to remind the reader: it is the amount of the investment in life effort). However, just as matter-of-factly, aesthetic judgment is the result of a subjective evaluation process of empirical facts and hence, the same signal can be rated differently. Humans assess what has been perceived according to personal criteria and against the background of personal interests, without however – and this distinguishes the naturalism of sociobiology from the post-modern attitude of anything goes – losing themselves in a rationally unapproachable arbitrariness. Aesthetic judgment is a subjective act of an objectively species-specific adaptation. "Beauty is in the eye of the beholder" is only half true for this reason. Symons (1995) expressed it more correctly as

"Beauty is in the adaptations of the beholder". Adaptations are information-processing mechanisms for solving biological problems of life and reproduction, which of course also process subjective data and, therefore can generate variable output with the same input. There are various life problems – for reasons of biological individuality – with different decision-making processes and therefore also different tastes, even though the Darwinian algorithm of aesthetic judgment must be thought of as a biologically evolved species-specific universal.

Summary

Adaptive interpretations of artificial aesthetics have previously been so dissatisfying, because they have sought utility where there is none, and thus failed to appropriately take the signal character of beauty into account. This paper explores the possibility of whether the functional logic of costly signaling theory could perhaps explain the evolution of aesthetics in the world of artifacts. There is some confirming evidence for this view. First of all: only what is costly is perceived as being beautiful. This applies, of course, only when taking the biological currency of costs into consideration, namely life effort. Whenever resources, vitality and time are spent, life and reproductive opportunities are being drawn upon, and this triad of biological costs constitutes the beauty of artifacts.

Second, beauty communicates the sociobiological qualities of those who have invested in the production of beauty. Here, we are dealing with a triad of messages, which at first glance appear to have little to do with each other. The "I am fit" comes from sexuality, the "I am strong" from the competition for power, and the "I am good" from morality. What these messages have in common is that their core information is not so obviously recognizable. "Good genes" cannot be seen; the power and strength of political protagonists can only occasionally be perceived and moral integrity cannot only not be seen, but it is even unlikely a priori, in a world of personal benefit maximizers. Therefore, all three messages require proof of their substance and provide costly signals – which alone due to their mere existence document hidden qualities of the signaler.

Signalers have a personal interest in reaching an audience to whom they can show their hidden qualities. Simple, economically rare emblems with messages, such as "I am fit, strong and morally good" were unable to evolve because a naive, untested acceptance of socially motivated information would be highly risky for signalees, since the interests of signaler and signalee are not identical. After all, signalees would have to expect to be exploited. On the other hand, they also have a vital interest in learning about the hidden qualities of their social partners, because they are seeking the best possible sexual mates, highly potent "Machiavellists" as patrons and solidarity partners who are reliable. Therefore, a detector which checks the signaler's personal statements about his or her quality is advantageous for signalees and precisely this function is fulfilled by aesthetic judgment. It assesses "honest signals" about hidden qualities. In the world of artifacts, aesthetic preferences have the same function that they have always had in a world of natural features, namely as aids to orientation for decisions in sexual and social affairs.

References

Aiken NE (1998) Power through art. In: Falger VSE, Meyer P, Van der Dennen JMG (eds) Sociobiology and politics. JAI, Stanford, pp 215–228
Baukus P (1996) Biologie der ästhetischen Wahrnehmung. In: Riedl R, Delpos M (eds) Die Evolutionäre Erkenntnistheorie im Spiegel der Wissenschaften. WUV, Wien, pp 239–261
Baumann C (1998) Naturwissenschaft und Kunst – Versuche der Begegnung. Nova Acta Leopoldina 77:193–204
Brooks R (2000) Negative genetic correlation between male sexual attractiveness and survival. Nature 406:67–70
Buchanan KL (2000) Stress and the evolution of condition-dependent signals. Trends Ecol Evol 15:156–160
Buss DM (1999) Evolutionary psychology – the new science of the mind. Allyn and Bacon, Boston
Coe K (1992) Art: the replicable unit – an inquiry into the possible origin of art as a social behavior. J Soc Evol Syst 15:217–234
Cooke B, Turner F (eds) (1999) Biopoetics – evolutionary explorations in the art. ICUS, Lexington
Darwin C (1871) The descent of man and selection in relation to sex. Murray, London
Dissanayake E (1992) Homo aestheticus: where art comes from and why. Free Press, New York
Dissanayake E (1995) Chimera, spandrel, or adaptation: conceptualizing art in human evolution. Human Nat 6:99–117
Dissanayake E (2000) Antecedents of the temporal arts in early mother–infant interaction. In: Wallin NL, Merker B, Brown S (eds) The origins of music. MIT Press, Cambridge, MA, pp 389–410
Eibl-Eibesfeldt I (1988) The biological foundation of aesthetics. In: Rentschler I, Herzberger B, Epstein D (eds) Beauty and the brain – biological aspects of aesthetics. Birkhäuser, Basel, pp 29–68
Folstad I, Karter AJ (1992) Parasites, bright males, and the immunocompetence handicap. Am Nat 139:603–622
Frisch M (1975) Montauk – Eine Erzählung. Suhrkamp, Frankfurt am Main
Grafen A (1990) Biological signals as handicaps. J Theor Biol 144:517–546
Green CD (1995) All that glitters: a review of psychological research on the aesthetics of the golden section. Perception 24:937–968
Grinde B (1996) The biology of visual aesthetics. J Soc Evol Syst 19:31–40
Henss R (1992) "Spieglein, Spieglein an der Wand..." – Geschlecht, Alter und psychische Attraktivität. Psychologie Verlags Union, Weinheim

Iwasa Y, Pomiankowski A, Nee S (1991) The evolution of costly mate preferences II. The "Handicap" principle. Evolution 45:1431–1442

Johnstone RA (1995) Sexual selection, honest advertisement and the handicap principle: reviewing the evidence. Biol Rev 70:1–65

Knight C (1998) Ritual/speech coevolution: a solution to the problem of deception. In: Hurford JR, Studdert-Kennedy M, Knight C (eds) Approaches to the evolution of language – social and cognitive bases. Cambridge University Press, Cambridge, pp 68–91

Lindström KM, Krakower D, Lundström JO, Silverin B (2001) The effects of testosterone on a viral infection in greenfinches (*Carduelis chloris*): an experimental test of the immunocompetence-handidcap hypothesis. Proc R Soc Lond Ser B 268:207–211

Low BS (1988) Pathogen stress and polygyny in humans. In: Betzig L, Borgerhoff Mulder M, Turke P (eds) Human reproductive behaviour – a Darwinian perspective. Cambridge University Press, Cambridge, pp 115–127

Lumsden CJ (1991) Aesthetics. In: Maxwell M (ed) The sociobiological imagination. SUNY Press, Albany, pp 253–268

Marquard C (1899/1984) The tattooing of both sexes in Samoa. McMillan, Papahure

Maynard Smith J (1991) Theories of sexual selection. Trends Ecol Evol 6:146–151

Miller GF (1999) Sexual selection for cultural displays. In: Dunbar R, Knight C, Power C (eds) The evolution of culture – an interdisciplinary view. Edinburgh University Press, Edinburgh, pp 71–91

Miller G (2000) Evolution of human music through sexual selection. In: Wallin NL, Merker B, Brown S (eds) The origins of music. MIT Press, Cambridge, MA, pp 329–360

Møller AP, Alatalo RV (1999) Good-genes effects in sexual selection. Proc R Soc Lond Ser B 266:85–91

Neiman FD (1997) Conspicuous consumption as wasteful advertising: a darwinian perspective on spatial patterns in classic Maya terminal monument dates. In: Barton CM, Clark GA (eds) Rediscovering Darwin: evolutionary theory and archeological explanation. American Anthropological Association, Arlington, pp 267–290

Olson VA, Owens IPF (1998) Costly sexual signals: are carotenoids rare, risky or required? Trends Ecol Evol 13:510–514

Orians GH (2001) An evolutionary perspective on aesthetics. Bull Psychol Arts 2:25–29

Owens IPF, Short RV (1995) Hormonal basis of sexual dimorphism in birds: implication for new theories of sexual selection. Trends Ecol Evol 10:44–47

Petrie M (1994) Improved growth and survival of offspring of peacocks with more elaborate trains. Nature 371:598–599

Pinker S (1999) How the mind works. Norton, New York

Power C (2000) 'Beauty magic': the origins of art. In: Dunbar R, Knight C, Power C (eds) The evolution of culture – an interdisciplinary view. Edinburgh University Press, Edinburgh, pp 92–112

Ralevski E (2000) Aesthetics and art from an evolutionary perspective. Evol Cognition 6:84–103

Rensch B (1984) Psychologische Grundlagen der Wertung bildender Kunst. Die Blaue Eule, Essen

Richter K (1999) Die Herkunft des Schönen – Grundzüge der evolutionären Ästhetik. Von Zabern, Mainz

Rowanchilde R (1996) Male genital modification – a sexual selection interpretation. Human Nat 7:189–215

Singh D, Bronstad PM (1997) Sex differences in the anatomical locations of human body scarification and tattooing as a function of pathogen prevalence. Evol Human Behav 18:403–416

Sitte P (1999) Bioästhetik – Biologie zwischen Erkennen und Erleben. In: Sitte P (ed) Jahrhundertwissenschaft Biologie – Die großen Themen. Beck, München, pp 407–425

Sütterlin C (1998a) Art and indoctrination – from the Biblia Pauperum to the third Reich. In: Eibl-Eibesfeldt I, Salter F (eds) Indoctrinability, ideology and warfare – evolutionary perspectives. Berghahn, New York, pp 279–300

Sütterlin C (1998b) Grenzen der Komplexität – Die Kunst als Bild der Wirklichkeit. Nova Acta Leopoldina 77:167–188

Symons D (1995) Beauty is in the adaptations of the beholder: the evolutionary psychology of human female sexual attractiveness. In: Abramson PR, Pinkerton SD (eds) Sexual nature – sexual culture. University of Chicago Press, Chicago, pp 80–118

Thornhill R, Gangestad SW (1999) Facial attractiveness. Trends Cognitive Sci 3:452–460

Veblen T (1899/1953) The theory of the leisure class. The New American Library, New York

Watts I (1999) The origin of symbolic culture. In: Dunbar R, Knight C, Power C (eds) The evolution of culture – an interdisciplinary view. Edinburgh University Press, Edinburgh, pp 113–146

Wedekind C, Folstad I (1994) Adaptive or nonadaptive immunosuppression by sex hormones? Am Nat 143:936–938

Wiessner P (1983) Style and social information in Kalahari San projectile points. Am Antiquity 48:253–276

Zahavi A (1975) Mate selection – a selection for a handicap. J Theor Biol 53:205–214

Zahavi A (1978) Decorative patterns and the evolution of art. New Sci 80:182–184

Zahavi A, Zahavi A (1997) The handicap principle – a missing piece of Darwin's puzzle. Oxford University Press, New York

Zuk M (1994) Immunology and the evolution of behavior. In: Real LA (ed): Behavioral mechanisms in evolutionary ecology. University of Chicago Press, Chicago, pp 354–368

Handaxes: The First Aesthetic Artefacts

STEVEN MITHEN

Introduction

The first stone artefacts made by our human ancestors appear 2.5 million years ago and are referred to as the Oldowan culture. These are nodules of basalt, chert and limestone that have been struck with a hammerstone to remove flakes. Both the flakes and the remnants of the nodules, referred to as cores, were used for the processing of animal carcasses; they may have also been used for other activities such as cracking nuts, chopping plants, removing bark and throwing at prey (Schick and Toth 1993). Although such artefacts have traditionally been associated with *Homo habilis*, it is possible that several early hominids including australopithecines produced such tools (Susman 1991). With regard to aesthetics, Oldowan tools have little appeal. There was no intentionally imposed design with the resulting cores simply being a product of the original shape of the nodule and the number of flakes removed. By 1.4 million years ago, however, a new type of artefact appears in the archaeological record – the handaxe. This appears quite different to anything in the Oldowan culture due to a deliberately imposed form that often exhibits considerable symmetry.

Handaxes were considerably more difficult to make than Oldowan tools as they involved a bifacial knapping method, as explained below. After first appearing in Africa (Asfaw et al. 1992), handaxes remain as a pervasive element of the archaeological record for more than 1 million years. The final handaxes were made in the late Mousterian cultures of Europe a mere 50,000 years ago. As such, handaxes are associated with a range of hominid species, including those assigned to *Homo ergaster, H erectus, H. heidelbergensis* and *H. neanderthalensis.* Four handaxes are illustrated in Fig. 1. These all come from sites in southern England between 500,000 and 100,000 years ago and show some of the variation in shape and degree of symmetry that can be found in handaxes from Africa, western Asia, southern Asia and Europe.

Many handaxes have a strong aesthetic appeal on account of their symmetry. In many specimens this is likely to be a mere by-product of the bifacial knapping method used in their production. However, in others, symmetry appears to be deliberately imposed by the careful removal of

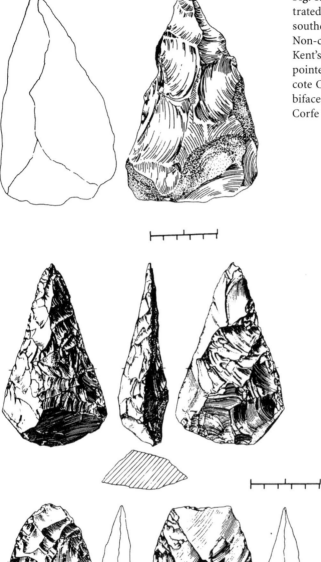

Fig. 1. Handaxes, as illustrated by examples from southern England. *Above* Non-classic biface from Kent's Cavern; *middle* classic pointed biface from Wolvercote Channel; *below* ovate bifaces from Cogdean Pit, Corfe Mullen. (Roe 1981)

flakes so that this is simultaneously achieved in three dimensions – in plan, in profile and in section (Wynn 1989). Moreover, the aesthetic appeal of handaxes might be supplemented by the use of finely grained raw materials, such as high quality flint, and – in rare instances – the presence of fossils left intact within the stone of the artefact (Oakley 1981). While archaeologists have frequently commented upon such aesthetic qualities, suggesting that such artefacts might lie at the root of art and mathematics (e.g. Gowlett 1984), little explanation has been offered with regard to two key questions: (1) why symmetry in artefacts appears pleasing to modern eyes; (2) why early humans invested time and effort in creating symmetrical handaxes to use in tasks such as animal butchery for which symmetry was quite unnecessary. Answers to these questions can be found by approaching these artefacts from the perspective provided by Voland (this Vol.): "In the world of artifacts, aesthetic preferences have the same function that they have always had in a world of natural features, namely as aids to orientation for decisions in sexual and social affairs". In other words, once handaxes are recognised as being artefacts for use in the social world in addition to that of tasks such as animal butchery, their aesthetic appeal can be explained (Kohn and Mithen 1999). Handaxes provide an excellent illustration of Voland's proposition demonstrating that it holds not only for modern humans, but also for our prehistoric ancestors.

Handaxe Manufacture and Use

To make a fine symmetrical handaxe, one must first secure appropriate raw material – not just the nodule of stone, but also suitable hammers, possibly including those of bone and antler as well as of stone. The costs of acquiring these materials, in time and energy, will be highly variable, depending upon one's local environment and mobility pattern. In many situations, raw material acquisition may have been the most costly part of the handaxe's manufacture. Once the raw material is secured, a range of different knapping actions are required, most of which are applied by working the artefact in a bifacial manner (Inzian et al. 1992; Pelegrin 1993; Schick and Toth 1993).

At first, relatively large cortical flakes must be removed, requiring use of a hard hammer. When the approximate shape has been created, other types of flakes are detached, notably thinning flakes, which travel across the surface of the artefact. These are struck with an antler, bone or wooden hammer at quite different angles, and with different degrees of force, to those initial hard hammer removals. To detach the thinning flakes, preparatory flakes may need to be removed to create a striking platform. Throughout the manufacturing process, the edge of the artefact may need to be slightly ground to remove irregularities that might otherwise deflect the force of the strike.

In light of the required planning of knapping actions (Gowlett 1984), mental rotations (Wynn 1989), and the range of hammers and striking methods, there can be little doubt that in the majority of artefacts a specific symmetrical form was imposed, even though raw materials may have constrained the options available and influenced the result (Ashton and McNabb 1994).

Many thousands of handaxes manufactured in this fashion have been excavated from sites in Africa, Europe and Asia, and then subjected to detailed metrical studies (e.g. Isaac 1977; Roe 1981; Villa 1983; Wynn and Tierson 1990). Archaeologists have undertaken microwear analysis, detailed re-fitting of debitage, and experimental studies concerned with manufacture and use (e.g. Jones 1980; Keeley 1980; Bergman and Roberts 1988; Austin 1994). Handaxes have also been at the centre of research regarding the evolution of human intelligence (e.g. Wynn 1979, 1989, 1993, 1995; Mithen 1996). Recent studies have challenged notions of chronological patterning for handaxe types, and placed emphasis on raw material and function, rather than culture and style, when explaining handaxe morphology (e.g. Ashton and McNabb 1994; Bosinski 1996). There has also been a greater recognition of the considerable variability in handaxes, ranging from the classic, highly symmetrical bifaces, to non-classic or atypical bifaces, which may lack a clearly imposed form (Ashton and McNabb 1994).

Handaxes are most likely to have been general purpose artefacts, being used for the butchery of animals, cutting wood, slicing meat and chopping vegetables. Direct evidence, however, is quite scarce. There are a few cases where microwear studies have been undertaken, such as on artefacts from Koobi Fora in Africa (Keeley and Toth 1981) and Hoxne in England (Keeley 1980). Both samples showed a range of wear traces, indicating they had been used for a variety of tasks. Experimental work appears to confirm this, as handaxes are clearly effective for a range of activities (Jones 1980, 1981; Schick and Toth 1993). It is conceivable that proteins or even DNA may survive trapped in microcracks of handaxes to provide direct evidence for their past function (Shanks et al. 2001). Handaxes may also have functioned as a source of flakes, having been carried around the landscape as curated artefacts (Hayden 1979), or as implements for throwing at game (Calvin 1993). The symmetry of many handaxes, especially those which Ashton and McNabb (1994) describe as non-classic, is most likely an unintentional product of either the shape of the original nodule of stone and/or the bifacial knapping technique. However, in others very small flakes have been intentionally removed to make exquisitely symmetrical artefacts which have no functional payoff in terms of the effectiveness of the artefact as a butchery or plant processing tool (Barker 1998). To explain the imposition of such symmetry, the role of such handaxes in social interaction must be considered.

The Social Context of Manufacture and Use

As handaxes were made by a variety of hominid types in numerous different geographical areas with different resources, the social context of manufacture and use is likely to have been variable. Nevertheless, the two common characteristics of the hominid species – large brains and habitual bipedalism – implies some common social characteristics. As Aiello and Dunbar (Aiello and Dunbar 1993; Dunbar 1993) have argued, large brains implies large groups. Archaeologists suspect that these groups were highly competitive, requiring individuals to adopt a range of Machiavellian social tactics to survive and prosper (cf. Byrne and Whiten 1988; Whiten and Byrne 1997). Even chimpanzees, with 50% of hominine brain size at most, live in socially complex societies in which friendships and alliances are constantly being adjusted (De Waal 1982). It seems likely that handaxe-making hominids would have had an advanced 'theory of mind' (Mithen 1996) and that deceptive behaviours would have been rife within their societies.

As large brains are metabolically expensive organs (Aiello and Wheeler 1995), the need for a high quality diet most likely involved substantial meat consumption, which in turn required co-operation in its acquisition by either hunting or scavenging. This dependency on animal carcasses is likely to have favoured large groups, due to the opportunities for food sharing and/or tolerated theft; another factor would have been the risk from carnivore predation in Pleistocene environments (Mithen 1994).

While handaxe making hominids are likely to have lived in complex social environments, had considerable technical skills and engaged in big-game hunting, it is unlikely that they had fully modern language and symbolic thought. The linguistic skills of human ancestors has been substantially discussed in recent publications (e.g. Dunbar 1993, 1996; Bickerton 1996; Mithen 1996; Deacon 1997) and remains open to much debate. Anatomical evidence suggests that there was a considerable development of vocal abilities between *Homo ergaster* and *H. neanderthalensis* with the encaphalisation between 600,000 and 250,000 years ago (Ruff et al. 1997) directly related to the evolution of language ability. However, species such as *H. ergaster, H. erectus* and *H. heidelbergensis* were most likely quite limited in both the extent of their lexicons and complexity of their grammar.

The absence of visual symbols in the archaeological record is frequently drawn upon to support the presence of only limited linguistic skills in such species. Prior to the cave art of the late Pleistocene, the most persuasive examples of visual symbols are the bones from Bilzingsleben, Germany, which have sets of parallel lines (Mania and Mania 1988), and the incised stone from Berekhat Ram, Israel, which some believe to represent a female figurine (Marshack 1997). Both date to between 300,000 and 250,000 years ago and are highly ambiguous – many archaeologists reject the notion that these have symbolic value. There is no evidence, therefore that the hominid species that made handaxes possessed symbolic thought.

With an absence of modern language abilities and symbolic thought, the hominid species that made handaxes were quite unlike modern humans. Consequently, analogies between the production of handaxes with that of similarly elaborate and aesthetic artefacts amongst modern humans which were then used for trade, in religious practices or as items of prestige are quite ill-founded. Those artefacts only functioned in such roles because they were invested with symbolic meanings and we have no reason to believe that this was the case with handaxes. A more effective explanation for the aesthetic appeal of handaxes is found by focusing on the social and sexual relations within the hominid societies.

Reliable Indicators: Handaxes as a Social Technology

Within such hominid social groups there is likely to have been intense social competition between males for access to mates. Whether or not pair-bonding had arisen at this stage in human history, females are likely to have had considerable choice over which males to select as mating partners. Males will have needed to display to females, in much the same fashion as in those species which utilise handicapping traits such as extravagant tail fans, antlers or canines (Zahavi and Zahavi 1997). In this regard, the symmetry of handaxes, and indeed the time and skill required for their manufacture may have acted as a reliable indicator as to the fitness of the individual who made the artefacts – a highly symmetrical well-made handaxe was a sign of 'good genes'.

Those hominids (male or female, see below) who were able to make fine symmetrical handaxes may have been preferentially chosen by the opposite sex as a mating partner. Just as a peacock's tail may reliably indicate the ability of the peacock to fight off parasites, acquire a nutritious diet and escape from predators, so might the manufacture of a fine symmetrical handaxe have been a reliable indicator of the hominid's ability to secure food, find shelter, escape from predation and successfully compete within the social group. As such that hominid would have been an attractive mate. These abilities would have had a genetic component and, consequently, would have been inherited by any offspring.

Critical to this argument is the wide range of variability one finds within those artefacts categorised as handaxes (Ashton and McNabb 1994), ranging from the classic, highly symmetrical forms which required considerable skill to manufacture, to those non-classic handaxes which may be asymmetrical and lack continuous flaking across the surface. Such variation is essential to the process of selection.

Handaxes would have acted as reliable indicators for four specific dimensions of fitness: resource location abilities, planning ability, good health, and capacity to monitor other individuals within the group (Kohn and Mithen 1999).

Knowledge of Resource Distribution

The ability to make a fine symmetrical handaxe indicates the possession of environmental knowledge because such artefacts require high quality raw material. Raw material needs to be procured, and hence by producing a finely made handaxe the hominid will be indicating a knowledge of raw material distributions in the landscape. Knowledge of where good quality stone is located is likely to go hand in hand with that of where good quality plants, carcasses, shelter and water are located. The ability to comprehend and exploit the environment in this way would be attractive in a mate, as an indication of heritable perceptual and cognitive skills. Consequently, hominids possessing finely made handaxes may have been preferentially selected as reproductive partners.

Forming and Executing Plans

The ability to make a fine, symmetrical handaxe would have been a reliable indicator of those mental capacities required for their production; such capacities may have been of value in other domains of activity. Quite simply, classic handaxes were difficult to make, requiring a high degree of intelligence (Gowlett 1984; Wynn 1989). The production of a handaxe required the ability to conceive and successfully execute a plan, probably involving a continual modification of that plan as contingencies arose during knapping, such as unexpected flaws in the material and miss-hits. It would also require persistence and determination. Handaxes would have been a 'test of character', indicating behavioural disposition to potential mates. Planning would also have been essential in other activities, such as finding food, whether by hunting or scavenging, and building alliances and friendships within highly competitive social groups. Hominids who could make fine symmetrical handaxes would have been demonstrating their ability at such planning, which would have been attractive to potential mates keen to secure good quality genes for their offspring.

One might pursue this at a more detailed level. As Wynn (1989) argued, when making a handaxe one needs to keep in mind the effect of a removal on the morphology of all three dimensions of the handaxe, not simply the one in most immediate view. Similarly, when building a friendship with one individual within a group, one must bear in mind the effect of this friendship on one's other social relationships, including those of hominids not currently in view.

Good Health

Being able to produce a handaxe would have also been a reliable indicator of a third feature attractive to members of the opposite sex: good health. Strength is required, notably for the removal of extensive thinning flakes. Good eyesight and eye-hand co-ordination are also essential. Consequently, poor knapping results would have been effective indicators of illness or physical decline.

Social Awareness

By consistently making particularly good handaxes, hominids would have demonstrated their ability to successfully survive and compete in social groups in spite of having the handicap of attending to handaxe manufacture. Their success would have depended on their ability to monitor the behaviour of others, so as to avoid deception or loss of political advantage.

Handaxes as Aesthetic Displays

Reliable indications of environmental knowledge, planning abilities and good health might have been given by making all sorts of complex artefacts, none of which required a symmetrical form. However, artefacts of a symmetrical form may have been particularly attractive to members of the opposite sex because of an evolved perceptual bias toward symmetry. An individual's attention would have been attracted to objects which were symmetrical in shape because our "visual systems, like those of many other animals, even fish, are exquisitely sensitive to patterns with a vertical axis of symmetry"(Dennett 1991, p. 179). In this regard, the symmetry of handaxes may not have been a reliable indicator at all, but simply "play[ed] on the perceptual biases of receivers to attract attention, provoke excitement, and increase willingness to mate" (Miller 1997, p. 96). There are two possible reasons for the perceptual bias towards symmetry: social monitoring and mate choice.

Symmetry and Social Monitoring

Braitenberg (1984) has suggested that our perceptual biases towards symmetry arise because in our deep evolutionary past the only other entities in the world with vertical symmetry would have been other animals. Consequently, a perceptual bias towards symmetry would have acted as an alarm signal to indicate they were being looked at by another individual. The other individual may have been of another species, a potential predator, or the member of one's own social group. Dennett (1991) has favoured

Braitenberg's explanation when stressing the significance of such perceptual biases for understanding the evolution of human cognition.

Symmetry as an Indicator of Good Genes

A second source for the presence of a perceptual bias towards symmetry regards mate choice itself, as bodily and facial symmetry may be a direct indicator of good genes. Symmetry abounds in the morphology of living things. This is because single genes control the development of features on both sides of an organism. High levels of symmetry are, however, difficult to achieve. The presence of genetic mutations, pathogens or stress during development may lead to the presence of asymmetries in bilaterally distributed features (Parsons 1992). Consequently, the degree of symmetry is a good indicator of the degree of genetic and physical health of an individual. Members of the opposite sex can use this relationship when seeking a mate with 'good genes'. The relationship between 'good genes' and symmetry has been established in several species. Male swallows, for instance, are chosen by females on the basis of the length of their forked tails (Møller 1988). The length of the tail is a cue to the genetic and physical health of the individual. It has been shown that there is a positive relationship between tail length and its degree of symmetry, demonstrating that tail length is indeed an 'honest advertisement' (Møller 1990). A similar correlation between the size of a sexually selected feature and its degree of symmetry has been observed in a number of other species: examples include the canine teeth of primates (Manning and Chamberlain 1993), spurs on birds, horns on beetles (Møller 1992), the tail fans of peacocks (Manning and Hartley 1991) and antlers of deer (Goss 1983). In all of these cases, it is evident that symmetry is a reflection of an individual's genetic and physical health, as has indeed been argued from theoretical grounds alone (Parsons 1992).

While it is assumed that females select males on the basis of the size of traits such as peacock tail fans and swallow tail lengths, they could alternatively be selecting on the basis of the degree of symmetry (Manning and Hartley 1991). This certainly appears to be the case for modern humans. Both men and women make substantial use of the degree of symmetry in the faces and bodies of those of the opposite sex when selecting reproductive partners. Thornhill and Gangestad (1994, 1996) have measured men's 'fluctuating asymmetry' and examined how this is related to several measures of reproductive success. By 'fluctuating asymmetry', they mean the difference in a range of characteristics that are found on both the right and left side of the body, such as hand width, ear length and foot width. By measuring a sample of these, they calculate an index for the degree of bilateral asymmetry for each of their subjects. Women, they argue, seek mates with low degrees of asymmetry as this is an indicator of 'good genes'. Men who are most symmetrical were found to be more facially attractive, to have had greater numbers of sexual partners, and to have begun sexual intercourse earlier in their life history. These traits were interpreted as indicating a greater potential for reproductive success.

The examples described above show that symmetry provides a cue to the genetic and physical health of a potential reproductive partner, and that in modern humans at least, women make substantial use of this cue when selecting males for sex. Due to the pervasiveness of the symmetry cue among animal species, and its specific presence in modern humans, it seems very likely that the males and females of all hominid species would have also used symmetry as a cue when selecting mates.

Handaxe Symmetry

To summarise, there are two possibilities why modern humans, hominids and a variety of animals are likely to have a perceptual bias to symmetry: social monitoring and mate choice. These do not appear to be mutually exclusive and both may have played a role in the evolution of the mind. The makers of handaxes were keying into this perceptual bias and making artefacts that caught the attention, and were most probably attractive, to members of the opposite sex.

Cheating and Handaxe Abundance

Handaxes functioned in the social domain as reliable indicators of 'good genes', and by exploiting the perceptual biases of the opposite sex towards symmetry. In both of these regards, handaxes are very similar to the ornaments of other species, such as elaborate bird plumage, canine teeth and deer antlers. However, there is, of course, one fundamental difference: a handaxe is not attached to a body, and hence a set of genes. This creates a major problem for the signal receivers: the sender of the signal may be a cheat. Without effective countermeasures, an individual could avoid the costs of making a handaxe by acquiring one of a quality beyond the cheat's own abilities. This might be acquired by stealing from another individual or simply collecting one from those discarded at an old activity area.

This is a virtue rather than a flaw in the theory, since it explains one of the most puzzling features of the archaeological record of handaxes: their great abundance. As has been frequently noted (e.g. Isaac 1977; Roe 1981; Wymer 1983), a common characteristic of Acheulian sites is the abundance of handaxes, with many appearing to be in pristine condition. At Boxgrove, for instance, not one of the excavated handaxes shows signs of macroscopic damage (Roberts et al. 1997). Why should a hominid have invested time in making another handaxe, instead of simply picking one up from the ground? This puzzle is solved when we recognise that handaxes were involved in the process of mate choice. Simply observing the possession or use of a handaxe would have been of limited value, since the artefact might have been made by a different individual, so a potential mate could only give a reliable indication of quality by actually being seen to make a handaxe. We should not be surprised, therefore, to find such abundant numbers of handaxes on Acheulian sites; once a handaxe had been discarded it had no further value in the game of mate choice. Once it had been made, much of its job was done.

Which Sex Made Handaxes?

There has been a traditional attitude of associating handaxes with assumed male activities of acquiring meat by either hunting animals or scavenging from carcasses. Recently, however, by drawing on data regarding tool use and manufacture in chimpanzees, it has been suggested that knowledge of how to make handaxes is most likely to have been transmitted by adult females to their offspring (Dennell 1994). There is no direct evidence for whether handaxes were predominately made by either males or females, or equally by both sexes. The theory proposed in this paper suggests, however that the most elegant, symmetrical handaxes were made by males.

Hominid societies during the Middle Pleistocene are likely to have been competitive and dynamic in which sexual selection operated on both sexes. For the simple biological reason of pregnancy and caring for young offspring, females would have invested more than males in reproduction and consequently, are likely to have chosen mates more carefully (Trivers 1972). As a consequence, females would tend to be more interested in indicators of genetic quality, creating intense inter-male competition and their use of various methods of self-advertisement, including handaxe manufacture.

At the same time, however, the socio-ecological conditions during the Middle Pleistocene would not have been conducive to the establishment of durable relationships in which males provisioned females and developed strong pair-bonds. Males are also likely to have been keen to select mates with good genes and hence, females may have also utilised artefacts as a means to display to males.

A further factor to consider is the need for display to members of the same sex. Many of the activities undertaken by human ancestors will have required co-operation, necessitating the selection of partners in working groups. At least 400,000 years ago, , hominids were engaged in big game hunting as evident from the discovery of spears at Shöningen (Thieme 1997). Such activity is likely to have produced large payoffs in terms of meat provision, but to have involved high risks. This becomes particularly evident from the skeletal evidence of the Neanderthals in which bone fractures are common, most likely having arisen during hunting activity. Consequently, there would have been a premium on the selection of hunting partners who could be relied upon in dangerous situations. In this context, the production of a fine symmetrical handaxe (by males or females) may have acted as much as a reliable indicator of health, fitness and intelligence when partners are being selected for activities such as hunting as they did to members of the opposite sex when choosing mates.

Summary

While archaeologists have frequently recognised the aesthetic qualities of handaxes, they have struggled to explain this due to their focus on the use of handaxes for activities such as animal butchery. By placing the production and use of handaxes in their social context, and drawing on notions of sexual selection and evolved psychological propensities to value symmetry, their aesthetic qualities are more readily understood. As Voland has proposed for all artefacts with aesthetic qualities, handaxes were "aids to orientation for decisions in social and sexual affairs".

Although handaxes remained as a prevalent artefact for more than a million years, they were eventually replaced by a technology involving a much greater use of retouched flakes, referred to as the Mousterian in Europe and West Asia, and the Middle Stone Age in Africa. Artefacts of these traditions lack the aesthetic qualities of handaxes and suggest that stone technology was no longer being used as actively in the social sphere. As Kohn and Mithen (1999) have argued, this may relate to a major change in social relations with a greater prevalence of male–female pair bonding, involving male provisioning of both mates and offspring, arising due to the increased reproductive costs faced by females due to encephalisation. In this context, females would have been less concerned with simply obtaining 'good genes' as they now required prolonged male investment. As such technology appears to have become more efficient as a hunting tool, such as in the first development of hafted spear points, and lost its social display function – females no longer wanted the indicator of having good health, knowledge and skill, they wanted the material results in terms of food and protection.

Handaxes play a crucial role in understanding the evolutionary history of aesthetic objects and the evolution of aesthetics. They were the first artefact to have been created with an aesthetic quality and were done so by human ancestors that lacked modern language skills and visual symbols. They demonstrate that our modern aesthetic preferences emerged very early in human evolution and support the view that these can be understood by examining the role of artefacts in social and sexual affairs.

References

Aiello L, Dunbar RIM (1993) Neocortex size, group size and the evolution of language. Curr Anthropol 34:184–193

Aiello L, Wheeler P (1995) The expensive tissue hypothesis. Curr Anthropol 36:199–221

Asfaw B, Beyene Y, Suwa G, Walter RC, White T, Woldegabriel G, Yemane T (1992) The earliest Acheulian from Konso-Gardula, Ethiopia. Nature 360:732–735

Ashton NM, McNabb J (1994) Bifaces in perspective. In: Ashton N, David A (eds) Stories in stone. Lithics Studies Society Occasional Paper No 4, Lithic Studies Assoc, London, pp 182–191

Austin L (1994) The life and death of a Boxgrove biface. In: Ashton N, David A (eds) Stories in stone. Lithics Studies Society Occasional Paper No 4, Lithic Studies Assoc, London, pp 119–126

Barker A (1998) A quantitative study of symmetry in handaxes. MA Diss, University of Reading, Reading, UK

Bergman CA, Roberts MB (1988) Flaking technology at the Acheulian site of Boxgrove, West Sussex (England). Rev Archeol Picardie 1–2:105–113

Bickerton D (1996) Language and human behaviour. University College Press, London

Bosinski G (1996) Stone artefacts of the European Lower Palaeolithic. In: Roebroeks W, van Kolfschoten T (eds) The earliest occupation of Europe. Analecta Praehistoria Leidensia vol 27. University of Leiden, Leiden, pp 262–267

Braitenberg V (1984) Vehicles: experiments in synthetic psychology. MIT Press, Cambridge

Byrne R, Whiten A (1988) Machiavellian intelligence: social expertise and the evolution of intellect in monkeys, apes and humans. Clarendon Press, Oxford

Calvin W (1993) The unitary hypothesis: a common neural circuitry for novel manipulations, language, plan-ahead and throwing. In: Gibson KR, Ingold T (eds) Tools, language and cognition in human evolution. Cambridge University Press, Cambridge, pp 230–250

Deacon T (1997) The symbolic species. Allen Lane, London

Dennell R (1994) Comment on 'Technology and Society during the Middle Pleistocene' by S. Mithen. Cambridge Archaeol J 4:3–32

Dennett D (1991) Consciousness explained. Penguin, London

De Waal F (1982) Chimpanzee politics: power and sex among the apes. Jonathan Cape, London

Dunbar RIM (1993) Coevolution of neocortical size, group size and language in primates. Behav Brain Sci 16:681–735

Dunbar RIM (1996) Grooming, gossip and the evolution of language. Faber and Faber, London

Goss RJ (1983) Deer antlers: regeneration, function and evolution. Academic Press, London

Gowlett J (1984) Mental abilities of early man: a look at some hard evidence. In: Foley R (ed) Hominid evolution and community ecology. Academic Press, London, pp 167–192

Hayden B (1979) Palaeolithic reflections. Humanities Press, New Jersey

Inzian M-L, Roche H, Tixier J (1992) Technology of Knapped stone. CNRS, Paris

Isaac G (1977) Olorgesailie. Chicago University Press, Chicago

Jones P (1980) Experimental butchery with modern stone tools and its relevance for Palaeolithic archaeology. World Archaeol 12:153–165

Jones P (1981) Experiment implement manufacture and use: a case study from Olduvai Gorge. Philos Trans R Soc Lond Ser B 292:189–195

Keeley L (1980) Experimental determination of stone tool uses: a microwear analysis. Chicago University Press, Chicago

Keeley L, Toth N (1981) Microwear polishes on early stone tools from Koobi Fora, Kenya. Nature 203:464–465

Kohn M, Mithen S (1999) Handaxes: products of sexual selection? Antiquity 73:518–526

Mania D, Mania U (1988) Deliberate engravings on bone artefacts by *Homo erectus*. Rock Art Res 5:91–107

Manning JT, Chamberlain AT (1993) Fluctuating asymmetry, sexual selection and canine teeth in primates. Proc R Soc Lond Ser B 251:83–87

Manning JT, Hartley MA (1991) Symmetry and ornamentation are correlated in the peacock's train. Anim Behav 42:1020–1021

Marshack A (1997) The Berekhat Ram figurine: a late Acheulian carving from the Middle East. Antiquity 71:327–337

Miller GF (1997) How mate choice shaped human nature: a review of sexual selection and human evolution. In: Crawford C, Krebs DL (eds) Handbook of evolutionary psychology: ideas, issues and applications. Lawrence Erlbaum Associates, Maahwah, NJ, pp 87–129

Mithen S (1994) Technology and society during the middle Pleistocene. Cambridge Archaeol J 4:3–33

Mithen S (1996) The prehistory of the mind. Thames and Hudson, London

Møller AP (1988) Female choice selects for male sexual tail ornaments in the monogamous swallow. Nature 332:640–642

Møller AP (1990) Fluctuating asymmetry in male sexual ornaments may reliably reveal male quality. Anim Behav 40:1185–1187

Møller AP (1992) Patterns of fluctuating asymmetry in weapons: evidence for reliable signalling of quality in beetle horns and bird spurs. Proc R Soc Lond Ser B 245:1–5

Oakley KP (1981) Emergence of higher thought. Philosophical Trans R Soc Lond Ser B 292:205–211

Parsons PA (1992) Fluctuating asymmetry: a biological monitor of environmental and genomic stress. Heredity 68:361–364

Pelegrin J (1993) A framework for analysing prehistoric stone tool manufacture and a tentative application to some early stone industries. In: Berthelet A, Chavaillon J (eds) The use of tools by human and non-human primates. Clarendon Press, Oxford, pp 302–314

Roberts MB, Parfitt SA, Pope MI, Wenban-Smith FF (1997) Boxgrove, West Sussex: rescue excavations of a Lower Palaeolithic landsurface (Boxgrove Project B, 1989–91). Proc Prehist Soc 63:303–358

Roe D (1981) The lower and middle Palaeolithic periods in Britain. Routledge and Kegan Paul, London

Ruff CB, Trinkaus E, Holliday TW (1997) Body mass and encaphalization in Pleistocene *Homo*. Nature 387:173–176

Schick K, Toth N (1993) Making silent stones speak: human evolution and the dawn of technology. Simon and Schuster, New York

Shanks OC, Bonnichsen R, Vella AT, Ream W (2001) Recovery of proteins and DNA trapped in stone tool microcracks. J Archaeol Sci 28:965–972

Susman RL (1991) Who made the Oldowan tools? Fossil evidence for tool behaviour in Plio-Pleistocene hominids. J Anthropol Res 47:129–151

Thieme H (1997) Lower Palaeolithic hunting spears from Germany. Nature 385:807–810

Thornhill R, Gangestad S (1994) Human fluctuating asymmetry and sexual behaviour. Psychol Sci 5:297–302

Thornhill R, Gangestad S (1996) The evolution of human sexuality. Trends Ecol Evol 11:98–102

Trivers RL (1972) Parental investment and sexual selection. In: Campbell BG (ed) Sexual selection and the descent of man. Aldine, Chicago, pp 139–179

Villa P (1983) Terra Amata and the Middle Pleistocene record from Southern France. University of California Press, Berkeley

Whiten A, Byrne R (1997) Machiavellian intelligence II: extensions and evaluations. Cambridge University Press, Cambridge

Wymer J (1983) The Palaeolithic Age. Croom Helm, London

Wynn T (1979) The intelligence of later Acheulian hominids. Man 14:371–391

Wynn T (1989) The evolution of spatial competence. University of Illinois Press, Urbana

Wynn T (1993) Two developments in the mind of early *Homo*. J Anthropol Archaeol 12:299–322

Wynn T (1995) Handaxe enigmas. World Archaeol 27:10–23

Wynn T, Tierson F (1990) Regional comparison of the shapes of later Acheulian handaxes. Am Anthropol 92:73–84

Zahavi A, Zahavi A (1997) The handicap principle. Oxford University Press, New York

IV Modular Aesthetics

Human Habitat Preferences: A Generative Territory for Evolutionary Aesthetics Research

Bernhart Ruso, LeeAnn Renninger and Klaus Atzwanger

Human evolutionary aesthetics is in many ways the study of humble everyday life – preferences and feelings evoked by a stimulus without self-conscious thought, and yet prevalent on an almost daily basis. While a recent trend in this line of research has been to focus on preferences related to sexual selection – mate selection, body composition, facial symmetry, and body movement analysis, quietly sitting in the shadows awaits an equally important area for the understanding of evolved aesthetic preferences – our (natural selection mediated) response to our physical environment. The current paper argues that a large part of the everyday aesthetic experience for humans involves a behavioral and emotional response to landscape. Since the selection of habitat was crucial in our evolutionary history, research on human habitat preference and perception is a vital area for the further understanding of evolved aesthetic tastes. A review of the major evolutionary theories and empirical evidence underlying habitat preference theory is given, ending with a discussion on the current status of habitat preference research and suggestions for future research directions.

Habitat Selection: The Evolution of Affect

An important principle in psycho-evolutionary theory is the recognition that the function and adaptedness of any particular aspect of human activity cannot be understood based on the activity's current role, but rather on its former function during the Pleistocene, the epoch in which modern humans evolved (Lorenz 1973; Cosmides and Tooby 1987; Symons 1989, 1990). Although selection of a habitat is not a priority for current human survival, habitat selection was a vital part of everyday survival for our ancestors. During the lengthy hunter-gatherer stage of evolution, frequent moves throughout a landscape were necessary in order to attain reliable resources for the long periods of generation and offspring dependency. Those individuals who were able to detect and seek a habitat that offered protection from predators and weather, food, water, and other resources were more successful than those who were not able to seek and perceive these qualities about a location.

Since the selection of a habitat decisively influenced the survival and reproductive success of the organism, the relevant psychological mechanisms underlying habitat selection were subject to strong selection pressure. A Darwinian approach to habitat selection proposes that one consequence of selection pressures on habitat selection mechanisms was the development of emotional responses to species-specific features of the environment. This is based upon the consideration that emotional reactions act as motivators for human behavior (Orians 1980). Preferred environments have an adaptive significance in that they effectively elicit like – dislike feelings which in turn motivate approach–avoidance behaviors appropriate to ongoing well-being (Ulrich 1986). Said simply: a large portion of our current feelings and behavioral responses toward environmental forms may be considered as evolutionary remnants that helped us to originally seek good habitat locations.

Habitat Preference: Evolutionary Theories

The idea that there is a link between human evolution and current aesthetic preferences is not new. Over the past 20 years, several theories have provided a sturdy framework for the empirical testing of environmental aesthetics (e.g., Appleton 1975; Orians 1980; Woodcock 1984; Kaplan and Kaplan 1989). Among these works, two of the most discussed theories are the prospect–refuge theory by Appleton and the framework of prediction of preference by Kaplan and Kaplan (Kaplan and Kaplan 1982; Kaplan 1987, 1992).

Appleton's approach to environmental aesthetics began in 1975 with what he called "habitat theory". Appleton hypothesized that humans prefer landscapes which promise to satisfy our basic biological needs, i.e., that signal vital resources and invite further exploration. From habitat theory, Orians developed the well-known "savanna theory" of biotope preference. Savanna theory postulates that modern humans show an innate preference specifically for savanna biotopes, the environment in which our ancestors did most of their evolutionary flourishing. Implicit to the theory is the assumption that biotope imprinting must have taken place phylogenetically, as humans should develop mechanisms that allow them to recognize habitats that have previously promoted survival throughout our evolutionary history. Savanna theory has remained in the spotlight of habitat preference research for many years and has provided a superb background for much empirical testing. Ungratefully, not all findings supported the theory which had initiated it. The preference for a moderate level of maintenance and traces of human life, like houses or fields cannot easily be explained by the savanna theory (Coeterier 1996; Hagerhall 2000). Moreover, why would such complex mechanisms for the preference of habitat – a problem for all members of our evolutionary lineage – have evolved solely within one epoch? When it comes to environmental preferences, it must be likely that choice mechanisms were put under selective pressure long before (and after) the Pleistocene era.

Additions to environmental aesthetics theory were made in 1984 by Eibl-Eibesfeldt. Eibl-Eibesfeldt described the human aesthetic condition globally, as being "phytophilic", having a strong psychological and behavioral attraction to green plants, and "hydrophilic", having a strong psychological and behavioral attraction to water. Regarding landscape preferences, he stated that structures that provide an easy-to-survey vegetation were useful features to evaluate the varying quality of landscapes, and that habitat selection should be based largely on a need for security. A synthesis of the aforementioned theories into an overarching framework is provided by Appleton's "Prospect Refuge" theory (1975, 1984), a theory which assumes that natural selection favored the survival of those individuals who preferentially settled in areas providing prospect, the ability to survey the landscape, and refuge, the ability to hide when in danger. The presence of environmental structures that allow prospect views would have increased the likelihood of spotting resources such as water and food, and of recognizing approaching threats such as predators, hostile conspecifics, or changing weather, so that decisions could be made about the opportune time to move on and to set priorities for activities. Since both prospect and refuge were crucial for survival, it is hypothesized that our ancestors developed positive response patterns – emotions as motivators – to areas that exhibited these qualities and negative response patterns to environments that did not.

The other major theory of environmental aesthetics comes from Stephen and Rachel Kaplan (Kaplan and Kaplan 1982; Kaplan 1987, 1992). The Kaplans point out that successfully negotiating terrain requires skill and knowledge of the environment. Accordingly, landscapes that signal to the observer an opportunity to explore without losing orientation should be preferred over those that fail to satisfy or even hinder this need. Kaplan's preference matrix contains four predictors that describe these exploration and orientation possibilities: complexity (the number of independently perceived elements in a scene), coherence (unity), legibility (identifiability and patterning), and mystery (the promise of future information). In a spontaneous decision, visual complexity helps determine the exploration quality, while coherence aids in rapid understanding. If more time is available for decision-making, then legibility helps in reading the landscape and mystery helps to properly evaluate the exploration possibilities. An extension of Kaplan's 'complexity' factor has branched into what is called the biodiversity theory (Erlich and Erlich 1992). The biodiversity theory claims that moderate to high biodiversity within an area (i.e., moderate to high complexity) is a signal that a landscape is ecologically stable, and therefore predictive of future resources. Moderate to high biodiversity should then evoke ratings of high aesthetic preference within humans. It is likely, however that although high complexity wakes our interest, those factors that help maintain orientation (coherence and legibility) are also important in preference ratings, as orientation is essential to survival. Biodiversity cues alone will not be enough to evoke positive aesthetic preference in humans.

Despite differences in the focus, the aforementioned theories maintain one common ground: it is evident that behavioral reactions to different structural or qualitative components of an environment have developed within humans, and that these reactions are related to our basic biological needs. An important discussion that then follows regards the ability of the human perceptual system to detect such environments.

Perception

Environmental aesthetics research distinguishes two realms of aesthetic variables: (1) the structure of features, the so-called formal aesthetics, and (2) the content of the features, the so-called symbolic aesthetics (Lang 1988). The distinguishing characteristics of formal aesthetics are form, proportion, rhythm, complexity, spatial arrangement, incongruence and novelty. The distinguishing characteristics of symbolic aesthetics are style, material, resources, degree of naturalness, and the actual content items, for example, water and trees. (Lang 1988; Groat and Despres 1990). From an evolutionary perspective, the term "symbolic aesthetics" could be criticized as being inaccurate. We expect that individuals should prefer such things as trees and water for what they actually are, rather than for any symbolic meaning they might have. The term "content aesthetics" might be more appropriate.

In both formal aesthetics and symbolic/content aesthetic areas, an important base for research relies on information about what the human perceptual system is more capable of. From an evolutionary perspective, perception should be closely tied to evaluation, where identification of criterion attributes will most likely implicate values and valuation. The objective of perception is to present our brain with a coherent and meaningful picture of the outside world and to give each object its place in the organized whole (Coeterier 1996). When we interact with the world, it is impossible to register and process all incoming information. Instead, we must select, order and condense the information. Which elements of our environment – both formal and symbolic/content oriented – are important enough to be selected by us as the prominent, organizing features of a habitat? These elements are likely to be the ones which influence our emotional, affective responses.

Research has found that despite differences in the background of participants and differences in type of landscape stimuli used, the criteria used to evaluate landscapes tend to be similar (e.g., Coeterier 1996). The most prominently perceived components of an environment are the unity of the landscape as a whole, its function (use), maintenance level, naturalness, spaciousness, development in time, soil and water, and its sensory qualities such as color and smell. Further, landscape features tend to be perceived and assessed within a context of all other visible features, rather than assessed lexically, feature per feature. The physical element of water, for example, is often a preferred content item. Within a schemata, however, the attribute that water confers depends on the context. A winding and twisting river might enhance the naturalness of a landscape, while a straight canal might enhance the perceived spaciousness. Thus, the actual content item 'water' is perceived simultaneously as a content item and as an abstract, formal attribute (e.g., spaciousness), derived from the overall gestalt of the landscape.

Water and vegetation tend to have further perceptual qualities of their own. Using color slides of natural waterscapes, Herzog (1985) found that the most predominant evaluation criteria for a scene that contains water are unity (coherence), spaciousness, identifiability, complexity, mystery and texture – terms which overlap Kaplan's preference matrix and Coeterier's (1996) eight perceptual attributes (listed in the previous paragraph). For vegetation, perception research has found that individuals tend to distinguish between different vegetation forms based predominantly on vegetation density, height, and leaf color. Plant form, spatial arrangement, and texture are less important preference assessment (Misgav 2000).

Much theoretical and empirical work on the connection between perception and preferences repeats the following finding: the emotional responses that we have to a landscape are rapidly made, without much need for cognitive processing. These rapid responses are typically made to general, rather than specific, schematic features of the environment (Blum and Barbour 1979; Zajonc 1980). Zajonc (1980) has given the term 'preferenda' to such features. The structural properties of an environment combine with biases in the human perceptual system to convey quickly, and with very little processing, salient general characteristics of a setting. Gross depth cues, coherence, complexity, development in time, and certain classes of content such as water and vegetation are perceived very quickly as they provide useful information about the location's ability to meet human biological needs. These preferenda are still active in our modern aesthetic evaluation system. Modern technology can provide valuable insights about the ways in which current preferenda are perceived. Syneck (1998), for example, has found that computer analysis of fractal dimension replicates the way in which humans perceive complexity in a landscape. Such findings can greatly increase our understanding of the perceptual foundations of acsthetic preferences.

Landscape: A Consensus on Aesthetic Preferences?

As predicted by evolutionary theory, landscape choice studies have confirmed that modern humans exhibit a preference for savanna-like habitats (Orians 1980; Balling and Falk 1982; Ulrich 1983, 1986). Orians (1980) lists eight variables that exert a positive influence on landscape selection and that are likely to have stemmed from phylogenetic adaptations to savannas: a preference for water, large trees, focal points, semi-open spaces, changes in elevations, unobstructed view of the horizon, plant growth and moderate complexity. Ulrich's (1981) review of landscape preference literature provided similar results, including the following positive preference points: (1) complexity that is moderate to high; (2) the complexity has structural components that establish a focal point and other order and patterning is also present; (3) there is a moderate level of depth that can be perceived unambiguously; (4) the ground surface texture is even and homogeneous, conducive to movement; and (5) a deflected vista is present (prospect). Although some components are missing, for example, development in time, mystery and function of a landscape, this summary integrates much of the formal aesthetics components on landscape preferences.

Symbolic/content variables, on the other hand, are also important for aesthetic evaluation. Studies have revealed, for example that the presence of artificial contents such as poles, cables, signs and vehicles within a landscape tend to lower the aesthetic preference ratings of that landscape (Herzog et al. 1976; Appleyard 1981; Anderson et al. 1983; Nasar 1990, 1994). We have an overall preference for natural environments over artificial environments (Nasar 1983, 1984; Kaplan and Kaplan 1989); so much that the addition of natural material to urban environments tends to dramatically increase the area's aesthetic appeal (Taylor and Atwood 1978). This is complicated, however, because we also tend to prefer that nature has some indicators of human control and intervention. Slides showing well-maintained pastures with short grass, for example, receive higher ratings than wild nature scenes with a high complexity level and rough ground texture (Hagerhall 2000; see also Misgav 2000). Perceived landscape management is related to a feeling of safety (Hagerhall 2000), and a positive correlation between perceived safety and preference exists. This relationship has also been shown by Kuo et al. (1998). Residents of urban public housing facilities rated photographs of their own neighborhoods with manipulated tree densities, tree placements and grass maintenance. Although tree placement had little effect on preference, both tree density and grass maintenance had strong effects on preference. Higher tree density and higher grass maintenance resulted in increased feelings of safety and aesthetic pleasure.

Finally, one of the content variables that continuously emerges in findings is water. There is considerable evidence to support the claim of many researchers that water is a landscape property which nearly always enhances scenic quality (see Ulrich 1981). Although negative affective reactions can be elicited by some water phenomenon (e.g., a stormy sea), a consistent empirical finding is that water evokes interest, aesthetic pleasantness and positive feelings such as tranquility (e.g., Hubbard and Kimball 1967; Civco et al. 1978; Palmer 1978). The universal appeal of water is likely to be biologically based, therefore largely independent of cultural or learned associations.

Behavioural and Physiological Outcomes

If our environment can evoke specific feelings (e.g., safety or tranquility), and differential affective states in general (e.g., like, dislike), then it is likely that environmental stimuli will also affect other aspects of our behavioral outcomes and biological functioning. Habitat preference researchers have provided substantial evidence that the assessment of our environment and the emotional response has a strong impact on our physiological and psychological health. The addition of natural materials to an urban environment not only increases subjective feelings of 'refreshment' and 'relaxation', but can also actually speed recovery in stressful situations. Ulrich (1984) observed that patients recovered faster in a nature-like environment than otherwise. It was sufficient to have a view of a small stand of deciduous trees to reduce patients' post-operative hospital stays and to increase their well-being. Hospital stays were shortened an average of 8.5 days and consumption of analgesic drugs was curbed.

Access to nature also has an impact on physical development. Kindergarten children who played in more natural playgrounds showed considerable improvements in motor fitness, balance and co-ordination skills compared with children who played in less natural playgrounds (Fjortoft and Sageie 2000). Grahn's (1996) comparison of nursery schools found similar effects. Children who played at an outdoor, nature-like playground had better health records, concentration, social skills and increased creativity compared to children who played at an indoor playground. Studies show that children play in barren playgrounds half as much as they do in relatively green playground. Moreover, children's access to social interactions with adults significantly reduced when playgrounds have less vegetation (Taylor et al. 1998). Adults prefer to be in green spaces, too.

If learning in children is improved by plant-rich environment, it might also be expected that adult cognitive processes also become more efficient when in plant-rich areas. Oberzaucher and Grammer (2000) investigated this idea by documenting the effect that indoor plants had on test-takers in a driver's license examination room. Individuals who took tests when the examination room was equipped with plants achieved significantly more points per time-unit than individuals who took tests when the room was void of plants.

Not only learning is affected by the presence or absence of nature; human social behavior is also influenced. Access to nature and natural views is known to yield more positive relations and decrease aggression among residents in a neighborhood (Kuo et al. 1998), to decrease domestic violence in families (see Kuo et al. 1998), and to decrease the amount of graffiti and vandalism (Brunson et al. 2001). Ruso and Atzwanger (2001) utilized a small fountain within a shopping mall to measure the affect of water on social interaction (frequency of body contacts) and tactile exploration in commercial settings. Both rate of interaction and rate of exploration significantly increased when the fountain was filled with water compared to when the fountain was dry. The effect was even stronger when the water was in motion. Similar observations about the relationship between water presence and increased human sociality were also made by Pitt (1989), who found that group size and affiliation among river recreationists is higher compared to other outdoor recreationists. The authors suggest that water, like other natural components, is not only important for the assessment of environment quality, but also has an immediate impact on human behavior.

Since environment and environmental components still have an influence on our modern-day psychology, health, development, cognitive ability, and social behavior, it makes sense that strong aesthetic preferences still exist regarding habitat components and/or any items that contain symbolic references to outdoor habitats. As habitat has been so closely tied to survival throughout our evolutionary history, aesthetic preference and physiological outcome will also be inextricably tied to one another. Because of the directness of this connection, we argue that the study of habitat preferences, perhaps more than any other preference area, is one of the most potent areas in which to learn about fundamental foundations of human aesthetic preferences.

Habitat Preference: Individual Differences

If aesthetic preferences, affect, and behavioral outcomes toward physical environments are considered to be outgrowths of evolutionary adaptations, it should make sense that these aspects are displayed relatively universally, regardless of cultural or experiential differences. Is there evidence that individual differences exist in preferences and responses to environmental variables?

Criticism can be made that the apparent consensus on preference for habitat may actually be a sampling artifact, as samples tend to be taken from people of similar background and experiences. Indeed, Lyons (1983) found that landscape preferences differed according to age and gender, with older people expressing lower preference ratings in all categories compared to younger people, and females preferring more vegetation than males. However, no theories were offered as to why or how this might occur. Differences in landscape perception and preference according to age have been found in several other studies. Gibson (1979) and Heft (1988) found that there is a perceptual difference that varies with age, with children interpreting the landscape and terrain in terms of functions, and adults tending to interpret the landscape terrain in terms of forms. We might also expect that perceptual differences will occur especially within the feature of complexity or biodiversity, as the perception of this feature is not fixed, but will be modified by category learning as one gets older.

The fact that differences occur in the strength of preferences for savanna environments over the lifespan is well documented, a finding that is offered as evidence that innate predisposition/phenotypical imprinting occurs in humans. At age 8, children select the savanna landscape as their preferred place to live in and to visit. By age 15, this changes such that savanna landscape, deciduous forest, and coniferous forest are liked equally well (Balling and Falk 1982). Because none of the participants in this study had ever been in tropical savanna, the authors postulate that a developmental pattern occurs, where innately programmed responses may be later modified by experience in particular settings (in this case, the deciduous forests of eastern US). This line of thinking can be pushed further: clearly, the savanna habitat is not available across the world. If all humans followed the impulse to stay in savanna-like environments, resource competition would result. It would be more adaptive to remain flexible and to maintain a preference for the environment that one has grown up in. If an individual and an individual's parents were able to survive and reproduce in a certain environment, then that individual is also likely to have similar success. Syneck's (1998) work also supports this hypothesis. Analysis of virtual landscape pictures found that before puberty children tend to prefer low complexity landscapes, while after puberty they prefer the high complexity, mountain-like landscapes which they had been surrounded by during maturation. Experience with an environment can change one's aesthetic preferences.

As might be expected, landscape evaluation has also been found to vary according to one's occupation. Brush et al. (2000) tested six groups of individuals, three of whom earned their living from the land (farmers, loggers, and foresters). Three types of landscape were assessed, forest, farmland and the urban edge. All groups, except the farmer group, choose the forest landscape to be most enjoyable, followed by farmland and the urban edge. The farmers, however, preferred farmland above all others. Although Brush's research took place in western USA, similar results were also found for farmers in the Iberian Peninsula (Gomez-Limon and Fernandez 1999). Livestock farmers preferred more open landscapes, while the other groups (recreationists and managers) preferred landscapes with denser vegetation. Buhyoff et al. (1982) also found that experience with an environment can influence preferences. Tree damage from insects was shown to decrease preference among observers who were knowledgeable about forestry, but not always decrease preference within unknowledgeable individuals. From such research, we can conclude that occupational bias, experience with and knowledge of a respective landscape can have an impact on how a landscape is perceived.

If occupation and amount of experience in an environment can influence preference, what about culture? Cultural aspects of landscape preferences were measured by Yang and Brown (1992). Cross-cultural comparison showed that Korean groups and the Western groups preferred the other's landscape styles, with Koreans preferring Western style and the Westerners preferring the Korean style. Regardless of cultural background, however, ratings of the importance of four landscape elements were universal, with the presence of water as the most important, followed by vegetation, rock, and the layout of the landscape. It seems that the assessment of cultural elements of landscape style is influenced by the cultural background, but the assessment of the basic landscape elements is not influenced by cultural factors. While several researchers have stressed culture as the pre-eminent determinant of preference (e.g., Lowenthal 1968; Tuan 1974), the majority of researchers have provided support for strong cross-cultural similarity in aesthetic judgments of landscapes ranging from interior landscapes, to urban landscapes, to natural landscapes (e.g., Shafer and Tooby 1973; Kwok 1979; Zube and Pitt 1981; Ulrich et al. 1991).

Habitat Preference Research: Conclusion

Many of the factors that affect habitat preferences and perception have now been identified. Fairly consistent for positive aesthetic evaluation of landscapes within most studies is the main theme, naturalness. Commonly valued content items across most studies are water, plants, and focal points such as landmarks, mountains, lakes or large trees. Color, density and distribution of the these items are used to evaluate the important landscape attributes of spaciousness, complexity, development in time, and maintenance. Within habitat preference research we see that many of the findings support each other. It is the case, however that not all findings are consistent. There are areas in which habitat preference research can be improved.

One complication in habitat preference research is an ambiguity and overlap in terminology. One researcher talks about coherence while the other talks about unity, one talks about exploring potential and legibility while the other talks about mystery and identifiability. Further, it may often be the case that the findings are confounded within variables. The finding that maintained grass areas are more preferred than wild fields, for example, may have something more to do with legibility and complexity than the variable 'human intervention/maintenance' per se. Important for future development in the field will be following a set terminology and more standardized methodology. This is also important for maintaining consistency in interpretations. When, for example, Kuo et al. (1998) report that high tree densities have the highest preference ratings and provide the greatest sense of safety, it has to be taken into account that even the highest tree density shown in this study is considerably lower than the forest densities used in other studies, in which 'high' tree densities are typically not the most preferred. Standardization in operational definitions of what 'low', 'moderate' and 'high' means in attributes, as well as standardized use of the attributes themselves, would add clarity to collective research efforts within the field.

Criticism has also been expressed about the quintessential 'lumper' rather than 'splitter' focus in landscape preference research. Researchers have tended to homogamize findings rather than look for differences among and within groups, a tendency which stems chiefly from funding complications. The funding for most habitat preference research comes from federal agencies seeking an empirical basis for establishing standards of aesthetic quality for landscapes. This pressures researchers to average group differences, rather than seek to understand the basis for possible differences in preferences (Brush et al. 2000). Although overall preferences may look similar across groups or cultures, hidden within the presumed similarity may be large differences in the way in which environmental components are actually perceived and processed. A fundamental argument of this chapter is to propose that habitat preference research is not only a research area for environmental planners and environmental psychologists, but also a crucial area for further understanding of human aesthetic preferences in general, a lucrative 'research landscape' for all aspects of evolutionary aesthetics research. We hope to see habitat preference research is not only published in landscape and urban planning journals, but also more and more in evolutionary-focused journals.

Future Research Directions

Now that many elements that lead to positive assessment of habitat have been identified (e.g., content items such as water and formal attributes such as spaciousness), the next direction for habitat research to focus on is to look more closely at the context of the viewer. Relph (1981), in his analysis of the psychological experience of space, explained that "all places and landscapes are individually experienced, for we alone see them through the lens of our attitudes, experiences, and intentions, and from our own unique circumstances (p. 36)". While evolutionary theory acknowledges that the 'lens' of human experience is set within a framework of evolutionary developments and species-specific perceptual constraints, the idea that individual differences and unique contextual circumstances might influence habitat tastes is certainly not precluded. Even more to the contrary, theory and research in the area of group differences is fully essential, as it is an important way to learn about the subtleties of aesthetics and evolved strategies. Understanding the basis for individual and group differences in landscape preferences may lead us to further understand subtle differences in evolved survival strategies that influence aesthetic preferences, much the same way that social aesthetic researchers have discovered that long- vs. short-term mating strategies may influence the manner in which an observer determines physical attractiveness ratings of other individuals (e.g., Johnston et al. 2001).

Because a suitable habitat must provide resources for carrying out many different activities over varying time frames, evaluation of habitats is a complex process for organisms (Orians 1980). The current status of a landscape is important, but the organism must also evaluate future states. Being sensitive to predictive mechanisms and learned information about an environment that relate to one's current and future goals would be advantageous. It might be hypothesized, for example that aesthetic preferences will differ between men and women due to sexual differences in habitat use over evolutionary time (sexual dinichism). Differences in evolved psychologies should cause differences in drive state, thereby influencing the response to landscapes. Do females prefer a different vegetation density type than males? Do males, due to differential evolutionary roles, place more value in having focal points and prospect views? With age, do people tend more toward refuge-oriented attributes rather than prospect-oriented attributes? Does marital status affect aesthetic tastes, for example that individuals raising families orient more towards refuge-oriented attributes, where singles orient more towards mystery and prospect attributes? Does short- vs. long-term mating strategy have an effect on landscape preferences? Do female landscape preferences change according to menstrual cycle stages? Is there a correlation between testosterone levels and interest in prospect-oriented attributes? Such questions tie together the expression of aesthetic preferences with ontogeny and phylogeny behind them.

Findings from habitat preference research can also be extended into other areas of aesthetic preference. To what extent do habitat preferences carry over into artistic preferences, for example, the use of natural or artificial colors in paintings, shape choice, the layout of a building or various architectural shapes? Can we identify 'prospect'- or 'refuge'-oriented variables in interior decoration styles, with focal points, complexity, legibility and mystery elements? If aesthetics can be broken into two components, social (mate choice, baby schema) and environmental (habitat choice, eating preferences), to what extent can evolutionary variables account for the overlap in each, for example, in the design of an automobile? To what extent is habitat preference multi-sensory? Are some groups or types of individuals more inclined to utilize specific multi-modal attributes? How does music tie into habitat preference attributes? In which ways can a song create an ambience of prospect, refuge, spaciousness? Environmental terms and thinking can be utilized as a basis to help us understand the human aesthetic condition throughout many domains.

Evolutionary theory provides a rich variety of ideas for the testing of aesthetic preferences. In general, evolutionary aesthetics can greatly benefit from further emphasis on habitat preference research. For our evolutionary ancestors, the emotional response to landscapes was one of the most fundamental determinants of survival. Today, we carry these evolutionary remnants with us. The development of such research is, then, an invaluable resource for our further understanding of how human aesthetic tastes evolve. Because of the previous extremely close connection between habitat selection and everyday survival for our ancestors, the study of habitat preferences, perhaps more than any other preference area, is one of the most potent areas in which to learn about fundamental foundations of modern-day aesthetic preferences. Research is just beginning.

Summary

The current paper argues that a large part of the everyday aesthetic experience for humans involves a behavioral and emotional response to landscape. Since the selection of habitat was crucial in our evolutionary history, research on human habitat preference and perception is a vital area for the further understanding of evolved aesthetic tastes. The authors focus on landscape perception as a feature of evolutionary aesthetics and strengthen the point that human aesthetic values are rooted on both social and environmental components. They discuss evolutionary theories of habitat preference and compare them with empirical data of studies which were conducted in the last 20 years, with a strong emphasis on the most recent studies. The discussed papers cover the fields of perception, landscape preference, behavioral and physiological reaction to the perceived environment and individual differences in habitat preference. The authors examine the gap between the empirical data and the underlying theories which initiated the studies in the first place and argue that not all findings are consistent when compared on a bigger scale. However, they try to mediate between theory and empirical data and conclude with thoughts and future research directions to close the gap in our understanding of landscape perception.

References

Anderson LM, Mulligan BE, Goodman LS, Regen HZ (1983) Effects of sounds on preference for outdoor settings. Environ Behav 15:539–566
Appleton J (1975) The experience of landscape. Wiley, London
Appleton J (1984) Prospect and refuge re-visited. Landscape J 3:91–103
Appleyard D (1981) Livable streets. University of California Press, Berkeley
Balling JD, Falk JH (1982) Development of visual preference for natural environments. Environ Behav 14(1):5–28
Blum GS, Barbour J S (1979) Selective inattention to anxiety-linked stimuli. J Exp Psychol 108:182–224
Brunson L, Kuo FE, Sullivan WC (2001) Resident appropriation of defensible space in public housing: implications for safety and community. Environ Behav 33(5):626–652
Brush R, Chenoweth R, Barman T (2000) Group differences in the enjoyability of driving through rural landscapes. Landscape and Urban Planning 47:39–45
Buhyoff GJ, Wellmann JD, Daniel TC (1982) Predicting scenic quality for mountain pine beetle and western spruce budworm damaged forest vistas. For Sci 28(4):827–838
Civco DL, Kennard WC, Lefor MW (1978) A technique for evaluating inland wetland photointerpretation: the cell analytical method. Photogram Eng Remote Sensing 44(8):1045–1052
Coeterier JF (1996) Dominant attributes in the perception and evaluation of the Dutch landscape. Landscape Urban Planning 34(1):27–44
Cosmides L, Tooby J (1987) From evolution to behavior: evolutionary psychology as the missing link. In: Dupré J (ed) The latest on the best – essays on evolution and optimality. MIT Press, Cambridge, MA, pp 277–306
Eibl-Eibesfeldt I (1984, 1995) Die Biologie des menschlichen Verhaltens – Grundriß der Humanethologie, 3rd edn. Piper, München
Erlich PR, Erlich AH (1992) The value of biodiversity. Ambio 21(3):219–226
Fjortoft I, Sageie J (2000) The natural environment as a playground for children. Landscape description and analyses of a natural landscape. Landscape Urban Planning 48:83–97
Gibson JJ (1979) The ecological approach to visual perception. Houghton Mifflin, Boston
Gomez-Limon J, Fernandez H (1999) Changes in use and landscape preferences on the agricultural-livestock landscapes of the central Iberian Peninsula. Landscape Urban Planning 44:165–175
Grahn P (1996) Wild nature makes children healthy. Swed Building Res 4:16–18
Groat L, Despres C (1990) The significance of architectural theory for environmental design research. In: Zube EH, Moore GT (eds) Advances in environment, behavior and design, vol 4. Plenum Press, New York, pp 3–53
Hagerhall CM (2000) Clustering predictors of landscape preference in the traditional Swedish cultural landscape: prospect-refuge, mystery, age and management. J Environ Psychol 20:83–90
Heft H (1988) Affordances of children's environments: a functional approach to environmental description. Children's Environ Q 5(3):29–37
Herzog TR (1985) A cognitive analysis of preference for waterscapes. J Environ Psychol 5:225–241
Herzog TR, Kaplan S, Kaplan R (1976) The prediction of preference for familiar urban places. Environ Behav 8:627–625
Hubbard HV, Kimball T (1967) An introduction to the study of landscape design. Hubbard Educational Trust, Boston

Johnston VS, Hagel R, Franklin M, Fink B, Grammer K (2001) Male facial attractiveness: evidence for hormone-mediated adaptive design. Evol Human Behav 22(4):251–267

Kaplan R, Kaplan S (1989) The experience of nature: a psychological perspective. University Press, New York

Kaplan S (1987) Aesthetics, affect, and cognition: environmental preference from an evolutionary perspective. Environ Behav 19(1):3–32

Kaplan S (1992) Environmental preference in a knowledge-seeking, knowledge-using organism. In: Barkow JH, Cosmides L, Tooby J (eds) The adapted mind: evolutionary psychology and the generation of culture. Oxford University Press, New York, pp 561–600

Kaplan S, Kaplan R (1982) Cognition and environment: functioning in an uncertain world. Praeger, New York

Kuo FE, Bacaicoa M, Sullivan WC (1998) Transforming inner city landscapes. Trees, sense of safety, and preference. Environ Behav 30:28–59

Kwok K (1979) Semantic evaluation of perceived environment: a cross-cultural replication. Man Environ Syst 9:243–249

Lang J (1988) Symbolic aesthetics in architecture: toward a research agenda. In: Nasar JL (ed) Environmental aesthetics: theory, research and applications. Cambridge University Press, New York, pp 11–26

Lorenz K (1973) Die Rückseite des Spiegels. Versuch einer Naturgeschichte menschlichen Erkennens. Piper, München

Lowenthal D (1968) The American scene. Geogr Rev 58:61–88

Lyons E (1983) Demographic correlates of landscape preference. Environ Behav 15:487–511

Misgav A (2000) Visual preference of the public for vegetation groups in Israel. Landscape Urban Planning 48:143–159

Nasar JL (1983) Adult viewers' preferences in residential scenes: a study of the relationship of environmental attributes to preference. Environ Behav 15:589–614

Nasar JL (1984) Visual preference in urban street scenes: cross-cultural comparison between Japan and the United States. J Cross Cult Psychol 15:79–93

Nasar JL (1990) The evaluative image of the city. J Am Planning Assoc 56:41–53

Nasar JL (1994) Urban design aesthetics: the evaluative qualities of building exteriors. Environ Behav 3(26):377–401

Oberzaucher E, Grammer K (2000) Phytphilie – Pflanzen steigern die Effizienz von kognitiven Prozessen. Homo 51 Suppl:94

Orians GH (1980) Habitat selection: general theory and application to human behavior. In: Lockard JS (ed) The evolution of human social behavior. Elsevier, New York, pp 49–66

Palmer JF (1978) An investigation of the conceptual classification of landscapes and its application to landscape planning issues. In: Weidemann S, Anderson JR (eds) Priorities for environmental design research, part 1. Environmental Design Research Association, Washington, DC, pp 92–103

Pitt DG (1989) The attractiveness and use of aquatic environments as outdoor recreation places. In: Altman I, Zube E (eds) Public places and spaces. Plenum Press, New York, pp 217–253

Relph E (1981) Rational landscapes. Croomhelm, London

Ruso B, Atzwanger K (2001) Water-induced well-being in shopping malls. In: Schultz M, Atzwanger K, Bräuer G, Christiansen K, Forster J, Greil H, Henke W, Jaeger U, Niemitz C, Scheffler C, Schievenhövel W, Schröder I, Wiechmann I (eds) Homo – Unsere Herkunft und Zukunft. Cuvillier Verlag, Göttingen, pp 182–186

Shafer EL, Tooby M (1973) Landscape preferences: an international replication. J Leisure Res 5:60–65

Symons D (1989) A critique of darwinian anthropology. Ethol Sociobiol 10:131–144

Symons D (1990) Adaptiveness and adaption. Ethol Sociobiol 11:427–444

Syneck E (1998) Evolutionary aesthetics: visual complexity and the development of human landscape preferences. Diss, University of Vienna, Vienna

Taylor A, Wiley A, Kuo F, Sullivan W (1998) Growing up in the inner city: green spaces as places to grow. Environ Behav 30(1):3–27

Taylor RL, Atwood BG (1978) Plant complexity, and pleasure in urban and suburban environments. Environ Psychol Nonverbal Behav 3:67–76

Tuan Y (1974) Topophilia. Prentice Hall, New Jersey

Ulrich RS (1981) Natural versus urban spaces: some psychophysiological effects. Environ Behav 13(5):523–556

Ulrich RS (1983) Aesthetic and affective response to natural environment. In: Altman I, Wohlwill JF (eds) Behavior and the natural environment. Plenum Press, New York, pp 85–125

Ulrich RS (1984) View through a window may influence recovery from surgery. Science 224:420–421

Ulrich RS (1986) Human response to vegetation and landscapes. Landscape and Urban Planning 13:29–44

Ulrich RS, Simons RF, Losito BD, Fiotito E, Miles MA, Zelson M (1991) Stress recovering during exposure to natural and urban environments. J Environ Psychol 11:201–230

Woodcock DM (1984) A functionalist approach to landscape preference. Landscape Res 9(2):24–27

Yang B, Brown TJ (1992) A cross-cultural comparison of preferences for landscape styles and landscape elements. Environ Behav 24:471–507

Zajonc RB (1980) Feeling and thinking. Am Psychol 35(2):151–175

Zube EH, Pitt DG (1981) Cross-cultural perceptions of scenic and heritage landscapes. Landscape Planning 8:69–87

Bodies in Motion: A Window to the Soul

KARL GRAMMER, VIKTORIA KEKI, BEATE STRIEBEL,
MICHAELA ATZMÜLLER and BERNHARD FINK

"From the invisible atom to the celestial body lost in space. Everything is movement....it is the most apparent characteristic of life: it manifests itself in all functions. It is even the essence of several of them."
Etienne-Jules Marey (1839–1904), the pioneer of motion research

Biological Theories and the Empirical Assessment of Attractiveness

Evolutionary psychologists have widely accepted that there are biological reasons for body shape and size preferences in potential sexual partners. Female physical attractiveness is considered to be largely a reflection of her potential reproductive success. Recent research suggests that male physical attractiveness is also based upon the same criterion. Reproductive success is defined as the optimum (for a given environment) number of children surviving to reach sexual maturity and to become parents themselves. Buss (1989) suggests that there are cultural universals in desired body size and shape for intersexual attraction, and that these derive from the division of labor between males and females during the course of evolution, where males specialized in hunting activities and females in food gathering and child rearing. Natural and sexual selection are believed to have operated in a way that men and women whose bodies were best suited for these tasks were most attractive to potential mates. We find that females consider their physical appearance as "efficient" when they attract mates, males consider their body as efficient when it promises success in male–male competition (Erikson 1968; Lerner et al. 1976).

In the light of evolutionary theory, it seems that the notion that beauty is just in the eye of the beholder has to be re-thought and modified (see Symons 1995). It was once widely believed that standards of beauty were arbitrarily variable (Etcoff 1994). Recent research suggests, however, that people's conception of what is facially attractive is remarkably consistent, regardless of race, nationality or age (Perrett et al. 1994). Three basic characteristics have repeatedly been shown to influence human attractiveness judgments: youth (reflects absence of senescence), certain sexually dimorphic sex hormone markers such as chin size (reflects hormonal health) and symmetry of bilateral traits (reflects developmental health; see Thornhill and Gangestad 1999 for review). These characteristics all pertain to health, leading to the conclusion that humans have evolved to view certain bodily features as attractive because the features were displayed by healthy others (Grammer and Thornhill 1994; Thornhill and Gangestad 1999). For attractiveness ratings of static two-dimensional (2D) stimuli, Henss (1992) found a median inter-rater correlation of 0.45. This would explain about 20% of the variance in attractiveness ratings of faces. Seen absolutely, this is not much. On the whole, the concordance in attractiveness ratings of faces seems to be rather low. This low value for the explanation of variance in attractiveness ratings could come from many different intervening variables such as learning factors, theoretical shortcomings of current theories of attractiveness which often put only the face into the center of their considerations and the fact that simple static pictures do not reflect the whole picture of attractiveness assessments in general. In this article, we will review the current approaches to attractiveness research and introduce a new point of view. In this view we see the face and the body as a dynamic communicative device whose motions contribute to attractiveness assessments.

The Role of Averageness

Thornhill and Gangestad (1993) suggested that preference for average trait values in some facial features could have evolved because in heritable traits the average denotes genetic heterozygosity. Heterozygosity could signal an outbred mate or provide genetic diversity in defense against parasites. Studies indicate that average faces are attractive, but can be improved by the addition of specific nonaverage features. However, a recent study by Halberstadt and Rhodes (2000) has found a strong relationship between averageness and attractiveness also for stimuli like dogs and wrist-watches. It may be that humans have a general attraction to prototypical exemplars, and that their attraction to average faces is a reflection of this more general attraction. The contribution of averageness to attractiveness is still a matter of debate (Thornhill and Gangestad 1999). Exactly what features contribute to the averageness effect remains unclear. Most studies find that there are some faces in the tested samples that are considered more attractive than average faces (e.g., Alley and Cunningham 1991). The conclusion of studies focusing on the attractiveness of averaged faces is still questionable. It could be that these faces show nonaverage features, which are developed under the influence of sex hormones.

The Role of Symmetry

The hypothesis that attractiveness assessments are sensitive to facial symmetry has been tested in a number of studies. Bilateral symmetry of bilaterally represented traits is positively correlated with heterozygosity in many animals, including humans (see Thornhill and Gangestad 1999 for review). Thus facial symmetry, like facial averageness, may display underlying heterozygosity and parasite resistance (Thornhill and Gangestad 1993). In addition, body bilateral symmetry reflects overall quality of development, especially the ability to resist environmental perturbations during development. This implies that a symmetrical face displays developmental homeostasis (Thornhill and Gangestad 1993). Symmetry of the secondary sexual traits of the face is hypothesized to display immunocompetence because the construction of symmetrical – especially large symmetrical – facial secondary sex traits is expected to require a higher quantity of sex hormones (Thornhill and Gangestad 1993, 1999).

Whereas a lot of studies demonstrate the direct effects of symmetry on attractiveness (Grammer and Thornhill 1994), other research suggests that symmetry can be associated with attractiveness for reasons other than direct effects of symmetry per se (Scheib et al. 1999). Enquist and Arak (1994), as well as Johnstone (1994), offered an alternative account, arguing that symmetry is more readily perceived by the visual system. The preference for symmetry may thus be merely a by-product of the perceptual system.

In sum, up to now studies that have used a 2D approach demonstrated an influence of symmetry on attractiveness (see Thornhill and Gangestad 1999). Yet there are several confounding variables that have to be considered carefully. Average faces tend to be more symmetrical than nonaverage faces and skin texture itself could affect symmetry measures in 2D data (Grammer and Thornhill 1994; Fink et al. 2001).

The Role of Sex Hormone-markers

In many species, including humans, testosterone production and metabolism mobilize resources for the efforts of males to attract and compete for mates (Ellison 2001). Testosterone affects a number of facial features. A high testosterone (T) to estrogen (E) ratio in pubertal males facilitates a lateral growth of the cheekbones, growth of the mandibles and chin, and a lengthening of the lower facial bones (e.g., Farkas and Munro 1987). Testosterone to estrogen ratio is also important for female development. The signaling value of many female body features is linked to age and reproductive condition, both of which correspond to a woman's ratio of T to E (Symons 1995; Thornhill and Grammer 1999). Attractive signals correspond to high ratios. Estrogen promotes women's fertility, but it could be a handicapping sex hormone for women in a similar way as T acts for men (Dabbs 2000). Markers of high E may reliably signal that a female's immune system is of such high quality that it can deal with the toxic effects of high E (Thornhill and Møller 1997; Thornhill and Grammer 1999).

In this context, skin condition is presumed to reliably signal aspects of female mate value. Charles Darwin (1872) himself attributed to selection pressures the greater visibility of the human skin associated with relative hairlessness. It has been proposed that skin condition reliably signals aspects of female mate value (Symons 1979, 1995; Barber 1995). Although no full explanation has been offered, it is taken for granted that small differences in skin texture and facial hair can have effects on women's sexual attractiveness. The contribution of skin texture to attractiveness has only recently been shown (Fink et al. 2001). In addition, human males, universally, are expected to be most sexually attracted by female skin that is free of lesions, eruptions, warts, moles, cysts, tumors, acne, and hirsutism (Symons 1995). According to Morris (1967), flawless skin is the most universally desired human feature. Skin and hair, while very sexy when healthy, seem to be equally repelling when not (Etcoff 1994; Barber 1995).

Major Problems in Attractiveness Research

Symmetry, averageness, hormone-markers and skin texture are probably interacting factors in attractiveness ratings. If so, how might the features that influence attractiveness be related to one other? Research has focused mainly on the analysis of single features and their contribution to attractiveness. It may be, however, that this approach, which is still commonly used, has a lot of shortcomings as attractiveness may not be reduced to the analysis of a single feature (Grammer et al. 2002).

Cunningham et al. (1995, 1997) proposed the 'multiple fitness model', which states that attractiveness varies across multiple dimensions, rather than a single dimension. Each feature is considered to signal a different aspect of mate-value. This would correspond to the 'multiple message hypothesis' (Møller and Pomiankowski 1993), which argues that each ornament signals a specific, unique property of the condition of an individual. The question that arises then is which cognitive processes are used in the signal receiver in order to come to a coherent interpretation of the "attractiveness" of a given stimulus.

In a recent study, Grammer et al. (2002) showed that a second hypothesis – the 'redundant signal hypothesis' (Møller and Pomiankowski 1993) – might provide a solution to this problem. In this case, mate choosers should pay attention to several sexual ornaments because, in combination, they provide a better estimate of general condition than any single ornament. Thornhill and Grammer (1999) showed that independent attractiveness ratings in Austria and the USA of the same women in each of three poses (face, front nude with faces covered, and back nude) are significantly positively correlated, as predicted by the redundant signal hypothesis. Because of the connection of the attractiveness features of the face, back and front to E, the correlation between the ratings of the different pictures implies that women's faces and bodies comprise features which amount to a single ornament of honest mate value. The results of this approach show that it is possible to extract features that are related to meta-theoretical evolutionary explanations for trait values in attractiveness ratings (Grammer et al. 2002). Further, they indicate that faces (and bodies) have to be considered as a whole when it comes to attractiveness assessments.

The Quality of Movement and Communication

Across numerous studies expressive people appear to be more attractive than unexpressive people (Larrance and Zuckerman 1981; Riggio and Friedman 1986; Friedman et al. 1988). DePaulo (1992) speculates that this is the case because they simply are endowed with greater physical beauty. However, there may be more to it. The attempts of unattractive people to behave more expressively enhances their attractiveness and expressive people seem to know how to regulate their attractiveness so that they can manage to appear beautiful even under difficult conditions that make others look less physically appealing. Sabatelli and Rubin (1986) have shown that physically attractive people are better at communicating their emotions spontaneously and that they are more successful in controlling their nonverbal behaviors. DePaulo (1992) proposes that this is due to the fact that attractive people have more confidence and ability. We will extend this idea with the suggestion that expressiveness is movement and the quality of movement. The question becomes how movement is controlled and what the underlying mechanisms are.

Nonverbal communication is seen as being context-dependent, as a multi-channel and a multi-unit process (a string of many events interrelated in communicative space and time) and as being related to the function it serves (Grammer et al. 1999). In general, methodology of nonverbal behavior, however, does not adequately reflect the underlying theory. In this regard, the repertoire analysis – the traditionally used method for studying behavior – has several shortcomings. Identifying reliable behavior categories is rather difficult and may not even be useful because a non-semantic approach for studying nonverbal behavior seems more appropriate. A widely neglected dimension of nonverbal behavior is the quality of movement, and even though evidence for its importance exists, there is an enormous lack of a systematic description of movement quality (Walbott 1982).

What does movement depend on, and what are the possible variables related to attractiveness? Body movement primarily is constrained by the laws of physics and is modulated in an individual person by the particular configuration of the biomechanical linkages and the respective motor control strategies. Consequently, movement quality should be determined by genetic factors, i.e., the heritability of bone structures and muscles, which create the physical apparatus. In addition, there are either learned or possible innate motor control strategies on a neuronal level. Another modifier could be the current physiological state and the influence of neurotransmitters (Cioni et al. 1997). Thus, body movements possibly reflect genetic factors and transient factors like current physiological states.

Human evolutionary anatomy suggests that the signaling value of the human body represents adaptations to hunting in males and food gathering and child rearing in females. Therefore, we have to deal with the construction of the body in terms of energy consumption. Most of these features will be apparent and could be amplified by the way a body actually moves. Evidence comes from the study of primate locomotion. Witte et al. (1991) investigated the mechanical requirements of a bipedal walking primate and the possibilities to meet these requirements with a minimum of energy. They found that human average body proportions of the legs and arms were distributed in a way that enables a human to walk faster without additional input of energy. If the proportions of the arms and the mass distribution of the arms are comparable to those of the legs, both arms and legs can be considered as pendulums which can swing at the same frequency, thus allowing one to walk (and other possible gross body movements) at an optimum energy level compared to the output. These authors further point out that an elongated and slim trunk shape is necessary in order to provide a great mass moment and inertia and thus stability against leg movement. This suggests that there must be an optimal body shape for movement. In a recent study, Manning and Taylor (2001) used ability in football as a proxy for male physical competitiveness. They found that professional football players had lower second to fourth digit ratios (2D:4D) than controls. The 2D:4D ratio acts as a measure of fluctuating asymmetry, and correlates negatively with prenatal and adult testosterone. Football players in first team squads had lower 2D:4D than reserves or youth team players. Players who had represented their country had lower ratios than those who had not. The authors suggest that prenatal and adult testosterone, therefore, promotes the development and maintenance of physical traits, which are useful in sports and athletics disciplines and in male–male fighting.

There is enough evidence that motion alone can provide a lot of information about a person. When a series of moving lights are attached to a walker's major joints, it can easily be recognized that a human body is in motion (Johansson 1973). Moreover, observers can recognize the walker's gender from such a display (Cutting and Kozlowski 1977; Barclay et al. 1978; Cutting et al. 1978). Dittrich et al. (1996) extended this approach to the question of whether it is possible to judge the emotional state of a human body from motion information alone. An ability to make this kind of judgment may imply that people are able to perceive emotion from patterns of movement without having to compute the detailed shape first. In this study, subjects were shown brief video clips of two trained dancers (one male, one female). The dancers were aiming to convey the following emotions: fear, anger, grief, joy, surprise, and disgust. The video clips portrayed fully lit scenes and point-light scenes, with 13 small points of light attached to the body of each dancer. Full-body clips gave good recognition of emotionality (88% correct), but the results for upright biological motion displays were also significantly above chance (63% correct). Biological-motion displays, which convey no information while static, are able to give a rich description of the subject matter, including the ability to judge emotional state. This ability is disrupted when the image is inverted. Besides the recognition of gender and emotions from motion, Montepare et al. (1988) showed the impact of age-related gait qualities on trait impressions depicted in point-light displays. Younger walkers were perceived as more powerful and happier than older walkers. A composite of youthful gait qualities predicted trait impressions regardless of the walkers' masculine gait qualities, sex, and perceived age. In a second experiment, subjects observed young adult walkers depicted in point-light displays and rated their traits, gaits, and ages. Young adults with youthful gaits were perceived as more powerful and happier than peers with older gaits, irrespective of their masculine gait qualities, sex, or perceived age.

The ability to discriminate between and to identify others must have been paramount for humans in the environment of evolutionary adaptedness (EEA) and is still immanent in our perceptual system. The most obvious way in which this is accomplished is by means of face recognition. However, this is not the only method available. Stevenage et al. (1999) found that gait could also serve as a reliable cue to identity. They found that the human visual system was sophisticated enough to learn to identify six individuals on the basis of their gait signature under conditions of simulated daylight, simulated dusk and point-light displays. It thus appeared that gait-related judgments could be made, and furthermore that these judgments were possible without reliance on shape information. In other words, gait was used as a reliable means of discriminating between individuals.

On the basis of motion perception alone, people can identify two of the major aspects of mate-choice which are identified up to now: youth and gender. Gender identification could be identical with the identification of averageness – if an average male or female moves more gender-typically and an average body build might represent a general solution for energy consumption. Thus, averageness should be the one candidate for research on motion and beauty.

Averageness could reflect a sex-prototypical body build, because the biomechanical linkages of the body show considerable sex differences which will also have a general influence on body motion. For adult women the proportions of muscle strength to body height and weight remain much the same as for children. In adolescent boys, however, there appears to be an additional stimulus for muscle growth that is particularly noticeable in the muscles of the upper limb girdle. This increase in muscle mass is generated by testosterone during puberty (Jones and Round 1998). Under the influence of estrogen, the female pelvis matures to its adult form and females become relatively wider-hipped and shorter-legged than their male peers, with the difference especially marked in the shoulder/hip ratio (Johnston 1998). Thus, sex dimorphic traits will influence motion, and motion could be as sex-specific as body build.

Symmetry is the second candidate to find its expression in motion quality. Møller et al. (1999) showed that symmetric chickens show more coordinated and more efficient walking behavior. Thus, symmetry could not only be an indicator of developmental stability, it could also be an indicator of movement efficiency and thus bodily efficiency of an individual. However, since external developmental instability only reflects one part of developmental instability, effects of stress on the neural system might just as well account for the development and the expression of external asymmetries. Studies of animal behavior have indicated that the fractal dimension of repeated behavior, such as movement, differs between healthy and sick individuals (Alados et al. 1995). Fractal dimension allows us to measure the persistence of natural phenomena, this means that if a process has been increasing for a period, it is expected to continue for another period. This also means that the movement is self-repeating, or self-identical. Thus, the ability to repeat behavior in a consistent way may provide important information about condition and represent a behavioral equivalent of morphological developmental instability (Møller 1998). Deviations from perfect symmetry in paired traits such as ear size and nostril width also indicate developmental instability and/or short-term fluctuations in hormones. In both cases, symmetry is thought to be optimal and to indicate high phenotypic quality. Actually, the performance in middle-distance runners depends upon the factors above – the more symmetric they are, the higher is the performance of the runners (Manning and Pickup 1998).

Unfortunately, there are no results yet from motion analysis that have looked at this possible relation. It seems obvious to some extent that fluctuating asymmetry is not only present in width of earlobes, elbows, ankles or feet. The structures that underlie motion, i.e., muscles and bones, are also prone to asymmetry. This suggests that fluctuating asymmetry may be more visible in motion than in static appearance. We suppose that fluctuating asymmetry generally influences motion abilities of the body. Assessing fluctuating asymmetry of a moving body would thus be recognized more easily than from a static one. Asymmetries of any kind in the limbs would disturb the optimal proportions and mass distribution of the limbs, which are prerequisites for optimal energy consumption during walking. Facial symmetry might be one cue to the link between motion and attractiveness perception. If attractive faces were more symmetrical, it follows that their emotional expressions could also tend to be more symmetrical. Ekman (1986) has pointed out that felt emotions differ from staged emotions in their asymmetrical muscular performance. This simply means that we could judge symmetrical faces as being more "honest". Indeed, we ascribe honesty more often to attractive people than to unattractive people. "What is beautiful is good " is a common perception prototype (Dion et al. 1972).

The third candidate for possible influences on body motion is sex hormones. As already mentioned, sex hormones will change general body build and generate a sex-dimorphic body build. However, sex hormones also affect motor patterns directly. From studies by Hampson and Kimura (1988), we know that females show better performance on several measures of manual speed and coordination in midluteal phase of the cycle than in menstrual phase. The same authors report better performance on the same motor tasks as above during the preovulatory surge in estrogen compared with their performance during menstruation. This finding suggests that higher levels of sex hormones may be associated with improved manual performance in women. Stenn and Klinge (1972) found that two of seven women examined during several cycles showed a peak in spontaneous arm movement activity during late luteal phase.

Although these findings are restricted to hand and arm movements, Grammer et al. (1997) found that differences exist in female movement quality during cycle in a self-presentation task. Females were asked to turn around 360° in front of a video camera. Females showed slower and more complex movements when they were in the presence of a male experimenter and had high estrogen levels. A neural network analysis that was applied to movements could classify high estrogen levels in females from body movement with 100% reliability. In view of these results, sex hormone profiles should be visible from body movements. The research on attractiveness has shown that many structures that are proposed to be attractiveness signals are developed under sex-hormone influence during puberty. With respect to sex-hormone-dependent female body structures, some of them are size-related fat deposits in breasts and buttocks. These are structures which probably might hinder movement if they are too big, and thus can be regarded as movement handicaps (Zahavi 1975), or when moving themselves, even as amplifiers for certain body movements. With regard to body structures in males, male sex hormones are responsible for typical male muscle build and are used widely in athletics to increase performance. Thus, male sex-hormone levels should not only be visible in static pictures through hormone markers – they should be more prominent in male body movement.

Human Body Movement and Physical Attractiveness

As we have seen, the human face has been the center of attractiveness research for a long time, but only rarely has attention been focused on the body and never before has movement been taken into account. Until now attractiveness has always been seen as a static feature, yet onlookers rarely perceive the human body in a static posture. We hypothesize that movement is one of the key variables determining gender and attractiveness of a person. In other words, people perceive gender and attractiveness of others not only through static facial features, but also through dynamic features, e.g., their gestures, their way of walking, and especially their individual body movements. When viewed from a greater distance, the static features such as symmetry of the face or skin texture are hard to spot. Movements would provide the perfect means to signal gender and attractiveness over greater distances. Up to now, we could identify the following possible influences from body build on movement:

1. Individual and sex-specific genetic influences on the biomechanical linkages, which can change movement quality.
2. Sex hormone influences on muscle build and fat distribution. Both factors will change mass distribution and thus again movement quality.
3. Symmetry which can influence mass distribution and the biomechanical linkages
4. Influences of current hormonal states, which might change motor patterns and thereby change motion quality.

In addition to these direct movement-related parameters, we must also take into consideration that movement of a body changes the view an observer has of this body constantly. This fact is independent of the movement itself, but might provide additional three-dimensional (3D) information on a body or face. Previous perceptual studies have assumed that the brain analyses texture and color of a surface independently of its 3D shape and viewing geometry (Troost 1998). A recent study on the perception of 3D shape showed that texture and color perception is strongly influenced by 3D shape perception (Bloj et al. 1999). It seems that there is a top-down influence of surface recognition on texture and color perception. George and Graham (2000) came to the same conclusion. Their study on the role of spatial and surface cues in the age processing of unfamiliar faces reveals that the 3D shape of a face is important for its recognition. For example, the 2D spatial configuration of the internal features in a face is obviously intimately tied to the 3D shape of the head. Likewise, the 2D cue of skin texture is based on the fine scale 3D properties of the facial surfaces. This finding is consistent with the work of O'Toole et al. (1999), who were able to demonstrate that both the 3D shape and 2D surface reflectance information contribute substantially to human recognition performance, thus constraining theories of face representation to include both types of information. This work was extended by Lander et al. (1999) who showed that movement information contributes substantially to person recognition. Their results suggest that the dynamics of the motion provide additional information in face perception. If this is the case for person recognition, we can assume that this also might be the case for attractiveness ratings.

Almost all research on facial and bodily attractiveness has been done with 2D frontal views. Hence, researchers have been studying an artificial situation. In actual interactions where attractiveness might play a role, faces are in motion and can be seen from many different angles. Stimulus-specific effects in face recognition over changes in viewpoint have been clearly demonstrated by O'Toole et al. (1998). In addition, it is unclear if the actual profile of a person corresponds to the information that can be gathered from a frontal view. To our knowledge, the only substantial research on facial profiles that has been done has used line drawings as stimuli. Modern research should focus on an integrative approach to the measurement of attractiveness, an approach that does not omit information that is present (and necessary) in the real-world representation of the phenomenon.

Here, we report two studies that are the very first attempt to investigate the influence of dynamic movement displays on the perception of attractiveness.

Study 1: Human Gait as Signal – Gender Recognition

Some of the studies cited previously indicate that it is possible to recognize the gender and identity of persons just by watching them walk. Walking style could also be influenced by gender identification. In addition, walking might not only provide clues to gender identification, it also could be used as a feature signaling mating status. Attractiveness in this case would be a compound of both. Our goal was to find out if both components, gender identification or being gender-prototypical, and mating status are present as clues in walking.

Participants were videotaped from the side while walking up and down an 8-m-catwalk. The camera was stationary and the lights were arranged in such a way that no shadows were visible on the background wall. The sample consisted of 93 males (mean age 25.7, SD = 6.1) and 96 females (mean age 23.19, SD = 4.4). After walking, the subjects filled out a questionnaire which asked for basic demographic data and mating status (if the subject is in an ongoing steady relationship, or looking for a partner). As Schmitt and Atzwanger (1995) reported that male walking speed depends on male status, we also asked for monthly net income. Females were asked if they were on the pill or not; nonpill takers were asked to indicate the first day of the last menstruation. Using the tables provided by Jöchle (1973), these data were used to calculate the probability of conception.

All subjects wore trousers and their normal street shoes. As possible intervening variables body size, weight and the number of female and male spectators were recorded. For each person four video sequences each showing a few steps were digitized. Digitization started the moment the person was completely visible on the video screen and ended when the person arrived at the opposite side of the video screen. *Figure 1 gives an overview of the applied recording method and the digital movie analysis procedures.*

Motion-energy-detection (Grammer et al. 1997, 1999) was used for analyzing the walking sequences. This is a model-free yet coarse descrip-

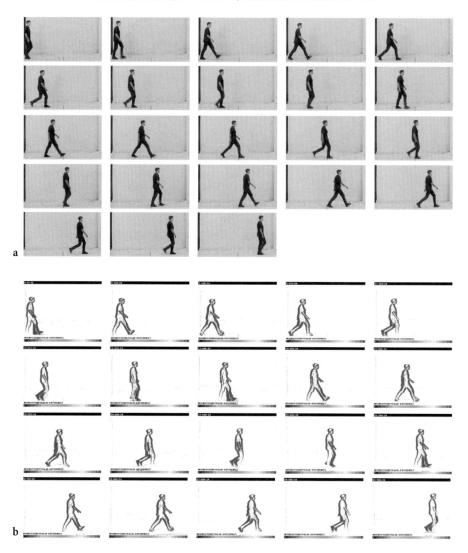

Fig. 1a, b. Motion energy in walking. **a** An original catwalk from study I. The movie clip consists of 46 frames (= 1.84 s) and every second frame with an odd number is shown. The walking consists of 1.5-stride cycles which equals three steps. **b** The difference in the pictures (first derivativee) from the same movie. This time every second even picture is shown. Gray density changes were calculated as mean gray density for the whole picture.

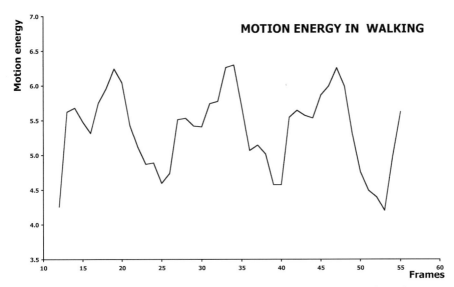

Fig. 1c. The resulting mean gray density changes over the whole movie. These data were used as input data for the neural network classification task

tion of general human body motion. Due to the fact that the pixels of every picture in a movie are a distribution of gray values from 0 to 255, we calculated the difference between two following pictures of the movie. This corresponds to the first derivate of the pictures in time. In the case of occurred movement, the gray values in one of the frames change. Hence, this method can detect the amount of motion and produce a time-based gray value distribution. We developed software that is able to compute physical parameters like speed, duration, emphasis, expressiveness and complexity from the gray value distribution over time (see Grammer et al. 1999 for a more detailed description of the software environment).

The parameters emphasis and expressiveness are related parameters, which include the amplitude of a burst of movement. They can, therefore, be seen as a measurement for the sweep of a movement. The complexity is defined by the number of minima and maxima of the curve. For the empirical analysis of the data, the four catwalks were then averaged. Note that the time we used here was extremely short. Average catwalks for females took 1.8 s (SD = 0.24), and for males 1.9 s (SD = 0.33).

In motion research several caveats are present. Basically, there is no theory of low-level motion detection, which can provide a bottom-up account of the perception of biological motion. Moreover, there is no theory of movement categorization, which can indicate what general properties of movement are crucial in perception, but evidence is beginning to accumulate that bidirectional processes are crucial to the recognition of human movement. Thus, we started the analysis with a classification task carried out by neural networks. This method provides two possibilities. First, if the neural network can reliably learn biological sex from motion, we know that this information is present in motion. However, even more important, if the neural network can learn and then classify sex, which is done on a continuum from 0 (female) to 1 (male), these data can be used to access gender information, for instance how high the classification as a female would be on the female scale. To do so, in the next step an artificial neural network was trained to classify these data according to the walker's sex.

We used the program SNNS 4.1 (Stuttgart Neural Network Simulator, University of Stuttgart http://www-ra.informatik.uni-tuebingen.de/SNNS) to simulate a time-delay neural network (TDNN) consisting of three layers (Waibel et al. 1989). This type of neural network is able to learn time structures in data and extract features translation invariant. The input layer entailed 75 units, the hidden layer 20 units, and the output layer 2 units. The organization of the input units allows the manipulation of the time span the network uses and the number of features it tries to extract. In our case, 15 frames were used and we tried to extract five features. This architecture also determines the overall length of the sequence which is 75 frames (3 s) in this case. The dataset was split randomly into a training set of 60 males and 60 females, a test set of 20 males and 20 females and a validation set of 10 males and 10 females. The training patterns were presented to the TDNN 5000–7000 times. Before training, the weights were set randomly, the learning rule used was time-delay back-propagation. After training, the neural network performance was tested on the independent test set. This was repeated until each subject had been classified ten times and then the median of the classification was calculated. For the assessment of correct classifications, we used a "winner-takes-all-rule", i.e., in the classification for a female, all values smaller than 0.5 are classified as males, all classification greater than 0.5 are classified as females.

The neural network was able to learn and it classified sex 68.2% correctly (males 67.9%, females 68.6%). This corresponds nicely to the accuracy of sex classification by human raters when confronted solely with point-light displays (see above) and underlines the validity of our method. Of the intervening variables (i.e., age, body size, body mass) body size and body mass showed a correlation with the size of classification as male or female. The greater the body size, the higher is the classification as male ($n = 169$, rs $= 0.33$, $p<0.001$) and the more the subject weighs, the higher is the classification as male ($n = 169$, rs $= 0.19$, $p<0.02$). Thus body size and mass were partialed out in the further analysis and the subjects were divided in six groups according to mating status (males mated/unmated) females (mated/unmated) and hormonal status (females pill/no pill). As the group of unmated females taking the pill had only four members, this group was dropped from the analysis.

For further analysis, a threshold method was applied (see Grammer et al. 1999) and single stride cycles were identified. These stride cycles were analyzed according to their number, their duration, their size (the amount of total change in one stride cycle), their complexity (the number of minima and maxima in one stride cycle, this indicates the presence and absence of additional body movements, like arms etc.), their speed (time to reach a maximum) present in time, their emphasis (size of the maximum of the curve in one cycle) and the expressiveness present in a stride cycle.

Within these variables, considerable sex-differences can be found. Males generate bigger changes per stride cycle (size, Kruskal-Wallis 1-Way ANOVA, $p = 0.03$), show significantly higher speed (Kruskal-Wallis 1-Way ANOVA, $p = 0.003$), higher expressiveness (Kruskal-Wallis 1-Way ANOVA, $p = 0.002$) and higher emphasis ($p = 0.0006$).

In the final analysis, we tried to find predictor variables for a classification of being male or female. When the effects of weight and body size are partialed out, we find negative correlations between speed, expression and emphasis and neural network classification as female in all groups with the exception of paired female nonpill takers (Table 1). The correlations between income as an indicator for status and walking quality show no consistent relations (Table 2). As for the correlations between conception

Table 1. Partial correlations of the neural networks' assessment of test subjects as females with the variables duration, information, expression, emphasis, and number of strides

Subjects	Df	Duration	Information	Expression	Emphasis	No. strides
Single males	25	0.124	−0.454*	−0.543**	−0.559**	0.110
Paired males	46	−0.044	−0.327*	−0.311*	−0.340*	0.065
Single female nonpill takers	22	0.193	−0.597**	−0.641***	−0.680***	0.155
Paired female nonpill takers	21	0.194	0.1766	0.154	0.182	−0.189
Paired female pill takers	23	0.024	−0.537**	−0.405*	−0.663***	0.033

* $p<0.05$; ** $p<0.01$; *** $p<0.001$.

Table 2. Partial correlations of test subjects' income with the variables duration, information, expression, emphasis, and number of strides

Subjects	Df	Duration	Information	Expression	Emphasis	No. strides
Single males	29	−0.314	−0.014	0.224	0.185	0.157
Paired males	48	0.085	−0.100	−0.140	−0.159	0.392*
Single female nonpill takers	23	−0.049	0.413*	0.412*	0.413*	−0.009
Paired female nonpill takers	20	−0.214	−0.194	−0.112	−0.182	−0.213
Paired female pill takers	27	0.271	−0.091	−0.224	−0.198	−0.432*

* $p<0.05$.

Table 3. Partial correlations of test subjects' probability of conceiving with the variables duration, information, expression, emphasis, and number of strides

Subjects	Df	Duration	Information	Expression	Emphasis	No. strides
Single female nonpill takers	23	0.232	0.256	0.227	0.199	−0.389
Paired female nonpill takers	22	−0.375	0.157	0.431*	0.349	0.083

* $p<0.05$.

probability and walking style, we find a correlation for expressiveness and walking in paired female nonpill takers (Table 3). This repeats the findings of Grammer et al. (1997) on hormonal influences of body movements. These results also show that sex recognition is possible from relatively gross measures of walking style, and that there is a hormonal influence on body movement.

Study 2: Attractiveness and Body Movements

In order to find out if a relation exists between attractiveness ratings and body movements, we asked 71 persons – 40 heterosexual women (mean age 25.1, SD = 6.2) and 31 men (mean age 23.5, SD = 3.3) at a club in Vienna/Austria to perform free dancing movements. The subjects were filmed in a static camera setting. The video sequences were digitized at 25 frames/s. Digitization starts with the beginning of the dancing movement and ends when the individual comes to a halt. As in the gait studies, the single dancing clips are very short. Their mean duration for females is 5.0 s (SD = 2.1) and for males it is 5.3 (SD = 2.4). Figure 2 shows the setup for this study.

In a rating experiment, dynamic quantised displays were created by applying a quantization (mosaic) function to the original movies. Each frame was specifically reduced to a series of blocks, 20 pixels horizontal and 20 pixels vertical. This procedure hides structural aspects of the original stimulus persons, but movement information can be detected through changes in color and lighting. Berry et al. (1991) introduced this method as a "noninvasive" alternative to the point-light-displays of Johansson (1973). They tested judgment accuracy of static versus dynamic displays and it was shown that only when raters viewed dynamic quantized displays, judgment accuracy was above chance level. Therefore, they concluded that this method successfully degrades individual information such as facial appearance, color and body shape and is a useful tool to investigate movement perception.

The 71 dynamic quantized displays were rated by 32 independent students (17 women, 15 men). The stimulus persons had to be judged on bipolar 7-point Likert scales in regard to the properties: attractive and erotic. The raters had no further information regarding the stimulus persons at their disposal (zero-acquaintance experiment) and their judgment relied only on the quantized displays).

Consensus of judgments was tested (Cronbach's alpha, inter-item correlation) and correlation between the movement parameters and the mean judgment was computed. The results were analyzed in four groups depending on who (men or women) rated whom (men or women). The consensus was low for attractiveness and eroticism (mean inter-item-correlations ranging from 0.13 to 0.19 and mean Cronbach's-alpha ranging from 0.69 to 0.79). An extremely low value was found when men rated other men for attractiveness (inter-item correlation = 0.08). See Table 4 for the mean inter-item correlations.

The concurrence was highest when raters rated stimulus persons of the same sex (men rating men and women rating women). The ratings for female stimulus persons regarding attractiveness were more concurring than for male stimulus persons. No significant correlation was found between the length of a dancing sequence and its rating. The length of the video sequences, therefore, did not influence the ratings. The mean rating for each stimulus person was correlated with its movement qualities (speed, emphasis, expressiveness and complexity). For this analysis a compound measure for attractiveness and eroticism was computed and the females were separated into two groups according to hormonal status: pill takers and nonpill takers.

At first, the same neural network classification task as in the gait study was carried out. In this case, the network was trained first with randomly selected 50% of the cases and then the other 50% percent were classified and vice versa. Astonishingly, the classification results are comparable to those of the gait study. Overall 61% was correctly classified. Females were classified correctly in 61% and males in 62% of the cases. Thus, the information on sex is not only in gait, but also in general movements like dancing.

The correlations for the whole body did not reach the level of significance, but certain trends were obvious (see Tables 5–7). Male stimulus persons were rated as more erotic and attractive, the greater the expressiveness and emphasis of their lower body movements were, while maintaining high complexity and high speed in the upper body. For the nonpill-taking females, increasing attractiveness and eroticism ratings were found to correlate with the high emphasis and high expressiveness and low speed of the upper body. Furthermore, nonpill-taking females were rated as more attractive and erotic when their movements were slower and less complex.

For female pill takers none of the correlations reach significance. Interestingly, when the correlations between classification as female by the neural network and the movement parameters are calculated, we find that the neural network sees low speed and high complexity in the lower body as "typical female".

Bodies in Motion: A Window to the Soul   313

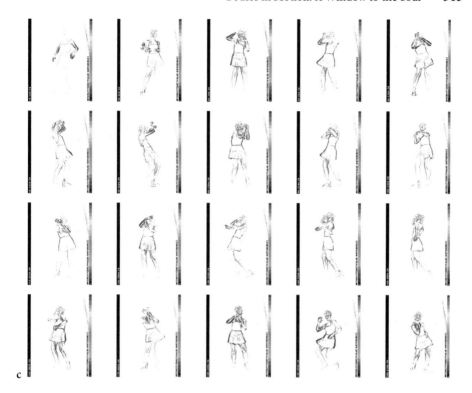

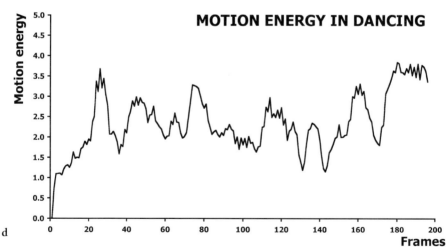

Fig. 2. Motion energy in dancing. **a** The original dancing movie; in this case, 200 frames (= 10 s). The figure shows every tenth frame from the sequence. Subjects danced without music, to their own beat. **b** Every tenth picture from the transformed movies. These quantized movies, which veil information about the person itself, but retain the body movements, were presented to the raters in the rating study. **c** The difference (first derivative) pictures for the movie above. From these pictures the mean gray density was calculated (**d**). In the last picture, the mean gray density curve over time is presented which was the basis for parameter calculation and neural network classification tasks

	Women rating men	Men rating women
Rater	n = 17	n = 15
Erotic	0.19	0.16
Attractive	0.16	0.15

Table 4. Mean inter-item-correlations of all raters for all stimulus persons

Table 5. Female and male raters rate female non-pill-taking stimulus persons, neural networks' classification of the stimulus person as male (n = 17, Spearman two-tailed)

Region of interest	Motion parameter	Female raters: compound attractiveness	Male raters: compound attractiveness	Classification as female
Whole body	Expressiveness	0.278	0.218	0.454
	Emphasis	0.345	0.198	0.267
	Speed	0.219	0.054	−0.272
	Complexity	−0.172	−0.026	0.255
Upper body	Expressiveness	0.466	0.427	0.369
	Emphasis	0.652*	0.634*	0.034
	Speed	0.465	0.397	−0.373
	Complexity	−0.467	−0.405	0.368
Lower Body	Expressiveness	0.200	0.057	0.169
	Emphasis	0.349	0.276	−0.145
	Speed	0.173	0.206	−0.488*
	Complexity	−0.151	−0.201	0.486*

* $p<0.05$.

Table 6. Female and male raters rate female pill-taking stimulus persons, neural networks' classification of the stimulus person as male (n = 23, Spearman two-tailed)

Region of interest	Motion parameter	Female raters: compound attractiveness	Male raters: compound attractiveness	Classification as female
Whole body	Expressiveness	0.131	0.178	0.101
	Emphasis	0.299	0.339	0.071
	Speed	0.235	0.229	0.127
	Complexity	−0.202	−0.194	−0.049
Upper body	Expressiveness	0.224	0.167	0.173
	Emphasis	0.316	0.158	−0.036
	Speed	0.152	−0.011	−0.046
	Complexity	−0.149	0.019	0.067
Lower body	Expressiveness	0.118	0.094	0.000
	Emphasis	0.277	0.197	0.027
	Speed	0.333	0.176	0.168
	Complexity	−0.333	−0.195	−0.135

Table 7. Female and male raters rate male stimulus persons, neural networks' classification of the stimulus person as male ($n = 31$, Spearman two-tailed)

Region of interest	Motion parameter	Female raters: compound attractiveness	Male raters: compound attractiveness	Classification as male
Whole Body	Expressiveness	0.133	0.388*	−0.077
	Emphasis	0.188	0.229	−0.311
	Speed	0.106	−0.144	−0.146
	Complexity	−0.102	0.173	0.175
Upper Body	Expressiveness	0.104	0.126	−0.077
	Emphasis	−0.039	−0.110	−0.089
	Speed	−0.223	−0.361*	−0.156
	Complexity	0.175	0.359*	0.139
Lower Body	Expressiveness	0.305	0.443*	0.137
	Emphasis	0.333	0.349	−0.003
	Speed	0.074	−0.170	−0.008
	Complexity	−0.053	0.202	0.069

* $p < 0.05$.

As a last analysis we tried to repeat the analysis of Alados et al. (1995) and calculated the fractal dimension for the time series of gray density changes. Hurst (1951) developed the rescaled range analysis, a statistical method to analyze long records of natural phenomena. We calculated an estimate of the Hurst coefficient using rescaled range analysis of the time series of gray value density in the following way (see Bassingthwaighte and Raymond 1994, 1995): let $x(j)$, $j = 1,2,...n$ be a series of N points. Subtract the mean from x so that the new mean is zero.

Over some portion of these data from $j = i,...i + kmax$, define VN $(i,k,kmax)$ = sum from $j = i$ to $i + k$ $(x(j))$, $k = 0$ to kmax. The mean from $j = i$ to $i+kmax$ is given by mean$(i,kmax)$ = VN$(i,kmax,kmax)/(kmax+1)$.

Define VR$(i,k,kmax)$ = sum from $j = i$ to $i+k$ $(x(j)-\text{mean}(i,kmax))$, $k = 0$ to kmax.

The difference between the maximum and minimum values of VN$(i,k,kmax)$ as k varies from 0 to kmax is the range, R$(i,kmax)$.

The difference between the maximum and minimum values of VR$(i,k,kmax)$ as k varies from 0 to kmax is also known as the range, R$(i,kmax)$.

The range of VR is called the trend-corrected range, because VR$(i,kmax,kmax) = 0.0$. The range, R$(i,kmax)$, in either case, is normalized by the standard deviation of the data, S, over the same x values. The rescaled range (R/S) is computed for the entire series.

The series is then halved. Each halved series has its R/S calculated. If the original series contained an odd number of points, the odd point is counted in the middle of the series and belongs to both half-length series. In this fashion, the series continues to be subdivided until the final pieces have at least six points.

The rescaled range is computed for all pieces of the series. The length (the number of points) of each piece is normalized by the total length of the series. Now the Y-on-X regression is between the logarithms of the normalized lengths and the logarithms of the rescaled ranges. The slope of the regression then is the Hurst coefficient and the relation between the Hurst exponent and the fractal dimension is simply $D=2-H$. A Hurst coefficient of $0.5<H<1$ corresponds to a profile-like curve showing persistent behavior. Indeed, the Hurst coefficient in all groups lie in this range [female pill-takers ($n=17$, $M=0.71$, $SD=0.008$), female nonpill-takers ($n=23$, $M=0.71$, $SD=0.20$); males ($n=31$, mean $=0.70$, $SD=0.16$)]. Thus, dancing can be classified as a persistent behavior in all groups, but we find only in the group of female nonpill-takers that fractal dimension correlates with ratings of attractiveness by males ($n=23$, $r=0.46$, $p=0.03$) and females ($n=23$, $r=0.45$, $p=0.03$). Attractive dancing by these females shows higher Hurst coefficients and thus lower fractal dimension.

Body Movements – The Hallmark of Attractiveness Assessments?

The results of the studies described above reveal that movement quality is related to sex and attractiveness, and movement can be used as a standalone indicator for attractiveness. Moreover, we were also able to show that gender classification goes together with movement parameters. This is possible from very short time slices. However, both studies are basically explorative and suffer from many shortcomings. First, the number of cases is small and thus statistical power not very high. Second, our motion description parameters might not exactly include the parameters that are used by observers for attractiveness assessments. Nevertheless, this first approach shows that there is attractiveness information in body movements.

In study II inter-rater correlations for movement and attractiveness seem to be quite low. The reason could be that the way we presented body movements is a highly abstract nonbiological stimulus. Further work should try to explore the missing 80% of variance in attractiveness assessments, and we will hypothesize that 3D information and movement will provide a substantial amount of additional variance explanation.

Our work showed that the way we are proceeding is feasible, but it leaves many open questions, although it is common place that body movements can or cannot be considered attractive. Surprisingly, a neural network is able to detect and classify "sex" as correctly as human observers can – independent from the source of body motion. Even more interesting is the fact that the classification as male or female shows connections to hormonal status and conception probability. We can speculate that the classification of the neural network might correspond to the gender identification of the subject.

If attractiveness to the opposite sex expresses itself in being gender-prototypical, the approach via movement analysis is feasible. Movements seem to be modulated by hormonal status, and the signal value thus would be the signaling of current hormonal state. General signaling of mate-value could be achieved with movement changes according to the general physical build. This second type of information could represent genetic factors and basic reactions of an organism to environmental influences. In addition, we could assume that hormonal influences during gestation and puberty have an impact of their own on the resulting body and thus its movement capabilities.

Regarding the correlations of movement qualities and ratings, one obvious gender difference was found: men seem to transfer information through movements of the lower body, whereas women seem to transfer information through movements of the upper body. Men and women were judged as more attractive and erotic the bigger the sweep of their movements was. In addition, women appeared more attractive and erotic, the slower and less complex their bodies moved. A study on nonverbal courtship communication (Grammer et al. 1999) showed that female interest in a communication partner also correlated with slow and less complex movements, although in this case smaller movements. The results on fractal dimension are difficult to interpret. It seems that persistence is higher in attractive dancing, at least in nonpill-taking females.

The results of the present studies reveal that the patterns of movement qualities contain information about attractiveness of an individual. Due to a lack of comparable studies, the discussion of these results remains difficult, but it could be shown that the former empirically neglected dimension of nonverbal behavior – the quality of movements – is worth a closer look. In addition, we want to emphasize that research on attractiveness must take into consideration that a more global and until now unnoted dynamic feature underlies the well-studied static features.

These results have to be put into a more realistic environment of human interaction and person perception than given by a mere 2D representation. Usually, we see a person at first from some distance and then the person moves towards us. The facial features cannot be assessed very well at this point. As Bente et al. (2001) pointed out, there is some hierarchical process in information processing. If the face is not available as a source of information, attention will shift to the next level, in this case to body form and especially body movements. Finally, the person is at a normal interaction distance (80–100 cm; Hall 1966). Bruce et al. (1996) assume that at this point information from the face is extracted. Invariant features like identity, gender, and youth are extracted as well as variant features like emotions or speech. The face and the head are mobile, thus we probably also identify structural information from a face that remains invariant despite the transformations in pose and expression. However, the dynamic patterns also hold information. The information in these body movements is the sex prototypicality of a person, probably their internal affective states and the current sex-hormonal state. Moreover, the movements will hold general genetic information and information about developmental stability and possible immunohandicaps.

In addition to these basic processes, motion provides 3D information. Bruce et al. (1993) showed that there is indeed a multiple determination of face sex by a range of different cues. When faces are displayed as 3D surfaces without their usual pigmented features and texture, performance at a sex-decision task was considerably above chance, at least when angled viewpoints were shown. There is no reason to assume that this type of processing sex in faces is different in perception of attractiveness, we can assume that 3D information also plays a prominent role in this domain. However, 3D perception is only possible on a constantly moving target. Although there is a rich body of literature about how faces are possibly categorized and mapped in the process of face recognition, there is basically no theory which could be applied to motion recognition. This has changed recently with the discovery of "mirror neurons".

Mirror neurons are a class of visuomotor neurons that have been discovered in the monkey's premotor cortex. They exist in primates without imitative and 'theory of mind' abilities. Mirror neurons respond both when the recorded monkey performs a particular action and when the same action, performed by another individual, is observed. Mirror neurons appear to form a cortical system matching observation and execution of goal-related motor actions. Experimental evidence suggests that a similar matching system also exists in humans (Gallese et al. 1996; Rizzolatti et al. 1996; Iacoboni et al. 1999). Recently, Buccino et al. (2001) have used functional magnetic resonance imaging (fMRI) to localize brain areas that are active during the observation of actions made by another individual. The observation of both object- and nonobject-related actions determined a somatotopically organized activation of premotor cortex. The somatotopic pattern was similar to that of the classical motor cortex homunculus. During the observation of object-related actions, activation, also somatotopically organized, was additionally found in the posterior parietal lobe. Thus, when individuals observe an action, an internal replica of that action is automatically generated in their premotor cortex. In the case of object-related actions, a further object-related analysis is performed in the parietal lobe, as if the subjects were indeed using those objects. These results bring the previous concept of an action observation/execution matching system (mirror system) into a broader perspective.

What might be the functional role of this matching system? One possible function is to enable an organism to detect certain mental states of observed conspecifics through their motion and the qualitative changes of body motions. Gallese and Goldman (1998) suggest that in humans the mirror neurons represent a primitive version of a simulation heuristics that might underlie mind-reading and, recently, Williams et al. (2001) have extended this by suggesting that sophisticated cortical neuronal systems have evolved in which mirror neurons function as key elements in order to become utilized to perform social cognitive functions. Humans thus might be able to use their own mental states to predict and explain mental processes of others from seeing their body movements.

We propose an extension to this simulation theory of mind. Besides psychological states, it could also well be possible to assess physiological states – in our case hormone levels – or even compare the movements to one's own movement capabilities. Hence, it would account for simple assessment of internal states and also attractiveness judgments. If we put this theory to the broader concept of attractiveness recognition, the perception of simple movements should generate internal states, which could be used to determine attractiveness and movement capabilities in a simple and uncomplicated way. We hypothesize that attractiveness ratings are ratings of the dynamic abilities of a body – which are also reflected in a static appearance and which are further accessible directly through simulation theory. Consequently, our brains might also be capable of not only reading minds, but also of reading bodies and their physiological states. Thus, body movements might indeed present a window to the soul.

Summary

Human physical attractiveness plays an important role in mate selection. Research suggests that people's conception of what is facially attractive is remarkably consistent, and the basic characteristics that seem to influence human attractiveness judgments have been repeatedly shown.

In this chapter, we review the current approaches to attractiveness research, averageness, symmetry, and sex-hormone markers, and discuss the main problems in attractiveness research, emphasizing the importance of nonstatic features.

We suggest a view which sees the face and the body as a dynamic communicative device whose motions contribute to attractiveness assessments. We try to support this assumption by presenting two recent studies which deal with gender recognition by human gait and the role of body movement in attractiveness assessments.

Our results suggest that movement quality is indeed related to sex and attractiveness, and movement can be used as a stand-alone indicator for attractiveness.

References

Alados CL, Escos J, Emlen JM (1995) Fluctuating asymmetry and fractal geometry of the saggital suture: two tools for detecting developmental instability caused by inbreeding depression in North African gazelles. Can J Zool 73:1967–1974

Alley TR, Cunningham MR (1991) Averaged faces are attractive, but very attractive faces are not average. Psychol Sci 2(2):123–125

Barber N (1995) The evolutionary psychology of physical attractiveness: sexual selection and human morphology. Ethol Sociobiol 16:395–424

Barclay CD, Cutting JE, Kozlowski LT (1978) Temporal and spatial factors in gait perception that influence gender recognition. Perception Psychophys 23:145–152

Bassingthwaighte JB, Raymond GM (1994) Evaluating rescaled range analysis for time series. Ann Biomed Eng 22:432–444

Bassingthwaighte JB, Raymond GM (1995) Evaluation of the dispersional analysis method for fractal time series. Ann Biomed Eng 23:491–505

Bente G, Krämer NC, Petersen A, de Ruiter JP (2001) Computer animated movement and person perception. Methodological advances in nonverbal behavior research. J Nonverbal Behav 25(3):151–166

Berry DS, Kean KJ, Misovich SJ, Baron RM (1991) Quantized displays of human movement: a methodological alternative to the point light display. J Nonverbal Behav 15:81–97

Bloj MG, Kersten D, Hurlbert AC (1999) Perception of three-dimensional shape influences colour perception through mutual illumination. Nature 402:877–879

Bruce V, Burton AM, Hanna E, Healey P, Mason O, Coombes A, Fright R, Linney A (1993) Sex discrimination: how do we tell the difference between male and female faces? Perception 22:131–152

Bruce V, Green PR, Georgeson MA (1996) Visual perception. Physiology, psychology, and ecology, 3rd edn. Psychology Press, Hove

Buccino G, Binkofski F, Fink GR, Fadiga L, Fogassi L, Gallese V, Seitz RJ, Zilles K, Rizzolatti G, Freund H-J (2001) Action observation activates premotor and parietal areas in a somatotopic manner: An fMRI study. Eur J Neurosci 13:400–405

Buss DM (1989) Sex differences in human mate preferences: evolutionary hypotheses tested in 37 cultures. Behav Brain Sci 12:1–49

Cioni M, Richards CL, Malouin F, Bedard PJ, Lemieux R (1997) Characteristics of the electromyographic patterns of lower limb muscles during gait in patients with Parkinson's disease when OFF and ON L-Dopa treatment. Ital J Neurol Sci 18:195–208

Cunningham MR, Roberts AR, Wu CH, Barbee AP, Druen PB (1995) Their ideas of beauty are, on the whole, the same as ours: consistency and variability in the cross-cultural perception of female attractiveness. J Personality Social Psychol 68:261–279

Cunningham MR, Druen PB, Barbee AP (1997) Angels, mentors and friends: trade-offs among evolutionary, social, and individual variables in physical appearance. In: Simpson JA, Kenrick DT (eds) Evolutionary social psychology. Erlbaum, Mahwah, pp 109–140

Cutting JE, Kozlowski LT (1977) Recognizing friends by their walk: gait perception without familiarity cues. Bull Psychonomic Soc 9:353–356

Cutting JE, Proffitt DR, Kozlowski LT (1978) A biomechanical invariant for gait perception. J Exp Psychol Human Perception Performance 4:357–72

Dabbs JM (2000) Heroes, rogues, and lovers. McGraw-Hill, New York

Darwin C (1872) The descent of man and selection in relation to sex. John Murray, London

DePaulo B (1992) Nonverbal behavior and self-presentation. Psychol Bull 111(2):203–243

Dion KK, Berscheid E, Walster E (1972) What is beautiful is good. J Personality Social Psychol 24:285–290

Dittrich WH, Troscianko T, Lea SE, Morgan D (1996) Perception of emotion from dynamic point-light displays represented in dance. Perception 25:727–738

Ekman P (1986) Telling lies. Norton, New York

Ellison PT (1999) Reproductive ecology and reproductive cancers. In: Panter-Brick C, Worthman C (eds) Hormones, health and behavior. Cambridge University Press, Cambridge, pp 184–209

Ellison PT (2001) On fertile ground: a natural history of human reproduction. Harvard University Press, Harvard

Enquist M, Arak A (1994) Symmetry, beauty and evolution. Nature 372:169–172

Erikson E (1968) Identity: youth and crisis. WW Norton, New York

Etcoff NL (1994) Beauty and the beholder. Nature 368:186–187

Farkas LA, Munro IR (1987) Anthropometric facial proportions in medicine. Charles C Thomas, Springfield, IL

Fink B, Grammer K, Thornhill R (2001) Human (*Homo sapiens*) facial attractiveness in relation to skin texture and color. J Comp Psychol 115(1):92–99

Friedman HS, Riggio RE, Casella DF (1988) Nonverbal skill, personal charisma, and initial attraction. Pers Soc Psychol Bull 14(1):203–211

Gallese V, Goldman A (1998) Mirror neurons and the simulation theory of mind-reading. Trends Cognitive Sci 2:493–501

Gallese V, Fadiga L, Fogassi L, Rizzolatti G (1996) Action recognition in the premotor cortex. Brain 119:593–609

George PA, Graham JH (2000) The role of spatial and surface cues in the age-processing of unfamiliar faces. Visual Cognition 7:485–510

Grammer K, Thornhill R (1994) Human (*Homo sapiens*) facial attractiveness and sexual selection: the role of symmetry and averageness. J Comp Psychol 108:233–242

Grammer K, Fieder M, Filova V (1997) The communication paradox and possible solutions. In: Schmitt A, Atzwanger K, Grammer K, Schäfer K (eds) New aspects of human ethology. Plenum Press, New York, pp 91–120

Grammer K, Honda R, Juette A, Schmitt A (1999) Fuzziness of nonverbal courtship communication: unblurred by automatic movie analysis. J Pers Soc Psychol 77(3):487–508

Grammer K, Fink B, Juette A, Ronzal G, Thornhill R (2002) Female faces and bodies: N-dimensional feature space and attractiveness. In: Rhodes G, Zebrowitz L (eds) Advances in visual cognition, vol 1: facial attractiveness – evolutionary, cognitive, cultural and motivational perspectives. Ablex, Westport, CT, pp 91–125

Halberstadt J, Rhodes G (2000) The attractiveness of nonface averages: implications for an evolutionary explanation of the attractiveness of average faces. Psychol Sci 11:285–289

Hall ET (1966) The hidden dimension. Doubleday, New York

Hampson E, Kimura D (1988) Reciprocal effects of hormonal fluctuations on human motor and perceptual-spatial skills. Behav Neurosci 102:456–459

Henss R (1992) "Spieglein, Spieglein an der Wand ..." Geschlecht, Alter und physische Attraktivität ("Mirror, mirror on the wall ..." Sex, age, and physical attractiveness). Psychologie Verlags Union, München

Hurst HE (1951) Long-term storage capacity of reservoirs. Trans Am Soc Civil Eng 116:770–799

Iacoboni M, Woods RP, Brass M, Bekkering H, Mazziotta JC, Rizzolatti G (1999) Cortical mechanisms of human imitation. Science 286:2526–2528

Jöchle W (1973) Coitus induced ovulation. Contraception 7:523–564

Johansson G (1973) Visual perception of biological motion and a model of its analysis. Perception Psychophys 14:201–211

Johnston FE (1998) Morphology. In: Ulijaszek SJ, Johnston FE, Preece MA (eds) The Cambridge encyclopedia of human growth and development. Cambridge University Press, Cambridge, pp 193–195

Johnstone RA (1994) Female preferences for symmetrical males as a by-product of selection for mate recognition. Nature 372:172–175

Jones DA, Round JM (1998) Human skeletal muscle across lifespan. In: Ulijaszek SJ, Johnston FE, Preece MA (eds) The Cambridge encyclopedia of human growth and development. Cambridge University Press, Cambridge, pp 202–205

Lander K, Christie F, Bruce V (1999) The role of movement in the recognition of famous faces. Mem Cognition 27:974–985

Larrance DT, Zuckerman N (1981) Facial attractiveness and vocal likeability as determinants of nonverbal sending skills. J Pers 49(4):349–362

Lerner RM, Orlos JB, Knapp JR (1976) Physical attractiveness, physical effectiveness, and self-concept in late adolescents. Adolescence 11(43):313–326

Manning JT, Pickup LJ (1998) Symmetry and performance in middle distance runners. Int J Sports Med 19(3):205–209

Manning JT, Taylor RP (2001) Second to fourth digit ratio and male ability in sport: implications for sexual selection in humans. Evol Human Behav 22(1):61–69

Møller AP (1998) Developmental instability as a general measure of stress. Adv Stud Behav 27:181–213

Møller AP, Pomiankowski A (1993) Why have birds got multiple sexual ornaments? Behav Ecol Sociobiol 32:167–176

Møller AP, Sanotra GS, Vestergaard KS (1999) Developmental instability and light regime in chickens (*Gallus gallus*). Appl Anim Behav Sci 62:57–71

Montepare JM, Zebrowitz M, McArthur L (1988) Impressions of people created by age-related qualities of their gaits. J Pers Soc Psychol 55:547–556

Morris D (1967) The naked ape: a zoologist's study of the human animal. McGraw-Hill, New York

O'Toole A, Edelman S, Buelthoff HH (1998) Stimulus specific effects in face recognition over changes in viewpoint. Vision Res 38:2351–2363

O'Toole A, Vetter T, Blanz V (1999) Three-dimensional shape and two-dimensional surface reflectance contributions to face recognition: an application of three-dimensional morphing. Vision Res 39:3145–3155

Perrett DI, May KA, Yoshikawa S (1994) Facial shape and judgement of female attractiveness. Nature 386:239–242

Riggio RE, Friedman HS (1986) Impression formation: the role of expressive behavior. J Pers Soc Psychol 50(2):421–427

Rizzolatti G, Fadiga L, Gallese V, Fogassi L (1996) Premotor cortex and the recognition of motor actions. Cognitive Brain Res 3:131–141

Sabatelli RM, Rubin M (1986) Nonverbal expressiveness and physical attractiveness as mediators of interpersonal perceptions. J Nonverbal Behav 10:120–133

Scheib JE, Gangestad SW, Thornhill R (1999) Facial attractiveness, symmetry and cues of good genes. Proc R Soc Lond Ser B Biol Sci 266:1913–1917

Schmitt A, Atzwanger K (1995) Walking fast: ranking high: a sociobiological perspective on pace. Ethol Sociobiol 16:451–462

Stenn PG, Klinge V (1972) Relationship between the menstrual cycle and bodily activity in humans. Hormones Behav 3:297–305

Stevenage SV, Nixon MS, Vince K (1999) Visual analysis of gait as a cue to identity. Appl Cognitive Psychol 13:513–526

Symons D (1979) The evolution of human sexuality. Oxford University Press, Oxford

Symons D (1995) Beauty is in the adaptations of the beholder: the evolutionary psychology of human female sexual attractiveness. In: Abramson PR, Pinker SD (eds) Sexual nature/sexual culture. University of Chicago Press, Chicago, pp 80–118

Thornhill R, Gangestad SW (1993) Human facial beauty: averageness, symmetry, and parasite resistance. Human Nat 4:237–269

Thornhill R, Gangestad SW (1999) Facial attractiveness. Trends Cognitive Sci 3:452–460

Thornhill R, Grammer K (1999) The body and face of woman: one ornament that signals quality? Evol Human Behav 20:105–120

Thornhill R, Møller AP (1997) Developmental stability, disease and medicine. Biol Rev 72:497–548

Troost JM (1998) Empirical studies in color constancy. In: Walsh V, Kulikowski J (eds) Perceptual constancy. Why things look as they do. Cambridge University Press, Cambridge, pp 262–282

Waibel A, Hanazawa T, Hinton G, Shikano K, Lang K (1989) "Phoneme recognition using time-delay neural networks," IEEE Trans Acoustics Speech and Signal Processing 37(3):328–339

Walbott HG (1982) Bewegungsstil und Bewegungsqualität: Untersuchungen zum Ausdruck und Eindruck gestischen Verhaltens. Beltz, Weinheim

Williams JHG, Whiten A, Suddendorf T, Perrett DI (2001) Imitation, mirror neurons and autism. Neurosci Biobehav Rev 25:287–295

Witte H, Preuschoft H, Recknagel S (1991) Human body proportions explained on the basis of biomechanical principles. Z Morphol Anthropol 78(3):407–423

Zahavi A (1975) Mate selection – a selection for a handicap. J Theor Biol 53:205–214

Perfumes

MANFRED MILINSKI

"Fragrant is the scent of your anointing oils, and your name is like those oils poured out; that is why maidens love you" (Song of Solomon 1, 3).

Introduction

All cultures are known to place great importance on artificial body odour, suggesting a deep-seated psychological awareness that human bodies should smell and perfumes have been used since the earliest times of recorded history (Stoddart 1986). Fragrances have been used for at least 5000 years and all traditional scents are found in modern perfumes. The purpose of using perfumes seems to be to enhance a person's sexual attractiveness, which is apparent from the advertising campaigns that usually accompany the marketing strategies of the perfume industry and which is testified also by the bible (see above). Perfumery is not only regarded as an art, it is also one of the earliest crafts and the basic techniques of today's perfumers are essentially the same as those of their Egyptian predecessors 4000 years ago (Dodd 1991). Many of today's perfume ingredients such as cassia, cinnamon, sandalwood, styrax, benzoin, jasmine, rose, etc. were already used for incense by ancient Chinese, Indian or Egyptian cultures 5000 years ago (Stoddart 1991). Detailed incense and perfume recipes, e.g., myrrh, labdanum, galbanum, olibanum in specified quantities can be found in the bible (Exodus 30:34–36) and are still used in some modern perfumes. Also, many of today's best-selling brands are rather old, e.g., 'Mitsouko' (Guerlain) from 1919, 'Chanel No. 5' from 1921, 'L'Air du Temps' (Ricci) from 1948, and many of the new perfumes are similar to established ones (H. & R. Fragrance Guide 1995), suggesting that fashion is not very important for human perfume preferences. Because of the long history of humans to select and use artificial scents, perfumes may have become part of our biology. It is, however, not obvious which part they play.

Humans can recognise at least 10,000 different odours (Prasad and Reed 1999) and olfactory communication is natural in humans (e.g. Doty et al. 1975; Doty 1981; Stoddart 1990; Gangestad and Thornhill 1998; Stern and McClintock 1998; Rikowski and Grammer 1999). Yet perfumologists are undecided about the biological significance of perfume use and of the observed great individual differences in preference for fragrances which is mirrored by the broad spectrum of perfumes on offer everywhere in the world (Van Toller and Dodd 1991).

When do people use perfumes? The Sun King, Louis XIV, is well known for his habit of using much perfume to conceal his hygiene problems and to make his presence more agreeable to the court and his feminine entourage (Le Norcy 1991). If, however, the masking of unpleasant body odour was the primary function of using perfumes, we would not expect so many different sophisticated mixtures with subtle bouquets of scents to exist on the market. Instead, one or a few strong scents would have been developed that are suited best for disguising bad smells. Daly and White (1930) noted that the functional significance of perfume may not be for the purposes of disguising or masking natural body odour, as other authors assumed, but to heighten and fortify natural odour. As Pratt (1942) put it, perfumes "unconsciously reveal what consciously they aim to hide". Thus, there are good reasons to dismiss the masking hypothesis, at least for a moment, in favour of Daly and White's hypothesis, which would also be a good candidate for explaining the existence of individual differences in preferences for perfumes. If there is anything to reveal, it should be something which is not the same for everybody.

If the choice of a personal perfume helps to "fortify natural odour", we need to understand (1) which kind of message is sent by natural odour, (2) who is the receiver of this message, (3) whether the receiver can be manipulated by mixing natural odour with fragrances, and (4) whether the sender could profit from manipulating the receiver. Assuming that the use of perfumes enhances sexual attractiveness (see above), we should start with understanding the advantage of sexual reproduction and mate choice.

The Evolution of Sex and Mate Choice

Evolutionary biologists have a big problem that is still unsolved. It seems to be a well-guarded secret of biology: we cannot explain why the great majority of animals and plants reproduce sexually. More precisely, we do not understand why males exist. We would have no problem if the world consisted only of asexual females. Following Crow and Kimura (1965; but see Kondrashov and Kondrashov 2001), the conventional textbook explanation states that sexual populations can react more quickly to environmental changes than asexual populations. If two new mutations A and B would be advantageous to cope with recent environmental change and these occur as usual in different individuals, sexuals can combine them within one generation. However, A and B have to occur sequentially in the same line in an asexual population, which takes time. The problem is that mixed populations containing both reproductive modes will turn very quickly into asexual ones because of the following (Fig. 1).

Let us imagine two females, a sexual and an asexual one. They have two offspring each. The sexual one has a son and a daughter, each with only

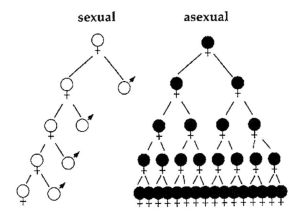

Fig. 1. The disadvantage of sexual versus asexual reproduction. See text for details

half of the mother's genome, whereas the asexual one has two daughters that are identical to their mother.

Each daughter (and only daughters) again produces two offspring. In this way, the asexual female has twice as many grandchildren as the sexual one, because sons neither lay eggs nor give birth, although they cost as much as daughters to be produced. Sons seem to be a waste. This difference becomes impressive when we look at further generations (Fig. 1). Models have shown that because of this double advantage of asexual reproduction, sexuals are outcompeted within several generations (e.g. Maynard Smith 1976). It needs a strong short-term advantage of sexual reproduction to survive this competition. If one thinks of tracking climatic changes, the correlations between selectively relevant features of the environment need to change direction between generations in order to support sexual reproduction. "But it is hard to believe that the world is like that" wrote Maynard Smith (1976) and, after he had evaluated the existing hypotheses for the evolution of sex, he concluded: "I fear that the reader may find these models insubstantial and unsatisfactory. But they are the best we have."

If we think that it might not be too bad and probably also enjoyable if we would be asexually reproducing females, after sexuals have been outcompeted, a fatal evolutionary mechanism would start to work and would drive us slowly, but definitely to extinction: Muller's ratchet (Muller 1964). We would assemble deleterious mutations in our germline, about one mutation per generation, until the genome is so degenerated after several hundred generations that the population becomes extinct. Deleterious mutations can only be cleaned away by sexual reproduction. However, Muller's ratchet is too slow to help the sexuals in their competition with asexuals.

Whatever the short-term advantage of sexual reproduction might be, if the selection pressure conferring this advantage should be relaxed or, even worse, if we use anything artificial so that sexual reproduction no longer confers the yet unknown advantage, we would turn into asexuals quickly and go extinct in the long run.

Sex is an inefficient way to reproduce: a sexual female throws away half of her genes (during meiosis), tries to find a suitable male who has done similarly (mate choice), and fills up what she has thrown away with what she gets from him. The short-term advantage of sexual reproduction should be such that the combination of the new half of genes with the half that she has retained should be more than twice as fit as if she had made just a copy of herself.

Infectious Diseases May Keep Us Reproducing Sexually

Now we have at least a hypothesis that might work. It is called the red queen hypothesis and it has several fathers (see Hamilton et al. 1990 for a review). Infectious diseases can change so rapidly between generations that hosts have to run as quickly as the red queen with Alice in Wonderland to be able to keep their position. If this hypothesis is true, the human species will turn into a parthenogenetic one if medicine becomes so successful that all diseases can be cured.

The red queen hypothesis (e.g., Hamilton 1980) assumes that infectious diseases change so rapidly that in every generation new combinations of genes for resistance are required to cope with the currently dominating parasites. A sexually reproducing female should select the male that has the right resistance genes that would, in combination with her genes, provide the offspring with the optimal immune response against the current infectious disease. This requires that (1) there is an enormous variation of resistance genes in the population, otherwise the female would not be able to choose, (2) she knows which immuno-genes she has herself and (3) she can "read" the immunogenetics of potential mates. These assumptions appeared to be too demanding that they could be fulfilled under natural conditions.

Everybody knows that it is extremely difficult to find a genetically suitable donor for someone who needs transplantation, e.g. a liver. This is because our immuno-genes, i.e. the major histocompatibility complex (MHC), contain the most polymorphic gene loci known in vertebrates (Apanius et al. 1997). This is bad for transplantation medicine, but good news for the red queen hypothesis: there are several hundred different immuno-alleles on the population market in various combinations of a few per individual that a female can choose from, if she can manage to choose among such alleles – but can she? Yes, indeed, she can. More than a dozen studies have shown that mice prefer as mates those males that have MHC alleles that differ from their own alleles and they can choose by smell only (e.g. Yamazaki et al. 1976, 1978; Egid and Brown 1989; Potts et al. 1991). This preference would result in offspring that are heterozygous in their immuno-genes which should enable them to fight off a broader spectrum of infectious diseases. Each MHC-allele codes for a specific MHC-molecule which can detect and present a specific class of foreign peptides to T-lymphocytes; T-lymphocytes kill infected cells and so eventually defeat the infection.

Fragrant Immuno-Genes in Humans

Immuno-genes "smell" (actually their gene products or specific ligands smell) and can thus be literally sniffed out by a choosy female. The fact that mice prefer mates that have MHC-alleles that are dissimilar to their own alleles proves that the mouse "knows" which alleles she (and males do the same) has herself. Therefore, all conditions required for the red queen hypothesis to work seem to be fulfilled, at least for mice. However, all vertebrates studied so far have a very similar MHC system so the findings from mice may be valid also for, e.g. fish and humans.

Evidence exists for MHC-dissimilar odour (Wedekind et al. 1995; Wedekind and Füri 1997) and mating preferences in humans (Ober et al. 1997). Wedekind et al. (1995) found in a double-blind study that women (not using the contraceptive pill) prefer the odour of T-shirts worn by MHC-dissimilar men to those with a more similar MHC genotype. The preference was reversed in women who took oral contraceptives. Note that the pill in those days mimicked a pregnancy and a pregnant woman would prefer the company of relatives to that of strangers to raise her children, after all, relatives share more genes than non-relatives. Supporting this view, it has been shown (Manning et al. 1992) that pregnant mice prefer to nest with MHC-similar individuals. Pregnant women change their taste for particular smells; they dislike both the smell of their partners and that of their particular perfumes (Hudson et al. 1996).

In the experiment, odours of MHC-dissimilar men reminded the women significantly more often of their own, real-life partners. Moreover, most of these women stated that they had not been on the pill at the time when they matched up. This observation supports the thesis that MHC-dependent body odours may play a role in human mate choice even under the reality of our smelly civilisation. These results have been replicated and extended to men whose preference was similar to that of women not taking contraceptives, although the reversed preference in women taking oral contraceptives was replicated only as a non-significant trend (Wedekind and Füri 1997).

MHC-correlated odour preferences in mice and humans are perhaps the key for understanding the function of using artificial scents in humans. One might take the view that a perfume might well be chosen to enhance one's own personal MHC scent. If this is right then there should never be a market leader among perfumes. Diversity in MHC would explain diversity in perfumes (Vollrath and Milinski 1995). There is evidence from studies with mice that perfumes interact with natural signals in mate–mate interactions: pregnancy failed in a high proportion of mice when the recently mated female was housed with a strange male, i.e., the "Bruce effect" (Bruce 1960); the female may prepare for a new pregnancy as quickly as possible after her mate has been removed by a new male. Interestingly, the Bruce effect is released also by anointing the mouse's original partner with a commercial perfume (Archunan and Dominic 1990). Perfume odorants are detected by mouse vomeronasal neurons which had been thought to detect only pheromones (Sam et al. 2001). Therefore, perfumes appear to mimic odour signals that are functional as pheromones in mice. What about humans?

Immunogenetics Influences Perfume Selection: An Experiment

Usually, people take a long time to find their perfume and then stay with it for many years. As a rule, customers try and buy perfumes for their *own* use as has been the habit in the past (Le Norcy 1991). Therefore, if perfumes are used to amplify in any way one's *own* body odours, the preference for perfume ingredients should correlate with one's own MHC type. However, no pronounced correlation is predicted between one's own MHC type and preference for perfume ingredients to be used by potential partners: for MHC-disassortative mating preferences, any other allele which is different from one's own alleles would do. To put it simply, you would be much more selective in what you like as your own perfume than what you like to smell on potential partners.

We (Milinski and Wedekind 2001) bought 36 high quality perfume ingredients mostly of natural origin, i.e. civet, patchouli, olibanum, benzoin, vanilla, oakmoss, tonka, leather, opoponax, heliotrope, myrrh, jasmine, lilac, carnation, caryophyllen, neroli, rosewood, ylang ylang, musk, sandal, amber, vetiver, cedar, moss, cardamom, tolu, cinnamon, styrax, castoreum, labdanum, bergamot, violet, tuberose, rose, orris and geranium. We put two drops of each on a smelling strip, sealed these strips in small numbered glass vials and sent them to the 137 MHC-typed persons of Wedekind's former projects in 1996. The package contained a questionnaire with a line scale between "unpleasant" and "pleasant" for each of the 36 scents. The test persons were asked (in German). "As how pleasant do you perceive this scent as a potential ingredient of a perfume/aftershave that you would like to use for yourself? Would you like to smell like that yourself? (mark between unpleasant and pleasant)."

An overall analysis revealed a statistically significant correlation between the MHC and the scorings of the scents. In a detailed analysis, we found that the two most common MHC alleles in our study group, HLA-A2 and -A1, for which we had the highest statistical power, each interacted significantly with the ratings of the 36 scents. This shows that one's own perfume may reveal one's own MHC alleles, at least the most common ones. An analogous analysis calculated with "dummy" alleles that were randomly assigned to the test subjects and that had the frequencies of the most common MHC-alleles did not result in any significant interaction term that involves one of these "dummy" alleles.

We did not dare publish these results before we had repeated the study. Therefore, we tested both for repeatability of the results with the same persons and for the second prediction, namely that we would expect no significant correlation between a person's MHC and the appreciation of the scents to be used by a potential partner. Two years later, each person was again sent a package with 36 coded glass vials, this time two sets of 18 vials each, each set containing the same 18 perfume ingredients (the last 18 listed above).

One set was to be scored as perfume ingredient used by self (numbers 1–18 in red for female, in blue for male smellers), the other set to be scored as ingredient to be used by potential partner (letters A–S in blue for female, in red for male smellers). Each smeller received one letter (in German), e.g., a female smeller starting with the set for potential partner: "Please assess the scents marked blue (A–S) as potential ingredients of a perfume that your partner uses. Would you like your partner to smell like that?" – "Please send your first set of assessments within the first envelope and wait at least for two days until you begin assessing the second set." – "Then please assess the scents marked red (1–18) as potential ingredients of a perfume that you would use yourself. Would you like to smell like that yourself?" – "Please send your second set of assessments within the second envelope."

Once again, the two most common MHC alleles interacted significantly with the ratings of the scents, but only when evaluated for "self". Figure 2 shows that the mean effects of two most common HLAs were significantly correlated between the first study and the ratings for "self" in the second study. Can we distinguish between persons with HLA-A2 and HLA-A1?

There are some hints. Persons with HLA-A1 appear to hate both No. 1 which is musk and 3 which is patchouli, and No 7 which is cardamom, whereas the persons with HLA-A2 like these ingredients in their personal perfume. However, we have to be cautious, this analysis is post hoc and descriptive and needs further study.

MHC genotype and ratings of the scents "for partner" did not correlate. This agrees with our second hypothesis. However, there was a statisti-

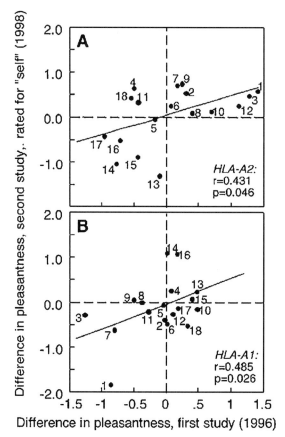

Fig. 2. Comparison of the effects the MHC appeared to have on the scorings of 18 scents that were presented in both studies, shown for the two most common HLAs A2 (**A**) and A1 (**B**). The figure shows the differences in the scorings of scent pleasantness (mean scoring of persons that possessed a given HLA or antigen combination minus mean scoring of persons that did not possess the HLA or antigen combination), i.e. positive values for a given scent indicate a preference for the scent by those that possess a given MHC specificity relative to those who do not possess it, while negative values indicate the contrary. The *numbers* correspond to the last 18 perfume ingredients listed on p. 330, beginning with musk (= 1)

cally significant correlation between the MHC and the scorings of the scents "for self" in both tests. This result suggests that persons who share one of these alleles have a similar preference for any of the perfume ingredients. This supports the hypothesis that perfumes are selected "for self" to amplify in any way body odours that reveal a person's immunogenetics. Individual perfume preferences were consistent even after 2 years. This finding agrees with the hypothesis that these preferences are individual traits that are immune to fashion to some extent, which would hardly be expected under the null hypothesis.

We found significant effects mainly for those antigens for which we had the highest statistical power to find any effect, if it existed. Many of the results from other antigens, for which we had less statistical power, were close to being statistically significant, which suggests that significant interactions of scent preferences with other antigens may exist and may become detectable with larger sample sizes.

Choice of Perfumes: Evolved Aesthetics?

If the function of perfume use is to signal the possession of specific MHC alleles, then, if anything, the ability to choose specific mixtures of scents that mimic or fortify one's own odours could have been developed. It would be extremely difficult for a person to find the mixture from the hundreds of brands that comes closest to the ideal fit with one's own odour. In agreement with this is that it seems to be a difficult and long-lasting process during which a person finds "her" or "his" perfume, usually affording many visits to a perfume shop. According to our hypothesis, the ideal perfume would be individually composed for any person as actually happened in former times: "Our great-grandmothers would consult with their perfumer. He would make, exclusively for them, a special perfume. The formula was secret and our great-grandmothers would never tell anyone what their delightful perfume was, *so much was it part of their personality.*" (Le Norcy 1991). Maybe we find a perfume most delightful or aesthetic that fits our biology, i.e. our immunogenetics best.

Do perfume ingredients resemble body odours? The scents of, e.g. jasmine and rose, apparently differ from human body odours. However, a natural flower oil may contain over 400 different odorants (Dodd 1991) and there are more than 270 known constituents in rose oil (Ohloff 1978). In 1951, the perfumer Jellinek described subscents ("Beigerüche") of many classic ingredients of natural origin as reminiscent of human body odours. These subscents do not need to have the same structure as body odours because structurally unrelated odorants can have similar odour (Beets and Theimer 1970). Starting as a "sensory bias" (e.g., Ryan et al. 1990), preferences for such subscents may have been developed. It may be because of their subscents that specific ingredients have a long tradition of being used for perfumes. Stoddart (1990) wonders why only about a hundred plant species have been used traditionally, which is an insignificant proportion of the total number of plant species in the world. "This suggests there may be something very special about them" as Stoddard says. This may be the possession of specific subscents. However, the subscent-hypothesis is only one among various potential mechanisms for the correlation between individual preferences for perfume ingredients and MHC-type.

Why perfumes? (1) Perfumes may be used as amplifiers of MHC-correlated body odours to overcome the smelly noise of human civilisation (Vollrath and Milinski 1995). If perfumes are used only to enhance body odours, an alternative strategy would be to enhance body odour itself or the perception of it. We would have to assume that constraints exist that render this alternative strategy less feasible. (2) Perfumes may be used strategically to signal the possession of specific MHC alleles when specific infectious diseases are around. Infected mice as well as ill humans are perceived as such by their changed odour which may become masked with the "healthy" perfume signal (Milinski and Wedekind 2001).

A Bit of History

The idea is not new: the painter of "Les Fleurs animées" obviously knew 150 years ago that humans prefer other perfumes for themselves than on their sexual partner (Fig. 3), and he knew that the mother passes on her perfume to her daughter (Fig. 4). I found in the museum of perfume in Paris that the daughter later on exchanged about half of the handed down ingredients for different ones, which seems to make sense, genetically.

Some Future Prospects

Recent research on MHC-dependent mate choice in female sticklebacks, a tiny, but wide-spread fish, showed that these fish are more clever than mice and humans (or researchers had not been alert enough): they count ("sniff out") the number of different MHC-alleles that a potential partner possesses and then go for the partner with which they would have the optimal number of different alleles in combination with their own alleles (Reusch et al. 2001). A recent study in humans has found similar results (Jacob et al. 2002): humans also seem to achieve an optimal similarity in immuno-genes. The optimum comes about by having enough different immuno-genes for fighting off infections and not having too many of them to stimulate auto-immune responses (Nowak et al. 1992). We need some imagination for understanding how this new finding can be translated in having shaped our combined preference for perfumes that we like on our partner and on ourselves – if we are as clever as sticklebacks.

Synthetic Perfumes – Lacking Biological Function?

It is not only the challenge of new materials that persuaded perfumers to use synthetic ingredients for creating new perfumes (Calkin and Jellinek 1994), but also an economic inevitability. A perfume that contains mostly synthetic scents can be much cheaper than its competitor that is composed of traditional natural ingredients that have become extremely expensive because their production is labour-intensive. Although, for example, natural rose oil consists of more than 270 different constituents (Ohloff 1978), its synthetic surrogate imitates only the few components that are subjectively reminiscent of "rose". However, if natural rose oils have been used for perfumes for 5000 years because of one or several of its subscents, which may not be perceived consciously, the synthetic mimic of rose may fail to fulfil its biological function of eventually signalling its bearers

Fig. 3. Painting by J.J. Granville "les fleurs animées" (1850), which suggests that humans are attracted by mates who use a different perfume from the one they use themselves

immunogenetics. I guess that customers who try to find "their" perfume may not accept completely synthetic perfumes. We may be facing a dilemma: only perfumes made of natural ingredients are aesthetic, but may become so expensive that most people can no longer afford to buy them, but cheap synthetic perfumes may not, as I assume, fulfil their biological purpose. There may be a way out of this dilemma. If researchers happen to detect the nature of the molecules with which vertebrates such as mice, fish and humans signal their immunogenetics by their body odour, it may be possible to synthesise them and add individual mixtures to any synthetic perfume and thereby restore its biological function; this would end a 5000-year-old culture and pave the way into a "brave new world" of synthetic sexual signalling. However by using further synthetic ingredients, the biological function of these new perfumes might be rescued and, thus, they might become as aesthetically appealing as their traditional precursors.

Summary

Sexual reproduction is still an evolutionary puzzle. A sexual female throws away half of her genes (during meiosis), and fills up what she lost with genes from a male. Sexual reproduction is only successful if the offspring with the new mixture of genes should be more than twice as fit as if she had just made a copy of herself. This increase in quality could be achieved by selectively "smelling out" suitable immuno-genes in potential partners which, in combination with the female's genes, offer optimal resistance against quickly changing infectious diseases. Mice, fish and humans exert this kind of mate choice. Humans, however, have potentially interfered with this adaptive strategy by using artificial scents, i.e., perfumes, in order to boost their sexual attractiveness, for at least the last 5000 years. It can be shown that an individual's preference for her/his personal perfume correlates with the person's immunogenetics. When evaluating perfumes of potential partners she/he changes her/his preference in a way that is expected if the use of perfumes amplifies the natural signalling of one's immunogenetics. Modern synthetic perfumes may not fulfil this biological function.

Fig. 4. Painting by J.J. Granville "les fleurs animées" (1850), which suggests a tradition of perfume ingredients being passed down from mother to daughter

References

Apanius V, Penn D, Slev PR, Ruff LR, Potts WK (1997) The nature of selection on the major histocompatibility complex. Crit Rev Immunol 17:179–224

Archunan G, Dominic CJ (1990) Stud male-induced protection of implantation in food-deprived mice: masking effect of an artificial scent on pheromonal odour. Ind J Exp Biol 28:371–372

Beets MGJ, Theimer ET (1970) Odour similarity between structurally unrelated odorants. In: Wolstenholme GEW, Knight J (eds) Taste and smell in vertebrates. Churchill, London, pp 313–321

Bruce MH (1960) A block to pregnancy in the mouse caused by proximity of strange males. J Reprod Fertility 1:96–103

Calkin RR, Jellinek JS (1994) Perfumery, practice and principles. Wiley, New York

Crow JF, Kimura M (1965) Evolution in sexual and asexual populations. Am Nat 99:439–450

Daly CD, White RS (1930) Psychic reactions to olfactory stimuli. Br J Med Psychol 10:70–87

Dodd GH (1991) The molecular dimension in perfumery. In: Van Toller S, Dodd GH (eds) Perfumery: the psychology and biology of fragrance. Chapman and Hall, London, pp 19–46

Doty R (1981) Olfactory communication in humans. Chem Senses 6:351–376

Doty RL, Ford M, Preti G, Huggins G (1975) Human vaginal odors change in pleasantness and intensity during the menstrual cycle. Science 190:45–60

Egid K, Brown JL (1989) The major histocompatibility complex and female mating preferences in mice. Anim Behav 38:548–550

Gangestad SW, Thornhill R (1998) Menstrual cycle variation in woman's preferences for the scent of symmetrical men. Proc R Soc Lond Ser B 265:927–933

Hamilton WD (1980) Sex versus non-sex parasite. Oikos 35:282–290

Hamilton WD, Axelrod R, Tanese R (1990) Sexual reproduction as an adaptation to resist parasites (a review). Proc Natl Acad Sci USA 87:3566–3573

H & R Fragrance Guide (1995) 3rd edn. Glöss and Co, Hamburg

Hudson R, Koch B, Heid B, Laska M (1996) Veränderungen der Geruchswahrnehmung während der Schwangerschaft: eine Längsschnittstudie. In: Brähler E, Unger U (eds) Schwangerschaft, Geburt und der Übergang zur Elternschaft. Westdeutscher Verlag, Opladen, pp 174–191

Jacob S, McClintock MK, Zelano B, Ober C (2002) Paternally inherited HLA alleles are associated with women's choice of male odor. Nat Genet 30:175–179

Jellinek P (1951) Die psychologischen Grundlagen der Parfümerie. Hüthig Verlag, Heidelberg

Kondrashov FA, Kondrashov AS (2001) Multidimensional epistasis and the disadvantage of sex. Proc Natl Acad Sci USA 98:12089–12092

Le Norcy S (1991) Selling perfume: a technique or an art? In: Van Toller S, Dodd GH (eds) Perfumery: the psychology and biology of fragrance. Chapman and Hall, London, pp 217–226

Manning CJ, Wakeland EK, Potts WK (1992) Communal nesting patterns in mice implicate MHC genes in kin recognition. Nature 360:581–583

Maynard Smith J (1976) The evolution of sex. Cambridge University Press, Cambridge

Milinski M, Wedekind C (2001) Evidence for MHC-correlated perfume preferences in humans. Behav Ecol 12:140–149

Muller HJ (1964) The relation of recombination to mutational advance. Mutat Res 1:2–9

Nowak MA, Tarczy-Hornoch K, Austyn JM (1992) The optimal number of major histocompatibility complex molecules in an individual. Proc Natl Acad Sci USA 89:10896–10899

Ober C, Weitkamp LR, Cox N, Dytch H, Kostyu D, Elias S (1997) HLA and mate choice in humans. Am J Human Genet 61:497–504

Ohloff G (1978) The importance of minor components in flavors and fragrances. Perfum Flavor 3:11–22

Potts WK, Manning CJ, Wakeland EK (1991) Mating patterns in seminatural populations of mice influenced by MHC genotype. Nature 352:619–621

Prasad BC, Reed RR (1999) Chemosensation: molecular mechanisms in worms and mammals. Trends Genet 15:150–153

Pratt J (1942) Notes on the unconscious significance of perfume. Int J Psychoanal 23:80–83

Reusch TB, Häberli MA, Aeschlimann PB, Milinski M (2001) Female sticklebacks count alleles in a strategy of sexual selection explaining MHC polymorphism. Nature 414:300–302

Rikowski A, Grammer K (1999) Human body odour, symmetry and attractiveness. Proc R Soc Lond Ser B 266:869–874

Ryan MJ, Fox JH, Wilczynski W, Rand AS (1990) Sexual selection for sensory exploitation in the frog *Physalaemus pustolosus*. Nature 343:66–67

Sam M, Vora S, Malnic B, Ma W, Novotny MV, Buck LB (2001) Odorants may arouse instinctive behaviours. Nature 412:142

Stern K, McClintock MK (1998) Regulation of ovulation by human pheromones. Nature 392:177–179

Stoddart DM (1986) The role of olfaction in the evolution of human sexuality: an hypothesis. Man 21:514–520

Stoddart DM (1990) The scented ape: the biology and culture of human odour. Cambridge University Press, Cambridge

Stoddart DM (1991) Human culture: a zoological perspective. In: Van Toller S, Dodd GH (eds) Perfumery: the psychology and biology of fragrance. Chapman and Hall, London, pp 3–17

Van Toller S, Dodd GH (eds) (1991) Perfumery: the psychology and biology of fragrance. Chapman and Hall, London

Vollrath F, Milinski M (1995) Fragrant genes help Damenwahl. Trends Ecol Evol 10:307–308

Wedekind C, Füri S (1997) Body odour preferences in men and women: do they aim for specific MHC combinations or simply heterozygosity? Proc R Soc Lond Ser B 264:1471–1479

Wedekind C, Seebeck T, Bettens F, Paepke AJ (1995) MHC-dependent mate preferences in humans. Proc R Soc Lond Ser B 260:245–249

Yamazaki K, Boyes EA, Thaler HT, Mathieson BJ, Abbott J, Boyes J, Zayas ZA (1976) Control of mating preferences in mice by genes in the major histocompatibility complex. J Exp Med 144:1324–1335

Yamazaki K, Yamaguchi M, Andrews PW, Peake B, Boyes EA (1978) Mating preferences of F2 segregants of crosses between MHC-congenic mouse strains. Immunogenetics 6:253–259

Do Women Have Evolved Adaptation for Extra-Pair Copulation?

RANDY THORNHILL and STEVEN W. GANGESTAD

Do women have special-purpose evolved adaptation that functions in pursuing copulations with men other than the main romantic partner, just as they have specialized adaptation for seeing color, estimating object distance, digesting fat, responding to stress, and a multitude of other problems that gave rise to successful selection for functional traits in human evolutionary history? This question can be asked in a variety of other ways without change in its conceptual content: do women have a feature(s) that is functionally designed/organized to accomplish extra-pair copulation (EPC)? Do women have a trait(s) that has the evolutionary purpose of extra-pair copulation? Do women have a bodily feature(s) that resulted in net reproductive success (RS) during human evolutionary history because its female bearers were conditionally unfaithful in their romantic relationships? Did human female ancestors become ancestors (i.e., out-reproduce the females that failed to become human ancestors) in part because they were infidels in romantic relationships?

Because an evolved adaptation is a product of past direct selection for a purpose or function, in this case the function or purpose of EPC, a question that is distinct from those above is: is EPC by women currently adaptive? Or, does women's EPC currently advance women's RS? An evolved adaptation may be currently nonadaptive and even maladaptive because the current ecological setting in which it occurs differs from the evolutionary historical setting that was the selection favoring it. An adaptation will be adaptive currently when the current ecological setting is the selection pressure that molded it (Williams 1966; Thornhill 1990, 1997; Symons 1992).

Reeve and Sherman (1993) define an adaptation as any trait that is currently adaptive, i.e., when variation in the trait co-varies with RS in a contemporary population. Their concept of adaptation is not the one used here. Here, we use the concept of an evolved adaptation and hence, the current adaptiveness of a trait is not a criterion for identifying it as an adaptation. The only criterion for identifying evolved adaptation is functional design (Williams 1966; Thornhill 1990, 1997; Symons 1992).

Martins (2000) suggests that an adaptation is a trait that persists in phylogenetic lines of species and higher taxonomic categories. However, persistence may occur because the trait is linked to another trait that is directly favored by selection – in which case the persistence is not a product of past direct selection, but instead of past *indirect* selection. Taxonomically conserved traits can be by-products of adaptations.

Women, just like all other organisms, are historical documents in that they possess the features they do as a result of the evolutionary process in the past. In terms of ultimate causation, each organism's features are of three types: adaptations, traits that are the product of direct selection for a function; by-products or incidental effects, traits that are the product of indirect selection and exist because direct selection favored a trait that had side-effects; and traits that result from mutation alone. The third trait type is very uncommon in nature compared to the other two types when considering the many thousands of traits of an individual woman. Mutation is a rare event and, because typically mutations are detrimental to performance and hence RS, they are selected against toward frequencies of zero. This negative selection is balanced by recurrent mutation, leading to mutations being at very low frequency in what is referred to as mutation--selection balance.

Incidental effects abound in the body and behavior of the individual. In humans they range from human behaviors such as piano playing, golf and reading to anatomical features such as the distance from the top of a person's left knee cap to the proximal end of her appendix. Incidental effects are far more common than adaptations in the totality of individual traits.

An adaptation is a piece of the individual's phenotype that exhibits functional or purposeful design and evolved by direct selection for the design because the design was a solution to an evolutionary historical problem that affected differential RS of individuals. Playing the piano has no evolutionary function because piano playing did not affect this evolutionary historical RS. Piano playing arises incidentally from adaptations directly selected because they served other functions. Similarly, the distance between a patella and the appendix lacks evolutionary function and arises from adaptations for body organization. Both incidental effects and adaptations are equally the product of the evolutionary process. Reading would not exist if certain psychological adaptations that function in human verbal language had not evolved (Pinker 1994).

The adaptationist research program, often called adaptationism, figuratively carves the organism's phenotype to separate empirically its adaptations from its by-products and traits in mutation–selection balance. Adaptations are those phenotypic mechanisms that exhibit functional design, which is identified by demonstration that a piece of the body is functionally organized to solve a problem. Said differently, the standard of evidence in identifying adaptation is showing fit or coordination between a trait and a problem. When the goodness of fit is sufficient to rule out chance association, one has evidence for adaptation. Williams (1966) emphasized that adaptation is an onerous concept in biology, and should be applied only when there is evidence of functional design. The presence of functional design in a trait eliminates incidental effect as an explanation and eliminates explanations based on chance causes such as mutation and drift. (Drift, like selection and mutation, is an evolutionary agent. Drift is differential RS of individuals due to chance. Selection is nonrandom differential RS because it arises from individuals' trait differences.)

The only natural process known to create functionally designed traits is Darwinian selection (natural, sexual and artificial). An adaptation then contains the evidence of its creator – the kind of historical selection that made it. Therefore, adaptations give the biologist information about the deep-time selection pressures that were actually effective in bringing about the evolutionary change that led to adaptation. Indeed, identifying functional design is the only means to infer these selection pressures (Williams 1966; Thornhill 1990, 1997; Symons 1992).

This chapter will critically review research on women's EPC and ask whether the standard of evidence for identifying women's EPC adaptation is met. We conclude that the evidence for EPC adaptation in women is considerable. It does not appear that women's EPC can be explained as a by-product due to direct selection of female sexual adaptation that functions in obtaining desirable long-term mates or in assuring pair-bond mate's insemination. Nor does it appear that women's EPC is an incidental effect of sexual adaptation in men for pursuing high partner variety. Nor does it appear that women's EPC functions to reduce the likelihood that males will injure or kill a female's offspring. We identify possibly productive avenues of future research that may be useful additional tests of the hypothesis that women have EPC adaptation. Women's EPC behavior involves olfactory and visual judgments of potential partners. Olfactory and visual aesthetic judgments are central to Darwinian aesthetics as well as traditional aesthetics (see Thornhill, this Vol.).

A central issue in this chapter is the method by which functional design can be demonstrated. We discuss this method further in the next section.

Functional Design: Fit of Phenotype to Problem

Any trait of the individual that has complex organization raises the possibility that the trait may be an adaptation. Complexity is highly unlikely to be the result of simple mutation alone. Moreover, complexity implies costs to survival and hence RS of the bearer and therefore rules out drift as the maker of the trait. The lateral line of certain fishes is a useful example. Its observed complexity suggested to biologists that it is an adaptation, but its function was not initially apparent. Eventually, the lateral line was shown to have functional design for processing of sound waves (Williams 1966). Evidentiary proof of adaptation is not found in complexity itself. Instead, functional design is the necessary and the sufficient evidence.

The concept of functional design is unambiguously illustrated by certain adaptations for camouflage. The red sea dragon is a species of sea horse that lives and hides among certain red seaweed. Not only is this fish the color of its home, it has specialized morphological projections from its body that look identical to branches of the seaweed. The dragon has a habitat preference for the red seaweed. Its habitat preference reflects specialized psychological adaptation for camouflage in the seaweed.

Aspects of this animal's morphology, behavior, and psychology are phenotypic solutions to the evolutionary historical problem of hiding from visual predators and from visually astute prey items that live in the seaweed.

Another equally impressive example of camouflage adaptation is seen in the common walkingstick insect (*Diapheromera velii*) in the southwestern USA. It is a master of disguise among the branches of its food plant, the *Dalea* shrub. This walkingstick can be found only on *Dalea*, where it is extremely difficult to locate because of its cryptic color and behavior of mimicking a *Dalea* branch. The reason the camouflage evolved was selection pressure from visual predators.

Direct selection for camouflage is required – both necessary and sufficient – to explain the existence and functional properties of these adaptations. The fit between phenotype and environmental problem in these two examples is near complete. This coordination cannot be explained by chance arising from incidental matching, genetic drift, or mutation. Observations of either species would lead to the conclusion of tight fit between phenotype and problem. Observers would come to this conclusion without use of any statistical tests; in fact, most people would agree that it is virtually impossible for the associations to be due to chance. The functional design is also easily communicated to biologists and ordinary people. One need not have heard of Darwinism to see the alignment between the traits considered and cryptic function. Darwinism, specifically its component of evolution by selection, is the ultimate scientific explanation of why the coordinated functional design exists. At the same time, functionally organized camouflage identifies past selection for hiding from visual predators, and, in some species, past selection for ambushing prey, as actual historical selection pressures that were effective in bringing about phenotypic evolution. The existence of camouflage adaptations places these historical selection pressures on biologists' list of realities of the deep-time history of life on Earth.

In many adaptationist studies, tests of hypotheses about functional design are conducted by experiments or in other controlled test settings, rather than by observation. All findings, regardless of the method generating them, are evaluated in terms of whether the phenotype solves a problem and the solution is often defined in terms of statistical significance. Statistical significance is the typical standard for evaluating whether results reject or support hypotheses in all disciplines of the life sciences, not just in adaptationism.

The specialized habitat preference behaviors of the sea dragon and the walkingstick and the walkingstick's behavior of assuming a stick-mimicking posture reveal the presence of information-processing and decision-making neural adaptations functionally designed for camouflage. The evidentiary basis of inferring psychological adaptation is identical to that with nonpsychological adaptation: fit between phenotype and problem. In the case of psychological adaptation, the problem domain is information-processing, decision-making and motivation.

The adaptations discussed reveal a fundamental property of adaptation: special-purpose functional organization. As Symons (1987) has argued in detail, on theoretical grounds, specialized function is expected in adaptation because the problems that give rise to selection for their solutions are specific, thus specific solutions will best solve, or only solve, these problems. Moreover, on evidentiary grounds, adaptations are almost always specific in function, including human psychological adaptations, such as each of the many psychological adaptations with specialized purpose associated with vision. Each of the gross functional components of the human body – digestion, excretion, skeletal support, immunity, growth, sexuality, reproduction, internal regulation, information-processing, and so on – actually is comprised of many functionally specific adaptations.

Women's Extra-Pair Mating

Social Significance and Frequency

Of all events within romantic relationships that evoke strong emotional reactions, extra-pair sexual relations are perhaps those that can lead to the most conflict and destructive consequences. Discovered or suspected infidelity can cause mateships to dissolve. Male suspicion of a mate's infidelity is the leading cause of male violence against women, which too frequently leads to spousal homicide (Wilson and Daly 1992; Figueredo and McCloskey 1993; Buss 2000). Moreover, a man's reliability of paternity in an offspring born to his wife is an important predictor of his willingness to invest in the offspring. As a man's confidence of paternity climbs, so does his actual investment and interest in his putative children (Anderson et al. 1999). One cue men use to adjust investment in their putative offspring is physical resemblance (Daly and Wilson 1995; Burch and Gallup 2000).

The clear significance of EPCs to the lives of many people reinforces the value of systematic and theoretically oriented research examining the factors that give rise to them. Traditional research on human EPC was largely atheoretical (for a review on extramarital sex, see Thompson 1983; on nonmarital extradyadic sex, Hansen 1986). Human EPC began to be examined in evolutionary theoretical terms especially in the early 1990s with the research of Baker and Bellis on human sperm competition (reviewed in their volume of 1995), but with occasional earlier treatments (e.g., Benshoof and Thornhill 1979; Symons 1979; Hrdy 1981; Smith 1984). The paucity of theoretically informed research on EPC until recently cannot be attributed merely to the infrequency of women's EPC. Large surveys of married women in the USA estimate that between 15% (Laumann et al. 1994) and 70% (Hite 1987) have had extramarital sex, with a median estimate being about 30% (for a review, see Thompson 1983; also Kinsey et al. 1953). In our study of 201 romantically involved university women of average age 20 years who had been in their current relationship, on average, about 2 years, 13% reported having EPCed in the current relationship and 17% ever in a relationship. In addition, 6% of a sample of British women with one main partner reported their *last* act of intercourse to be an EPC (Bellis and Baker 1990).

While socially monogamous bird species have been studied in greater detail than humans in regard to extra-pair paternity, humans seem to have a pattern of extra-pair paternity similar to that in these birds. Across many bird species in which there is extended male–female pair bonding for offspring production, the median extra-pair paternity rate is 10–15%, with rates over 20% not uncommon (Birkhead and Møller 1995; Petrie and Kempanaers 1998). Nonpaternity rates among purported mother–father pairs vary across modern human populations, perhaps from 1–30% (MacIntyre and Sooman 1991). For instance, recent studies yielded rates of 1% in Switzerland (Sasse et al. 1994) and 12% in Monterey, Mexico (Cerda-Flores et al. 1999). In the latter study, the rate among a low socioeconomic status group was 20% (see also Baker and Bellis 1995; Hill and Hurtado 1996; Beckerman et al. 1998).

Men's Sexual Proprietariness

Men, but not women, can be duped about parentage as a result of EPC, leading to cuckoldry, the unknowing investment in another man's offspring. Men show a rich diversity of mate-guarding tactics ranging from sexual jealousy, vigilance, monopolizing a mate's time and mate spoiling to mate abuse. The emotional and motivational basis of mate guarding by men also supports cultural systems such as the ideology of female virginity and sexual modesty in general, and female circumcision and claustration (Dickemann 1981; Thornhill and Thornhill 1987; Wilson and Daly 1992; Buss 2000). There is also evidence that men's mate guarding includes adjustments of ejaculate size to defend against the mate's insemination by a competitor (Baker and Bellis 1995; Shackelford et al. 2002) and more intense guarding of mates of high-fertility status (young or not pregnant) than mates of low-fertility status (older or pregnant; Flinn 1987; Buss 2000).

This diversity of male interests and activities is thought to be a condition-dependent output of a sex-specific male psychological adaptation that has been referred to as men's sexual proprietariness. This label describes the machinery's hypothesized functional design of processing information and motivating feelings and behaviors that exclude other males from sexual access to one's mate, and even exclusion when the mate desires copulation with another (Wilson and Daly 1992).

Women, too, guard their mates with a range of behaviors, but the proprietariness involved is focused on exclusion of competitors from access to a mate's resources rather than monopolization of sexual access. Though both sexes express concern about a mate's infidelity, across a variety of cultures, men more than women are concerned about the *sexual* fidelity of their partners (Buss et al. 1999; Buss 2000).

The occurrence and frequency of EPC by women are not evidence of EPC adaptation in women. As we emphasized, adaptation is an onerous concept and is applied to traits when, and only when, there is evidence of functional design. That female EPC occurs generally in pair-bonding species also is not evidence that women have EPC adaptation. The cross-species data indicate that women behave similar in EPC to females of other species that live in social systems like that of humans. Even the research that indicates that female birds of some species may have adaptation for EPC provides no evidence that women have such adaptation. Evidence that women have EPC adaptation can only be uncovered by study of *women's* EPC.

There is considerable evidence of functional design and hence adaptation in men's sexual proprietariness. This documents the occurrence of female EPC in human evolutionary history. Specifically, it documents that female EPC was of sufficient importance to have generated selection in males in the deep-time past that created adaptation that functions to solve the problem of female EPC. Were there no recurrent negative effect of female EPC on male RS in human ancestral settings, men would not have evolved these often costly, functionally specialized sets of emotional responses (Buss 2000). The existence of men's sexual-proprietariness adaptation is not, however, evidence that women have EPC adaptation. Women's EPC adaptation would be demonstrated only by evidence of a trait(s) in women that is functionally designed for EPC.

From the standpoint of a pair-bonded man, female EPC may be forced or unforced. Either way, female EPC lowers male paternity confidence. There is evidence that women possess adaptation for reducing the likelihood of their rape and for dealing with the negative consequences of rape, such as divestment by pair-bond mate (see review in Chap. 4, Thornhill and Palmer 2000). This evidence is simultaneously evidence for the occurrence during human evolutionary history of rapes frequent enough to have given rise to effective selection that led to women's anti-rape adaptation. Although aspects of men's sexual proprietariness can be explained as a solution to rape of mate (e.g., insemination of a relatively large ejaculate in copulations following physical separation of mates), many features of it seem to function as solutions to a mate's consensual EPC (e.g., gift giving and other behaviors that display willingness to invest in a current mate and function to increase the woman's perception of his mate value over that of his competitors). Again, however, evidence of consensual EPC by human ancestral females is not evidence of EPC adaptation in women.

Evidence for Design in Women's EPC

Certain EPC behavior of women suggests a co-evolutionary history of sexually antagonistic selection and, therefore, female EPC adaptation and historically adaptive female EPC. Intersexual arms races imply adaptation in each sex to cope with that in the opposite sex. Sexually antagonistic selection leads to adaptations that benefit the sex of the adaptation's owner, at the reproductive expense of the opposite sex (e.g., Rice 1996; Rice and Holland 1998). Due to continuing co-evolution of such adaptations in each sex to counter those in the opposite sex, neither sex is expected to be fully adapted to the adaptations of the other. EPC may have been pursued by ancestral females in a manner to avoid detection by the pair-bond partner. Females may sometimes benefit reproductively from having mates other than their main partner sire offspring (e.g., to increase genetic quality or diversity of offspring; Jennions and Petrie 2000). Females sometimes also may benefit from pursuing EPC by getting resources in exchange for sex from males other than the primary partner. Regardless of the type of benefit a female receives from EPC, she may gain by maintaining the investment of the main partner, which is most likely if he does not detect her EPC. Selection may have favored female EPC adaptation that motivates subtle, hidden EPC when benefits of EPC are high and costs low. Consistent with this, we have found that there is a significant sex difference in knowledge of EPC behavior among romantically involved couples and that men are less likely to know (Gangestad and Thornhill, unpubl.).

However, this pattern may be a by-product of, for instance, women's reduced risk-proneness (if, in fact, women are less risk-prone than men; see Wilson and Daly 1985). More research is needed to determine if women's secrecy about EPC is replicable, and if so, whether it has design specific to maximizing benefits and lowering costs of EPC per se. For example, does secrecy positively covary with the costs of EPC to individual women (e.g., the degree of resource loss from the main partner they would incur if detected)?

Perhaps promoting women's secrecy in EPC is their greater ability, in comparison to men, to control facial and body expressions of emotions and shielding of feelings about relationships (i.e., selective avoidance of thoughts about potential mates). Although women appear to be more expressive than men, women have enhanced ability to regulate their feelings and their true feelings are better hidden (Bjorklund and Kipp 1996). Bjorklund and Kipp emphasize that these psychological features in women may be EPC adaptation, allowing secretive EPC when benefits exceed costs. If this is the case, the expectation is that the hiding of true feelings about potential extra-pair mates or actual EPC partners will be especially acute at high conception risk in the menstrual cycle.

Variable male mate guarding across their partner's menstrual cycle may be related to an evolutionary historical arms race between females deceptively seeking EPC and men trying to prevent it. Earlier research found that men direct more guarding, including behaviors that increase the man's mate value in the eyes of mate, when mates are not pregnant (as opposed to pregnant) and are young (compared to older mates; Flinn 1987; Buss 2000). We, in collaboration with Christine Garver, have found that individual pair-bonded women report more guarding by mates during the fertile time of the menstrual cycle than in relatively infertile times (Gangestad et al. 2002; also see below).

Women's fertility status was determined by tracking women across their cycles. The women involved were not using hormone-based contraception. Peak fertility was identified by using a commercially available kit that measures the luteinizing hormone surge, which occurs 24–48 h before ovulation, corresponding to peak conception probability (Jöchle 1973; also see Thornhill and Gangestad 1999a).

The largest effect of women's fertility status was on men's vigilance – monitoring their mate's whereabouts and activities, including looking for evidence of extra-pair sexual interest in their personal belonging. Other male mate guarding behaviors – monopolizing partner's time and giving into partner's wishes – however, also significantly increased prior to ovulation. The effects were strengthened for primary partners of women who did not claim that the relationship was sexually exclusive, and strengthened also as women approached ovulation and thus peak fertility. Men whose mate guarding increased most dramatically in the days preceding their partner's peak fertility had partners who especially expressed sexual attraction to and fantasy about other men (also see below) and, hence, perhaps had the most to worry about with respect to potential sexual competitors. There was no evidence that men engaged in increased attempts to be around partners at their peak fertility in their cycles because their partners were simply in a better, more prosocial mood. These results suggest that ancestral females may have clandestinely pursued EPC disproportionately more at peak conception probability because men's mate guarding is most intense at this time in the menstrual cycle. The fact that men's mate guarding at fertile cycle points is directed at control of consensual EPC is seen in the guarding tactics that are directed at increasing a male's attractiveness relative to other males and at detecting consensual EPC.

Evidence that men who are pair-bonded to women engage in more mate guarding at the high conception risk times of their mate's menstrual cycle does not imply that female fertility across the cycle has not been directly selected to be hidden from others.

Concealed fertility by females and fertility detection by males are expected to have co-evolved in an evolutionary arms race between the sexes (Benshoof and Thornhill 1979; Baker and Bellis 1995). Thus, neither party in the race is anticipated to have completely solved the problem it faces in the opposite sex.

The fact that fertility in women's cycle is not conspicuous is evidence that women have been directly selected for deception of mates. Benshoof and Thornhill (1979) suggested that such deception led to direct selection on females for self-deception about personal fertility. They suggested that concealed fertility in women is an EPC adaptation that functions to obtain a sire of high genetic quality without the primary mate's detection. They further suggested that the self-deception allows cryptic pursuit of EPC in which controlled conception is affected by the proximate factors of selective sperm transport (e.g., by the occurrence of copulatory orgasm) or induced ovulation. Research has verified the potential of female orgasm in selective sperm retention (Baker and Bellis 1995; also see below).

Additional studies are needed to better understand men's shifts in mate guarding across their partner's menstrual cycle. Gangestad et al.'s (2002) findings involved only 31 women with a primary partner. Moreover, these findings were based on women's reports of the partner's mate guarding, not on the partner's report. However, we have found separately in a study of 201 couples that romantically involved men and women's independent reports of the man's mate guarding show a moderate positive correlation (about 0.5; see Gangestad and Thornhill 1997a, b).

Further Evidence of Design in Women's EPC: What They Say Is Desired

Greiling and Buss (2000) studied what women state they want or would want in a one-night stand outside their main sexual relationship. The one-night stand is such a brief relationship that women ordinarily can receive few or no long-term material benefits. The women rated their wants on a 1–9 scale, with 9 being most desirable. Compared to what women want from their main partner (Buss and Schmidt 1993), these women's desires about the extra-pair partner focused on his sexiness, sensuality, physical attractiveness, and high desirability to the opposite sex with ratings above an average of 8.0 on each variable. Women's evaluation that they focus on or would focus on sexiness and male physical attractiveness during EPC is manifested in their actual EPC behavior because women seem to prefer symmetric men (attractive to many women, see below) as EPC partners (Gangestad and Thornhill 1997b). Given women's interest in sexy and sensuous EPC partners, it is not surprising that many women say that sexual gratification is an EPC motive. In particular, women's copulatory orgasms appear to be important as women rate them as more of a potential asset in extra-pair sex than merely experiencing sexual gratification (Greiling and Buss 2000).

In addition to wanting to have extra-pair sex that includes a climactic zenith with a physically attractive, sexy, sensuous man who is highly desired by many women, which is consistent with EPCing to get a sire with good genes, women claim that other desires influence their EPC behavior: obtaining a male protector or backup long-term mate, testing the potential for a new long-term relationship, and evaluation of personal mate value (Greiling and Buss 2000). All these motives could be functional designs of EPC adaptation in women.

Further Evidence: The Timing of EPC

If the benefits and costs of specific cognitive, emotional, and overt behavioral responses varied across contexts ancestrally, selection is expected to have acted to shape psychological adaptations that produce responses contingent on context – that is, these psychological adaptations will give condition-dependent outputs. Such adaptation is evidenced in the condition-dependent expression of men's mate guarding, discussed above. Suppose ancestral females could have secured a better genetic complement for their offspring from extra-pair males at the expense of the RS of their main partners, who impose a cost on such female behavior when they observe it (e.g., by divestment in the female or her offspring). Because women obtain genetic benefits from mating only when at fertile times in the menstrual cycle, but could pay the costs of EPC throughout the cycle, selection is expected to have shaped female interest in and pursuit of men who possess indicator traits of high genetic quality such that the interest and pursuit changes across the cycle to maximize benefits and minimize costs.

Consistent with this is the menstrual-cycle shift in the frequency of women's EPC behavior found in a large survey of UK women. The women's EPC (6% of copulations reported by women with primary pair-bond partners) occurred more often on high-fertility days, whereas the frequency of in-pair sex was more evenly distributed across the cycle (Bellis and Baker 1990). Indeed, in these women, the peak rate of EPCs (occurring about the tenth day of the cycle) is about 2.5 times the minimal rate of EPCs (occurring the last week of the cycle; Baker and Bellis 1995).

Also, a menstrual-cycle shift in women's EPC is suggested by Gangestad et al.'s (2002) research on men's mate guarding discussed above. Men appear to respond to peak conception risk in a partner by increasing mate guarding, especially men with polyandrous mates who report more sexual attraction to and fantasy about other men. Also suggestive is the finding that women with a primary partner are most likely to attend a singles-scene nightclub without the mate when at peak fertility in the menstrual cycle (K. Grammer, unpubl. data).

Strong evidence that functionally specific psychological adaptation in women affects the motivation underlying this pattern of women's greater pursuit of EPCs at fertile times in the cycle comes from Gangestad et al.'s (2002) study. Research participants were 51 young women not using hormone-based contraceptives. They filled out questionnaires about their sexual interests and fantasies at two points in their menstrual cycle: once within 5 days before a luteinizing hormone surge, which is a period of high fertility with the surge corresponding roughly to the peak (Jöchle 1973; Baker and Bellis 1995; Wilcox et al. 1995; Thornhill and Gangestad 1999a), and once during the luteal phase (very low conception risk). There were three salient findings: first, on average, the effect of fertility status on women's sexual attraction to and fantasy about *primary* partners was small and statistically insignificant. Second, the same women's attraction to and fantasy about men other than primary partners increased substantially during the fertile phase of the cycle. Third, the second effect showed a significant increase in strength over the course of days after menses until the time of peak fertility.

The timing of women's interest in and fantasy about nonpartner men as well as the timing of their actual EPC behavior is evidence that women have a psychological adaptation that assesses circumstances and motivates EPC when genetic benefits can be realized (at fertile cycle times).

Further Evidence: Symmetry and Extra-Pair Copulation

Also suggesting that EPC by women has been directly selected because it resulted in genetic benefits to offspring is evidence that women choose as EPC partners men who evidence developmental stability, by virtue of body bilateral symmetry. Given that symmetric men invest less in their romantic relationships than asymmetric men, these choices are probably not for increased male investment (Gangestad and Thornhill 1997a, b).

In humans, symmetry is probably involved in good-genes sexual selection. There is evidence that an expression of a preference for a symmetric male would tend to result in a female obtaining a mate with relatively few mutations that disrupt development, with relatively many genes associated with accrual and efficient use of resources that enhance development, and with immune-system competence (Møller and Swaddle 1997; Thornhill and Møller 1997; Gangestad and Thornhill 1999). In both sexes, relatively low asymmetry seems to be associated with increased genetic, physical and mental health, including cognitive skill and IQ (Furlow et al. 1997; Thornhill and Møller 1997; Yeo et al. 2000; Thoma et al. 2002). Also, symmetric men appear to be more muscular and vigorous (Gangestad and Thornhill 1997a), have a lower basal metabolic rate (Manning et al. 1997), and may be larger in body size (Manning 1995; Gangestad and Thornhill

1997a; but see Simpson et al. 1999) than asymmetric men. Symmetric men also have expressions of facial sex hormone (testosterone) markers that may indicate health in terms of reproductive hormones (Gangestad and Thornhill 2003). Symmetry is a major component of developmental health and overall condition and appears to be heritable (Gangestad and Thornhill 1999; see also Van Dongen 2000; Fuller and Houle 2003), as expected given that symmetry is condition-dependent and thus will capture the inevitable heritability of condition (Rowe and Houle 1996).

Beyond women's choice of symmetric men as EPC partners, women show other sexual preferences for symmetric men. In men, low asymmetry has been shown to predict components of mating success such as a relatively high number of sexual partners, earlier age of first sex, quicker sexual access to a new romantic partner (Thornhill and Gangestad 1994, 1999b; Baker 1997; Gangestad and Thornhill 1997a), facial attractiveness (Thornhill and Gangestad 1994, 1999b; Baker 1997; but see also Gangestad and Thornhill 1997b), and number of EPC partners (Gangestad and Thornhill 1997b). Moreover, men's symmetry predicts a relatively high frequency of their sexual partners' copulatory orgasms (Thornhill et al. 1995). Shackelford et al. (1999) also found higher rates of reported female copulatory orgasm in women paired to relatively attractive men compared to women paired to less attractive men when relationship satisfaction was statistically controlled.

Female copulatory orgasm may be a mechanism of cryptic female choice in contexts in which women mate with multiple partners (see Baker and Bellis 1995; Thornhill et al. 1995). Baker and Bellis (1995) provide evidence that female copulatory orgasm may be involved in selective sperm retention, and it may be involved in selective bonding as well (Thornhill et al. 1995). Either of these two selective mechanisms, or both, may give an advantage to the sperm of relatively symmetric men over that of asymmetric men when women have multiple mates. Women's pattern of orgasm appears to favor EPC partner's sperm over in-pair partner's sperm, according to a study of UK women by Baker and Bellis (1995). Women's preference for symmetric EPC partners, coupled with their greater likelihood of sperm-retaining copulatory orgasms with symmetric men, could lead to symmetric men's ejaculates having disproportionately higher fertilization success during women's EPC.

That women having ovulatory menstrual cycles (not using hormonal contraceptives) prefer the body scent of relatively symmetric men, but only during the fertile times of their menstrual cycles, has been found in four separate studies (Gangestad and Thornhill 1998; Rikowski and Grammer 1999; Thornhill and Gangestad 1999a; Thornhill et al. 2003). This effect was not seen across the cycles of women using hormonal contraceptives. Nor do men exhibit a preference for the scent of symmetric women. The research on the scent of men's symmetry indicates that fertile women's olfactory assessment changes across the cycle such that symmetric men may be favored as offspring's sires.

Female preference for the facial features of men with symmetric bodies also changes across the menstrual cycle. University of New Mexico men ($N = 12$) and women ($N = 60$) rated facial photos of 65 men from a remote village on the island of Dominica (West Indies) on a 10-point scale of physical attractiveness. Nine body features on each of the Dominican men were measured by Gangestad and the features' asymmetries were relativized for feature size and then summed into a composite asymmetry index (as in prior studies; see Gangestad and Thornhill 1999 for methods). Gangestad's measures are strongly correlated with measures made by a villager who was totally unfamiliar with the hypothesis that symmetric men are more sexually attractive to women than asymmetric men. Each sex of raters found the symmetric men's faces to be more attractive than the faces of asymmetric men (male raters, r with target's age partialed out $= 0.25$, $0.02 < p < 0.05$; female raters, $r = 0.24$, $0.05 < p < 0.1$). Among women who were not using hormone contraceptives, the strength of preference for symmetric men is significantly positively correlated with their risk of conception based on menstrual cycle point (partial $r = 0.38$, $p < 0.01$, $N = 45$; Fig. 1).

Ovulating women's menstrual-cycle shift from no or weak preference at low fertility for male pheromonal and facial markers of body symmetry to a relatively strong preference for these markers at high fertility suggests a psychological adaptation functionally designed to motivate obtaining a sire of high genetic quality. Since most reproductive-age women in traditional societies (e.g., hunter–gatherers) are pair bonded most of the time,

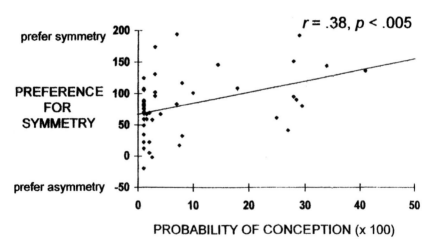

Fig. 1. The relationship between University of New Mexico women's facial attractiveness preference for Dominican men with symmetric bodies and the women's risk of conception across the menstrual cycle ($r = 38$, $p < 0.005$). Each point is the regression coefficient of a women's physical attractiveness ratings of the men's facial pictures regressed on the men's body asymmetry. Conception risk is based on actual data (Jöchle 1973). There were 65 men and 45 women (nonpill users) raters

this psychology is likely to have evolved in the context of adult females typically being pair-bonded. Thus, nonpaired university women's change in preference across the menstrual cycle may importantly reflect motivation pertaining to EPC adaptation.

Additional evidence that women have EPC adaptation, specifically adaptation that evaluates benefits and costs of EPC and motivates EPC when benefits are high and costs are low, would be demonstrated if women with low-mate-value main partners show a greater shift in preference for indicator traits of symmetry at the high-conception-risk cycle point than women who are in a pair-bond with a man of high mate value. Each of the major categories of men's mate value – physical attractiveness, including body scent, and possession of resources or status – are predicted to influence the extent to which a woman shifts preference across the cycle.

Further Evidence: Hormone Markers and Extra-Pair Copulation

Women say they favor men with physical strength and physical attractiveness for short-term relationships (Buss and Schmitt 1993). Similar criteria, including sexy, are prioritized according to women's judgments when they pursue or might pursue a one-night stand outside a main relationship (Greiling and Buss 2000). Physical strength, sexiness, and physical attractiveness in men are due, in part, to testosterone effects during a male's development, apparently especially adolescence and puberty (Thornhill and Gangestad 1999b). However, do women actually favor more testosteronized men for EPC?

Indicating that this is the case are studies finding that female preference for male hormone markers changes across the menstrual cycle in women who are not using hormone-based contraception. Such women prefer more masculine male facial features at fertile times in the cycle than at infertile cycle times. When infertile, such women prefer less masculinized or even slightly feminized male faces. Women using the contraceptive pill did not show this preference shift (Penton-Voak et al. 1999; Penton-Voak and Perrett 2000; Johnston et al. 2001). Also, ovulating women's (nonpill users) judgments of male faces across the cycle change to an attractiveness preference for more masculine at high-conception-risk times, but their judgments of male faces that receive attributions of "good dad," "dominant," "intelligent," "masculine," "average", and "androgenous" do not change across the cycle. This result indicates that only women's attractiveness judgments of men and not their other judgments of men change across the cycle, suggesting that fertile women's preference for more masculine faces is not the result of a general change in sensitivity or of enhanced ability to detect facial features in general at high-conception-risk times (Franklin and Johnston 2000; Johnston et al. 2001). These studies use computer graphics to manipulate the masculinity or femininity of a composite male face by exaggerating or reducing the shape differences between male and female average faces, thereby manipulating the sexually dimorphic features affected by testosterone and estrogen.

Further evidence that women's responses to male faces change across the menstrual cycle was found by Oliver-Rodriguez et al. (1999). They found that the magnitude of the P300 response of the evoked potential (a positive potential around 300 ms following presentation of a stimulus, which covaries with the emotional salience of the stimulus) of women in the fertile cycle phase correlated with their rating of male facial attractiveness, but not their ratings of female facial beauty. During the infertile phase, women's responses were undifferentiated and covaried with both male and female attractiveness judgments.

Men's facial masculinity is affected by androgen production during development, which, like symmetry, may be a signal of superior condition during development (Thornhill and Gangestad 1993, 1999b; Thornhill and Møller 1997). Indeed, male facial masculinity positively covaries with body symmetry and, hence, these traits tap a common underlying factor (Gangestad and Thornhill 2003). Our working hypothesis is that, ancestrally, this factor related to overall condition and heritable fitness and, accordingly, females express preferences for symmetry and masculinity.

The change in women's preference for male hormone markers across the cycle may be related to women having evolved to seek EPCs for genetic benefit rather than material benefits. First, Penton-Voak et al. (1999) found that individual women's preference for facial masculinity increased from the low-conception-risk cycle time to the high-conception-risk cycle time when judging men as potential short-term mates, but not when judging them as long-term mates. Second, these authors also reported the trend in their data ($0.05 < p < 0.1$), suggesting that women in pair-bond relationships underwent a greater shift in preference for facial masculinity than women without partners. Additional research is needed to determine if women paired to relatively low-mate-value men exhibit a greater cyclic change toward increased preference of more masculine faces at high conception risk. Third, relatively high facial masculinity in men is perceived by people as evidencing a man who will invest less in a mateship than a man with lower facial masculinity (Perrett et al. 1998; Johnston et al. 2001), and there is evidence that possibly related attributions (e.g., how agreeable or warm a man is) are accurate (Berry and Wero 1993; Graziano et al. 1997).

Women's preference for men with less masculinized or slightly feminized faces during the infertile period may function to secure certain material benefits from men who are willing to invest as a result of their relatively low genetic quality. Fertile women's preference for symmetrical men and men with highly masculine features indicate that women are willing to trade off between physical attractiveness (and thus heritable benefits from mate choice) and material benefits (and willingness to provide those benefits) in mate choice.

Further Evidence: Other Conditional Factors

Women's fertility variation across the cycle is a condition that affects shifts in their visual and olfactory mate preferences. Another conditional factor is whether women are pursuing long-term relationships or EPCs (one-night stands or other short-term sexual relationships). These two conditional factors combine to yield a shift at fertile cycle times that focuses on male physical features of symmetry, hormone markers and attractiveness in EPCs and short-term, sexual relations in general. This shift is consistent with the existence of female sexual adaptation that gives responses that ancestrally were adaptive by producing offspring of high genetic quality as a result of mating with men other than the main partner.

Another conditional factor suggesting EPC adaptation is association of women's EPC with romantic attachment style. Gangestad and Thornhill (1997b) found in a study of 201 romantically involved couples that a woman's attachment style is not related to the number of times she has been an EPC partner. Yet, women's number of EPC partners is related to their anxious and avoidant attachment styles. Whereas women's anxious attachment was positively associated with their number of EPC partners, women's avoidant attachment was negatively associated with their number of EPC partners. These effects persisted even when the total number of EPC partners was statistically controlled. It appears, then that women most likely to engage in EPCs are those who are open to intimacy (nonavoidant), yet fearful of abandonment (anxious). Possibly, these women have been exposed during their development to cues of noninvestment by men (cf. Belsky et al. 1991; Cashdan 1993), which lead them to engage in EPCs as hedges against mate abandonment and loss of mate's resources, a perception that may reflect an ancestral, positive correlation between being reared in an environment of noninvestment by males and likelihood of mate abandonment. The fact that these same women were not more likely to be the EPC partners of men is consistent with this scenario. That is, it is not apparent how *being* an EPC partner, as opposed to *having* an EPC partner, would hedge against a mate's abandonment.

A final condition is the association of women's EPC with socioeconomic level (SES). Low-SES women appear to have a higher rate of EPC than higher SES women. This is indicated by the data on cuckoldry across SES levels that we cited above. This pattern may be driven partly by motivation to secure resources from multiple men in settings in which most men lack sufficient resources to satisfy a woman's ancestral perception of the necessary resource level for successful reproduction. Alternatively, the pattern may reflect increased emphasis on female choice for sires with good genes when most men lack the material goods that can serve as a basis for resource-based mate choice. Future research might examine the cyclic preference changes of ovulating women of different SES levels. A pattern suggesting that women possess EPC adaptation that is sensitive to need for material resources from an EPC partner would be that low-SES women show little or no shift toward preference for masculine faces at fertile cycle times compared to higher SES women. However, if low-SES women experience less cost in terms of resources lost if their EPC is detected by their main partner, they may perceive EPC for genetic quality of high benefit in general.

Overview

Women's EPC is a fundamental behavioral component of the human mating system. That women's EPC may be a threat universally to men's paternity confidence is evidenced in men's sexual jealousy across all societies. This evidence ranges from men's emotion of jealousy itself to a rich diversity of cultural institutions, such as female sexual modesty and claustration (Dickemann 1981; Thornhill and Thornhill 1987; Wilson and Daly 1992). We have emphasized that the widespread occurrence of women's EPC is not evidence for EPC adaptation. Men's nipples and the human bellybutton also are widespread, but both are incidental effects, not adaptations. Men's sexual proprietariness is the phenotypic stamp of female EPC being a powerful and effective selective force on males in human evolutionary history, yet this evidence that females engaged in EPC during human evolutionary history is not evidence that woman possess EPC adaptation. Similarly, neither the widespread occurrence of EPC in birds and even in gibbons, which, like humans, are a great ape, nor the evidence that females of some species of pair-bonding birds have EPC adaptation is evidence that women have EPC adaptation.

The widespread occurrence of women's EPC now and historically, given its potentially high costs stemming from lowering paternity reliability of the main partner, leads to the theoretical expectation that past selection will have fashioned EPC adaptation with design that gives EPC behavior when its benefits exceed its costs to reproductive success of females in ancestral settings. This hypothesis about the expected past selection for historically adaptive female EPC predicts that women will possess EPC adaptation. Support for the hypothesized historical selection is evidence of a feature(s) in women that has specialized function for EPC. If functional specialization for EPC is not present in women, then EPC must be a by-product, an evolutionary outcome of indirect selection. Given the frequency of EPC in women, which rules out mutation alone as EPC's ultimate cause, and its complexity and other costs, which rules out drift as ultimate cause, women's EPC must ultimately be explained by direct or indirect selection. Functional specialization rules out by-product hypotheses (and drift and mutation alone) because functional specialization is a coordination between a trait's organization and achieved function. It is very unlikely that such coordination will occur by chance. Standard statistical procedures test whether hypothesized coordination is real, i.e., not explicable by chance.

One obvious incidental effect hypothesis for women's EPC is that it occurs as a by-product of women's sexual desire directly selected because the desire achieved mating with a long-term mate(s). This is soundly questioned by evidence that women's expressed desires for traits of short-term extra-pair partners focus on physical attractiveness, sexiness and sensuality, whereas women's expressed desires for more committal relationships target, to a much greater extent, a man's resources and status (Buss and Schmitt 1993; Buss 2000; Greiling and Buss 2000). Some of this variation in desire reflects stable individual differences in preference for short- vs. long-term partners (Gangestad and Simpson 1990), which may, in part, reflect developmentally triggered, strategic attachment styles (Belsky et al. 1991; Thornhill and Furlow 1998; Chisholm 1999), but much of it is due to conditional shifts within individual women (Greiling and Buss 2000). Thus, there is evidence that many women have different criteria when pursuing EPCs compared to long-term relationships and that female eroticism in response to male physical features is prioritized in EPC pursuits. Given the importance to women of male-provided resources, a facultative focus in women's EPC on male physicality, which may mark, in part, genetic quality, and on the female's eroticism and coital orgasm, which may retain the ejaculate of the EPC partner, provide evidence for EPC adaptation in women that is designed to obtain a sire of high genetic quality for offspring.

The nature of women's libidinous changes as reflected in their fantasy about and attraction to nonpartner men across the menstrual cycle also indicates that women's EPC is not an incidental effect of their desire for mating with their long-term partner (Gangestad et al. 2002). Women do not report a significant increase in sexual attraction to and fantasy about their primary partner when near maximum conception risk in the menstrual cycle. In general, however, women initiate more sex with their main partner at mid-cycle, though this pattern is not related to women's increased interest in and fantasy about men other than a current partner. Women's sexual attraction to and fantasy about their primary partner is marginally significantly related to the increased frequency of sex they initiated with him. Their sexual fantasy about and attraction to men other than the partner did not significantly predict sex initiated with their partner. From the EPC-adaptation hypothesis, one would predict that women who have a primary partner who possesses the traits that women desire when highly fertile (e.g., facial masculinity or body symmetry) to be more attracted to their partners at peak fertility in the menstrual cycle. Possibly, these women are responsible for the effect of fertility status on female-initiated sex. This issue should be explored in future research, which will require simultaneous assessment of women and their partners.

Thus, women's increased sexual interest in nonpartner men at mid-cycle does not seem to be accompanied, in general, by an expression of lust toward the primary partner. If this is confirmed in future research, it will seriously question the hypothesis that women's EPC is a by-product of a generally heightened libido that assures insemination by a main partner.

Another by-product hypothesis for women's EPC is that it occurs as a side-effect of men's sexually selected adaptation for obtaining many mates. According to this hypothesis, women engage in EPC for the same reason men do and the sexes have similar sexual psychological adaptation affecting EPC. Symons (1979) and Geary (1998) review the many lines of evidence demonstrating that women are not designed to seek partner variety per se. Moreover, women's changes in olfactory and visual mate preferences across the menstrual cycle and their actual EPC behavior of preferring symmetric EPC partners and enjoying high personal sexual arousal with these partners indicate that EPC in women is associated with a high degree of mate choice and apparently sire choice as well. These preferences and behaviors are inconsistent with women simply pursuing a variety of sex partners through EPC.

The expression of a high degree of female choice during women's EPC also is hard to reconcile with Hrdy's (1999) hypothesis that female primates' behavior of mating with multiple males functions to give many males in the social group a probability of paternity, thereby reducing the likelihood that subsequent offspring will be injured or killed by males. Another potential material benefit gained by EPCing females is food or protection for offspring from males with some paternity prospect. The hypothesis that women's EPC functions to give multiple males paternity as a strategy of getting male material benefits is an alternative to EPC functioning to obtain a sire of high genetic quality – that is, the two hypotheses envision different effective historical selection. Hrdy's hypothesis applied to women's EPC predicts either that women will pursue more EPCs at infertile than at fertile cycle points or that they pursue EPCs equally across the cycle. The former prediction assumes that females desire to give the main partner the highest paternity reliability. The latter prediction of uniform across-cycle pursuit of EPCs is expected under the assumption that sire genetic quality is unimportant in human female choice. Apparently, these predictions are not met and instead women's EPC is concentrated at fertile cycle times. This concentration, coupled with the male physical features that EPCing women value, implies that EPC functions to gain good genes for offspring. The prediction that EPCing women will target primarily more testosteronized or symmetric men to reduce such men's aggression toward offspring is a prediction of Hrdy's hypothesis. However, such men as EPC targets of *fertile* women is not. Such men as EPC targets of fertile women, coupled with women's secrecy and their partner's increased mate guarding at fertile times, implies that a sire of high genetic quality is an important goal of women's EPC.

Research to date indicates that women have EPC adaptation. This evidence is of the form of statistically significant coordinations between, on the one hand, certain aspects of women's sexual behavior and psychology and, on the other, high benefit and low cost EPC. One piece of this evidence is that women show preferences for potential markers of male genetic quality primarily or only when they are most likely to conceive, which may function to maximize the net benefit of EPC (benefit of getting good genes minus cost of lost investment if detected). Numerous studies reveal that women who are having ovulatory cycles and are at peak risk of conception prefer the scent of male symmetry, facial testosterone features, and possibly other facial traits correlated with body symmetry. Additional evidence of women's EPC adaptation is that women seem to have motivation to EPC secretly, which lowers the cost of EPC pertaining to its detection by the pair-bond mate. Women's fertility concealment in the menstrual cycle may be part of directly selected EPC adaptation for secret pursuit of a sire with good genes. Another piece of evidence is women's desire for EPC partners who are sexy, physically attractive and physically desired by many women, and women tend to show orgasmic patterns conducive to high sperm retention with EPC partners. The evidence that male physical attractiveness is a certification of hormonal and developmental health and overall condition (Thornhill and Møller 1997; Thornhill and Gangestad 1999b), when coupled with the likelihood that health and general condition are usually heritable (and thus were often heritable in human evolutionary historical generations), implies that women's EPC, at least in part, is designed to obtain sires of high genetic quality for offspring. This genetic benefit interpretation is reinforced by the evidence that men of high physical attractiveness and symmetry tend to invest less in romantic relationships than physically unattractive/asymmetric men.

This is not to say that women do not often seek material or direct benefits by EPCs. They apparently do. Hill and Hurtado (1996), in a study of Ache Indians, and Beckerman et al. (1998), in a study of the Barí, provide evidence that extramarital copulations by women enhance offspring survival, possibly as a result of the greater parental investment from multiple "fathers." Yet female preference for male-provided resources does not negate an important role for good genes sexual selection when fitness is heritable. Considerable genetic variation in fitness is seen in most natural populations (Burt 1996; Kirkpatrick 1996; Gangestad 1997).

It is conceivable that women's EPC functions to secure mates who are physical protectors of women instead of sires with good genes. Thornhill and Palmer (2000; also Smuts and Smuts 1993) have summarized ethnographic data indicating that protection by a pair-bond mate may reduce sexual coercion by other males. Our prior studies have shown that although symmetric men invest less with respect to certain types of investment, they perceive themselves and are perceived by their romantic partners as better protectors of the female partner (Gangestad and Thornhill 1997a). In addition, symmetric men engage in more fights with other men than asymmetric men (Furlow et al. 1998). Protection of the female is a form of male investment that least interferes with a man's access to partners other than his pair-bond mate because his protective ability is attractive to women in general and, compared to investment in the form of time, honesty, and sexual exclusivity, competes with male pursuit of additional partners to a lesser degree. Thus, female preference for male protection is not necessarily an explanation that is exclusive of good-genes mate choice. Symmetric men and highly testosteronized men may have more protection ability and better genes for offspring fitness. In addition, the physical protection benefit would be gained by females at all cycle times, but women show the preference for symmetry and testosterone-based features of males primarily at high-conception-risk cycle points. If protection is the primary or only motivation for women's EPC, then unpaired and thus unprotected women should exhibit the strongest shift at mid-cycle in preference for symmetry and testosterone markers, women paired to low-mate-value men should show intermediate shifts, and women paired to high-mate-value men should show the least preference change.

It is also conceivable that symmetric men and testosteronized men may be preferred by women at mid-cycle because of a preference for adequate sperm. Manning et al. (1998; also Baker 1997) have shown a positive correlation between ejaculate size, sperm quality and body symmetry in men. This, too, however, is not necessarily an alternative to good genes.

One possible good-genes mechanism is that symmetrical men may have particularly rare MHC (major histocompatibility) genotypes, making them attractive to many fertile women. MHC loci are involved in recognizing antigens and thereby initiating the immune response. It has been proposed by several workers that heterozygosity at MHC may defend against parasites (see Penn and Potts 1998 for review; also Wedekind et al. 1995; Wedekind and Füri 1997). Wedekind et al. (1995) and Wedekind and Füri (1997) found that normally ovulating women, but not women on the pill, prefer the scent of men who possess dissimilar MHC genotypes, and hence, men with rare MHC genotypes are expected to be attractive to relatively many women. If having many rare MHC alleles is associated with high parasite resistance, then symmetric men might have more rare MHC alleles (see Thornhill and Møller 1997 for review of human health, including infectious disease, in relation to symmetry). The MHC effect, however, works in both directions in that both sexes prefer the scent of MHC-dissimilar individuals (Ober et al. 1997; Wedekind and Füri 1997). Our results indicate that only women have the symmetry-scent preference (Gangestad and Thornhill 1998; Thornhill and Gangestad 1999a, Thornhill et al. 2003). Furthermore, the preference of women who are not using hormonal contraceptives for MHC-dissimilar men appears to be unrelated to their relative fertility across the menstrual cycle (Thornhill et al. 2003). This result suggests that the scent of men's symmetry and the scent of MHC-dissimilarity may be different pheromonal systems and that aspects of the genome other than MHC are responsible for the symmetry–scent preference. The symmetry–scent preference may reflect assessment of potential sires in terms of overall genetic quality pertaining to viability.

Additional studies could clarify the design of EPC indicated by research to date and test for some other design features of women's EPC. More research is needed to examine specifically whether women's motivation in EPC is guided by costs and benefits stemming from each woman's personal mateship status, which will require study of women and their main partner, as well as women's reports about their EPC partners. A critical prediction from the hypothesis that women's EPC functions to get good genes for offspring is that women with a main partner who is relatively low in physical mate value will exhibit, when at peak conception risk, stronger attraction to and fantasy about nonpartner men with physical markers of high genetic quality than women with a primary partner of high physical mate value. Additional research also is needed to explore the apparent connection in women between romantic relationship attachment style and EPC and between women's SES and EPC. Finally, additional research could determine if women's EPC reflects psychological EPC adaptation that functions to secure material benefits from EPC partners.

Summary

The adaptationist research program figuratively carves the organism's phenotype to separate empirically its adaptations from its by-products and traits in mutation–selection balance. The standard of evidence in identifying adaptation is showing fit or coordination between a trait and a problem. When the goodness of fit is sufficient to rule out chance association, one has evidence for adaptation. This paper critically reviews research on women's extra-pair copulation (EPC) and asks whether the standard of evidence for identifying EPC adaptation in women is met. We conclude that the evidence for EPC adaptation in women is considerable. This evidence ranges from (1) women's expressed desires about traits of EPC partners; (2) women's EPC behavior itself; (3) changes in women's preference for testosterone markers and certain other facial features of men across the menstrual cycle; (4) women's menstrual cycle shifts in olfactory preference for men's scent associated with developmental stability; (5) women's reports about their fantasy and attraction to nonpartner men across the cycle; (6) men's mate guarding across their partner's menstrual cycle; and (7) findings indicating that there was an evolutionary historical inter-sexual arms race in the context of female mateship infidelity, which implies relevant adaptation in both sexes. Women's EPC does not seem explicable as an incidental effect of other adaptation. A number of incidental effect hypotheses are considered. We also identify avenues of future research that may be useful additional tests of the hypothesis that women have EPC adaptation. A central issue in the paper is the method by which functional design can be demonstrated. Women's EPC behavior involves olfactory and visual judgments, salient areas of aesthetic valuation.

Acknowledgments. The paper benefited from criticisms by the participants in the Darwinian aesthetics workshop held at the Konrad Lorenz Institute, especially K. Bayertz, R. Coss, M. Cunningham, I. Eibl-Eibesfeldt, K. Grammer, and S. Mithen, and the participants in the evolutionary problems discussion group at the University of New Mexico, especially P. Andrews, C. Fincher, M. Franklin, P. Keil, and P. Watson.

References

Anderson KG, Kaplan H, Lancaster J (1999) Paternal care by genetic fathers and stepfathers I: reports from Albuquerque men. Evol Human Behav 20:405-431

Baker RR (1997) Copulation, masturbation and infidelity: state-of-the art. In: Schmitt A, Atzwanger A, Grammer K, Schafer K (eds) New aspects of human ethology. Plenum Press, New York, pp 163-188

Baker RR, Bellis MA (1995) Human sperm competition: copulation, masturbation and infidelity. Chapman and Hall, London

Beckerman S, Lizarralde R, Ballew C, Schroeder S, Fingelton E, Garrison A, Smith H (1998) The Bari partible paternity project: preliminary results. Curr Anthropol 39:164-167

Bellis MA, Baker RR (1990) Do females promote sperm competition? Data for humans. Anim Behav 40:997-999

Belsky J, Steinberg L, Draper P (1991) Childhood experience interpersonal development and reproductive strategy: an evolutionary theory of socialization. Child Dev 62:647-670

Benshoof L, Thornhill R (1979) The evolution of monogamy and loss of estrus in humans. J Soc Biol Struct 2:95-106

Berry DS, Wero JL (1993) Accuracy of face perception: a view from ecological psychology. J Personality 61:497-503

Birkhead TR, Møller AP (1995) Extrapair copulation and extra-pair paternity in birds. Anim Behav 49:843-848

Bjorklund DF, Kipp K (1996) Parental investment theory and gender differences in the evolution of inhibition mechanisms. Psychol Bull 120:163-188

Burch RL, Gallup GL Jr (2000) Perceptions of paternal resemblance predict family violence. Evol Human Behav 21:429-435

Burt A (1996) Perspective: the evolution of fitness. Evolution 49:1-8

Buss DM (2000) Dangerous passions. Free, New York

Buss DM, Schmitt DP (1993) Sexual strategies theory: a contextual evolutionary analysis of human mating. Psychol Rev 100:204-232

Buss DM, Shackelford TK, Kirkpatrick LA, Choe JC, Hasegawa M, Hasegawa T, Bennett K (1999) Jealousy and the nature of beliefs about infidelity: tests of competing hypotheses about sex differences in the United States, Korea and Japan. Personal Relationships 6:125–150

Cashdan E (1993) Attracting mates: effects of paternal investment on mate attraction strategies. Ethol Sociobiol 14:1–24

Cerda-Flores RM, Barton SA, Marty-Gonzalez LF, Rivas F, Chakraborty R (1999) Estimation of nonpaternity in the Mexican population of Nueveo Leon: a validation study with blood group markers. Am J Phys Anthropol 109:281–293

Chisholm JS (1999) Attachment and time preference: relations between early stress and sexual behavior in a sample of American university women. Human Nat 10:51–83

Daly M, Wilson M (1995) Discriminative parental solicitude and the relevance of evolutionary models to the analysis of motivational systems. In: Gazziniga MS (ed) The cognitive neurosciences. MIT Press, Cambridge, MA, pp 1269–1286

Dickemann M (1981) Paternal confidence and dowry competition: a biocultural analysis of purdah. In: Alexander R, Tinkle D (eds) Natural selection and social behavior. Chiron, New York

Figueredo AJ, McCloskey, LA (1993) Sex money and paternity: the evolutionary psychology of domestic violence. Ethol Sociobiol 14:353–379

Flinn M (1987) Mate guarding in a Caribbean village. Ethol Sociobiol 8:1–28

Franklin M, Johnston VS (2000) Abstract of paper presented at 2000 meeting of Evolution and Human Behavior conference. Amherst College, Amherst, MA

Fuller RC, Houle D (2003) Inheritance of developmental instability. In: Polak M (ed) Developmental instability: causes and consequences. Oxford University Press, New York (in press)

Furlow BF, Armijo-Prewitt T, Gangestad SW, Thornhill R (1997) Fluctuating asymmetry and psychometric intelligence. Proc R Soc Lond Ser B 264:823–829

Furlow B, Gangestad SW, Armijo-Prewitt T (1998) Developmental stability and human violence. Proc R Soc Lond Ser B 266:1–6

Gangestad SW (1997) Evolutionary psychology and genetic variation: non-adaptive fitness-related and adaptive. In: Bock GR, Cardew G (eds) Characterizing human psychological adaptations. Ciba Foundation Symposium 208. Wiley, New York, pp 212–223

Gangestad SW, Simpson JA (1990) Toward an evolutionary history of female sociosexual variation. J Personality 58:69–96

Gangestad SW, Thornhill R (1997a) Human sexual selection and developmental stability. In: Simpson JA, Kenrick, DT (eds) Evolutionary social psychology. Lawrence Erlbaum, Mahwah, NJ, pp 169–195

Gangestad SW, Thornhill R (1997b) The evolutionary psychology of extrapair sex: the role of fluctuating asymmetry. Evol Human Behav 18:69–88

Gangestad SW, Thornhill R (1998) Menstrual cycle variation in women's preferences for the scent of symmetrical men. Proc R Soc Lond Ser B 265:927–933

Gangestad SW, Thornhill R (1999) Individual differences in developmental precision and fluctuating asymmetry: a model and its implications. J Evol Biol 12:402–416

Gangestad SW, Thornhill R (2003) Facial masculinity and fluctuating asymmetry. Evol Human Behav (in press)

Gangestad SW, Thornhill R, Garver C (2002) Changes in women's sexual interests and their partners' mate retention tactics across the menstrual cycle: evidence for shifting conflicts of interest. Proc R Soc Lond Ser B 269:975–982

Geary DC (1998) Male, female: The evolution of human sex differences. American Psychological Association, Washington, DC

Graziano WG, Jensen-Campbell LA, Todd M, Finch JF (1997) Interpersonal attraction from an evolutionary perspective: women's reactions to dominant and prosocial men. In: Simpson JA, Kenrick DT (eds) Evolutionary social psychology. Erlbaum, Mahwah, NJ, pp 169-195

Greiling H, Buss DM (2000) Women's sexual strategies: the hidden dimension of short-term extra-pair mating. Personality Individual Diff 28:929-963

Hansen GL (1986) Extradyadic relations during courtship. J Sex Res 22:382-390

Hill K, Hurtado AM (1996) Ache life history. de Gruyter, New York

Hite S (1987) Women and love: a cultural revolution in progress. Grove Press, New York

Hrdy SB (1981) The woman that never evolved. Harvard University Press, Cambridge

Hrdy SB (1999) Mother nature: a history of mothers, infants and natural selection. Pantheon, New York

Jennions MD, Petrie M (2000) Why do females mate multiply? A review of the genetic benefits. Biol Rev 75:21-64

Jöchle W (1973) Coitus-induced ovulation. Contraception 7:523-564

Johnston VS, Hagel R, Franklin M, Fink B, Grammer K (2001) Male facial attractiveness: evidence for hormone-mediated adaptive design. Evol Human Behav 22:251-267

Kinsey AC, Pomeroy WB, Martin CE, Gebhard PH (1953) Sexual behavior in the human female. WB Saunders, Philadelphia

Kirkpatrick M (1996) Good genes and direct selection in the evolution of mating preferences. Evolution 50:2125-2140

Laumann EO, Gagnon JH, Michael RT, Michaels S (1994) The social organization of sexuality. University of Chicago Press, Chicago

MacIntyre S, Sooman A (1991) Nonpaternity and prenatal genetic screening. Lancet 338:869-871

Manning JT (1995) Fluctuating asymmetry and body weight in men and women: implications for sexual selection. Ethol Sociobiol 16:145-155

Manning JT, Koukourakis K, Brodie DA (1997) Fluctuating asymmetry, metabolic rate and sexual selection in human males. Evol Human Behav 18:15-21

Manning JT, Scutt D, Lewis-Jones DI (1998) Developmental stability, ejaculate size and sperm quality in men. Evol Human Behav 19:273-182

Martins EP (2000) Adaptation and the comparative method. Trends Ecol Evol 15:296-299

Møller AP, Swaddle JP (1997) Asymmetry developmental stability and evolution. Oxford University Press, Oxford, UK

Ober C, Weitkamp LR, Cox N, Dytch H, Kostyu D, Elias S (1997) HLA and mate choice in humans. Am J Human Genet 61:497-504

Oliver-Rodriguez JC, Guan Z, Johnston VS (1999) Gender differences in late positive components evoked by human faces. Psychophysiology 36:176-185

Penn D, Potts WK (1998) Chemical signals and parasite-mediated sexual selection. Trends Ecol Evol 13:391-396

Penton-Voak IS, Perrett DI (2000) Female preference for male faces changes cyclically: further evidence. Evol Human Behav 21:39-48

Penton-Voak IS, Perrett DI, Castles D, Burt M, Koyabashi T, Murray LK (1999) Female preference for male faces changes cyclically. Nature 399:741-742

Perrett DI, Lee KJ, Penton-Voak I, Rowland D, Yoshikawa S, Burt DM, Henzi SP, Castles DL, Akamatsu S (1998) Effects of sexual dimorphism on facial attractiveness. Nature 394:884-887

Petrie M, Kempenaers B (1998) Extra-pair paternity in birds: explaining variation between species and populations. Trends Ecol Evol 13:52–58

Pinker S (1994) The language instinct: how the mind creates language. William Morrow, New York

Reeve HK, Sherman PW (1993) Adaptation and the goals of evolutionary research. Q Rev Biol 68:1–32

Rice WR (1996) Sexually antagonistic male adaptation triggered by experimental arrest of female evolution. Nature 381:232–234

Rice WR, Holland B (1998) The enemies within: intragenomic conflict, interlocus contest evolution (ICE), and the intraspecific Red Queen. Behav Ecol Sociobiol 41:1–10

Rikowski A, Grammer K (1999) Human body odour, symmetry and attractiveness. Proc R Soc Lond Ser B 266:869–874

Rowe L, Houle D (1996) The lex paradox and the capture of genetic variance by condition-dependent traits. Proc R Soc Lond Ser B 263:1415–1421

Sasse G, Muller H, Chakraborty R, Ott J (1994) Estimating the frequency of nonpaternity in Switzerland. Human Hered 44:337–343

Shackelford TK, Weekes-Shackelford VA, LeBlanc GJ, Bleske Euler HA, Hoier S (1999) Female coital orgasm and male attractiveness. Human Nat 11:299–306

Shackelford TK, LeBlanc GJ, Weekes-Shakelford VA, Bleske-Rechek AL, Euler HA, Hoiers S (2002) Psychological adaptation to human sperm competition. Evol Human Behav 23:123–138

Simpson JA, Gangestad SW, Christensen PN, Leck K (1999) Fluctuating asymmetry, socio-sexuality and intrasexual competitive tactics. J Pers Soc Psychol 76:159–172

Smith RL (1984) Human sperm competition. In: Smith RL (ed) Sperm competition and the evolution of animal mating systems. Academic Press, London, pp 601–660

Smuts BB, Smuts RW (1993) Male aggression and sexual coercion of females in nonhuman primates and other mammals: evidence and theoretical implications. Adv Stud Behav 22:1–63

Symons D (1979) The evolution of human sexuality. Oxford University Press, Oxford

Symons D (1987) If we're all Darwinians, what's the fuss about? In: Crawford CB, Smith MF, Krebs DL (eds) Sociobiology and psychology: ideas, issues, and applications. Erlbaum, Hillsdale, NJ, pp 121–146

Symons D (1992) On the use and misuse of Darwinism in the study of human behavior. In: Barkow J, Cosmides L, Tooby J (eds) The adapted mind. Oxford University Press, Oxford, pp 137–159

Thoma RJ, Yeo RA, Gangestad SW, Lewine JD, Davis JT (2002) Fluctuating asymmetry and the human brain. Laterality 7:45–58

Thompson AP (1983) Extramarital sex: a review of the research literature. J Sex Res 19:1–22

Thornhill N, Thornhill R (1987) Evolutionary theory and rules of mating and marriage pertaining to relatives. In: Crawford C, Smith M, Krebs D (eds) Sociobiology and psychology: ideas, issues, and applications. Erlbaum, Hillsdale, NJ, pp 373–400

Thornhill R (1990) The study of adaptation. In: Bekoff M, Jamieson D (eds) Interpretation and explanation in the study of behavior. Westview, Boulder, CO, pp 31–62

Thornhill R (1997) The concept of an evolved adaptation. In: Daly M (ed) Characterizing human psychological adaptations. Wiley, London, pp 4–13

Thornhill R, Furlow B (1998) Stress and human reproductive behavior: attractiveness women's sexual development postpartum depression and baby's cry. Adv Stud Behav 27:319–369

Thornhill R, Gangestad SW (1993) Human facial beauty: averageness, symmetry, and parasite resistance. Human Nat 4:237–270
Thornhill R, Gangestad SW (1994) Fluctuating asymmetry and human sexual behavior. Psychol Sci 5:297–302
Thornhill R, Gangestad SW (1999a) The scent of symmetry: a human sex pheromone that signals fitness? Evol Human Behav 20:175–201
Thornhill R, Gangestad SW (1999b) Facial attractiveness. Trends Cognitive Sci 3:452–460
Thornhill R, Møller AP (1997) Developmental stability, disease and medicine. Biol Rev 72:497–548
Thornhill R, Palmer CT (2000) A natural history of rape: biological bases of sexual coercion. MIT Press, Cambridge, MA
Thornhill R, Gangestad SW, Comer R (1995) Human female orgasm and mate fluctuating asymmetry. Anim Behav 50:1601–1615
Thornhill R, Gangestad SW, Miller R, Scheyd G, McCollough JK, Franklin M (2003) Major histocompatibility genes, symmetry and body scent attractiveness in men and women. Behav Ecol (in press)
Van Dongen S (2000) The heritability of fluctuating asymmetry: a Bayesian hierarchical model. Ann Zool Fennici 37:15–23
Wedekind C, Füri S (1997) Body odor preference in men and women: do they aim for specific MHC combinations or simply heterozygosity? Proc R Soc Lond Ser B 264:1471–1479
Wedekind C, Seebeck T, Bettens F, Paepke AJ (1995) MHC-dependent mate preferences in humans. Proc R Soc Lond Ser B 260:245–249
Wilcox AJ, Weinberg CR, Baird BD (1995) Timing of sexual intercourse in relation to ovulation. N Engl J Med 333:1517–1521
Williams GC (1966) Adaptation and natural selection: a critique of some current evolutionary thought. Princeton University Press, Princeton, NJ
Wilson M, Daly M (1985) Competitiveness risk taking and violence: the young male syndrome. Ethol Sociobiol 6:59–73
Wilson M, Daly M (1992) The man who mistook his wife for a chattel. In: Barkow JH, Cosmides L, Tooby J (eds)The adapted mind: evolutionary psychology and the generation of culture. Oxford University Press, Oxford, pp 289–326
Yeo RA, Hill D, Campbell R, Vigil J, Brooks WM (2000) Developmental instability and working memory ability in children: a magnetic resonance spectroscopy investigation. Dev Neuropsychol 17:143–159

Subject Index

13C/14C ratio   81
abstraction   149
acacia tree *(Acacia nilotica)*   91
accentuation   135
activation patterns   47, 49, 53, 57
adaptationism   12, 342, 344
adaptations   15–16, 27, 42, 53, 145, 253, 255–256, 286, 341–342, 345, 347
adaptedness   191–192
adaptive value   175
adaptiveness   19
aesthetic adaptation   27
aesthetic appeal   201, 246, 261, 263
aesthetic artefacts   266
aesthetic attraction   110–112, 115
aesthetic cognition   255
aesthetic display   268
aesthetic experience   20
aesthetic expression   70, 71, 88, 99
aesthetic judgement   30, 239, 251, 254
aesthetic perception   133, 253, 255
aesthetic preferences   18, 71, 83, 85, 117, 133, 174, 180, 239, 248, 253–254, 263, 272, 280–283, 286, 290–291
aesthetic sense   174, 176, 191–192
aesthetic value   246
Aesthetica   133
aesthetical skills   30
aesthetics- artificial aesthetics   242
aesthetics- classical   23
aesthetics- content aesthetics   282
aesthetics- enviromental aesthetics   280–282, 291
aesthetics- evolution of aesthetics   39, 40–42, 44, 55, 59, 60, 65, 246
aesthetics- experimental aesthetics   11, 31
aesthetics- formal aesthetics   282
aesthetics- in the humanities   10
aesthetics- natural aesthetics   242
aesthetics- neuronal aesthetics   40, 46, 55, 62, 65
aesthetics- scientific aesthectics   10, 11
aesthetics- sensory aesthetics   133, 163
aesthetics- sexual aesthetics   11, 240
aesthetics- social aesthetics   291
aesthetics- symbol aesthetics   282
aesthetics- traditional aesthetics   9, 20, 25, 343
aesthetics- Western aesthetics   206
African rock python (*Python sebae*)   81, 106
African tiger lily (*Agapanthus africanus*)   91
aichmophobia   95, 111
alarm signal   268
amygdala   103
ancestral cue   18, 21, 27, 74
androgenetic alopecia   213
anorexia   201, 217, 220, 225
anti-rape adaptation   347
antisnake response   106
Ardipithecus ramidus kadabba   80
Ardipithecus ramidus ramidus   81
Aristotle   132, 136
art   22, 40, 44–45, 60, 71, 95, 131, 132, 137, 145, 159, 163, 202, 248, 250, 255, 263
artificial body odour   325
artificial enviroment   284
artificial intelligence   51
artistic expression   114, 116
asexual reproduction   326, 327
asymmetry   208, 269, 352, 362
athleticism   215, 223, 225, 230
Australian pine (*Pinus nigra*)   83

Australopithecus afarensis 82
Australopithecus anamensis 81, 82
autism 101
averageness 296

baboons (*Papio spp.*) 81
baby schema 158
baldness 213
barn swallows (*Hirundo rustica*) 173, 183
beard 184, 186, 222, 223
beauty 39, 40, 41, 44, 56, 59, 61-62, 65, 116, 131,139,158, 175-177, 180, 201-204, 231, 239, 242-244, 246, 248, 251-252, 255-256
beauty- artificial beauty 249
beauty- bias 24
beauty- concept of beauty 55
beauty- enhancement 209
beauty- essence beauty 21
beauty- inflationary devaluation of beauty 246
beauty- intellectual 10, 24, 29
beauty- myth 203
beauty- natural beauty 248
beauty- perception 174
beauty- physical beauty 83, 299
beauty- preference 59, 175, 184
beauty- production 174
beauty- signal qualtity 247
beauty- universal beauty 23
bifacial knapping method 261
biodiversity 281-282, 287
biological reductionism 39
biophilia 11, 16, 26, 28, 255
biotype preference 280
birth complications 188
black mamba (*Dendroaspis polylepis*) 108
blue draceana (*Draceana indivisa*) 91
BMI (body mass index) 189
bodily qualities 208
body composition 279
body deformation 244
body form 28
body movement 279, 299, 303, 307, 310, 316
body preference 295
body symmetry 354, 361

body weight 216
bodybuilding 223-224, 231
bonnet macaques (*Macaca radiata*) 101, 111
bonobos *(Pan paniscus)* 178
brain 39, 265
brandings 202
brownish gopher tortoise (*Gopherus polyphemus*) 106
bulimia 220
by-product 78, 253-255, 297, 342-343, 358, 360, 363

California ground squirrel (*Spermophilus beecheyi*) 75
camouflage 182, 344
capuchin monkeys *(Cebus apella)* 134
carrion crows (*Corvus corrone*) 134
categorical perception 145
celibacy 221
chase-away model 177
cheating signals 182
chimpanzee (*Pan troglodytes*) 106, 112, 141, 178, 265
chimpanzee painting 141
chopper 93
cobra (*Naja naja*) 106
coevolution 348
coherence 281, 283, 288
color perception 304
color preference 216
comic strips 223
Commiphora spp. 91
common walkingstick insect (*Diapheromera velii*) 344-345
communication 155, 182
communication- mother-child communication 252
communication- nonverbal communication 157, 299, 317
communication- prey/predator communication 247
communication- quality of communication 299
communication- visual communication 157
compariot 213
complexity 281, 283, 290, 343
condition-dependent outputs 351

conspecific recognition 99, 112
conspicuous leisure 245
constructivism 145, 201–202, 205, 230
co-optation 252
coral snake (*Micrurus spp.*) 106
cost-benefit-calculation 179, 182, 246
costly signal 240–241, 249, 251, 256
counteracting diversity 164
courtship displays 248
Crépe myrtle (*Lagerstroemia indica*) 91
Cro-Magnon 97
cross-cultural 26, 69, 70, 102, 116, 153, 157, 167, 189, 204–206, 209, 216, 225–226, 230–231, 288
cross-historical 209
cross-species 347
cultural adaptation 160, 229
cultural evolution 164
cultural filter 57
cultural history 248
cultural perception 201
culture-specific 159–160
curvaceousness 216

danger recognition 91
Darwin 2–4, 209–210, 297, 248
Darwinian algorithm 256
demonstration of ability 267
demonstrative waist 250
dendritic spine 76
determinism 54
developmental deprivation 75
developmental homeostasis 297
dot-pattern-selective cells 110
dress styles 214
Drosophila 72

ecological niche 45
ecological variations 214
economic competition 250
economic prosperity 231
EEA (enviroment of evolutionary adaptedness) 301
EEG (electroencephalography) 54, 86
encoding of reality 167
enculturation 42
entoptic phenomena 143
enviromental information 14

EPC (extra-pair copulation) 341, 345–346, 353, 355, 357–358, 360, 362–362
EPC- adaptation 343, 347–349, 351, 355, 361
EPC- in birds 358
EPC- in gibbons 358
eroticism 255, 311, 359
estrogen 186, 188, 297, 301, 303
ethnic group 221
evaluative perception 133
evil eye 101
evolution of aesthetic preferences 240
evolution of sex 327
evolutionary biology 59, 174, 239
evolutionary ethology 59
evolutionary history 286, 291, 347
evolutionary linguistics 255
evolutionary optimization 51
evolutionary psychology 13, 25–26, 69, 295
evolutionary theory 41, 58, 173, 280, 284, 290, 291, 295
expressional clues 153
expressive feature 207
eye-like schemata 99, 103, 117
eyespot patterns 99, 115, 155

face detecting 52
face perception 99, 115, 304
face processing 205
face recognition 73, 102–103, 301
facial asymmetry 226
facial attractiveness 186, 187, 230, 296
facial characteristics 211
facial expression recognition 76, 103–104
facial feature 208
facial hair 204, 222, 298, 230–231
facial patterns 90
facial qualities 208
facial symmetry 18, 26, 179, 186, 187, 254, 279, 296, 361
false gharial (*Tomistoma schlegelii*) 105
fashion industry 231
father presence 219
fatness 202
feelings 20, 22
felid spot evolution 110–111

female beauty   224
female breasts   188, 190, 207, 303
female choice   174
female copulatory orgasm   353
female mate competition   180
feminism   203
fingernails   211
Fisherian runaway process   177, 182–183, 185, 190–192
fitness   60, 207, 213, 226, 240, 272, 361–362
fitness maximization   173, 243, 256
fitness-promoting   239, 242
flakeblades   93
fMRI (functional magnetic resonance imaging)   71, 103, 108, 318
food sharing   265
fragrances   325
free-rider problem   251
functional design   13, 17, 19, 341, 343
functional organization   51–52
functional preference   254

Gaboon viper (*Bitis gabonica*)   106
gender identification   301
gender recognition   305
gene flow   12
gene frequency   12
gene-drift   12, 343–344
generalization of features   148
genetic abnormalities   208
genetic diversity   296
genetic fitness   173
genetic mutations   12, 269, 327, 342, 344
genetic quality   183
geometrical patterns   139
gestalt psychology   135, 144
giant civet-like viverrid (*Viverra leakeyi*)   80
giant hyena (*Pachycrocuta*)   81
glass snake (*Ophisaurus ventralis*)   106
GNP (gross national product)   218
good genes   177, 181, 186, 192, 240–241, 252, 256, 266, 269, 351–352, 358, 362
graphical symbolism   69
green monkeys (*Chlorocebus aethiops sabaeus*)   111
grooming   201, 202, 205, 207, 209–214, 217, 219, 220, 225, 229–230

group cohesiveness   69
group differences   289–290
group solidarity   251
guppy (*Poecilia reticulata*)   227

habitat perception   71, 79, 288, 291
habitat preference   14, 279–280, 285–289, 291, 344–345
habitat theory   280
hair color   212
hairstyle   212–213
handaxes   254, 261, 263–264, 266–272
handicap principle   177, 182–183, 186–187, 190–192, 240, 242–245, 247
harmony   132
hereditary diseases   186
heritability   15, 78, 177, 240, 296, 299, 361
heterozygosity   296, 329, 362
heterozygosity hypothesis   178, 181
hips   188, 207
Homo erectus   139, 265
Homo ergaster   81–82, 139, 261, 265
Homo habilis   261
Homo heidelbergensis   261, 265
Homo neanderthalensis   97, 261, 265, 272
honest signal   184, 189, 191, 240–242, 245, 247–248, 252–253
hormonal influence   310
hormonal status   317–318
host-parasit coevolution   177
human evolutionary history   341
humanities   10, 46
hydrophilia   281
hypermasculine body ideal   223–224

identifiability   289
ideology   25
illusory face expression   102, 104
image anthropology   65
immune response   181
immune system   208
immunocompetence handicap hypothesis   184
immuno-genes   328–330, 334
imprinting   107
incidental matching   344
Indian python (*Python molurus*)   106
indicator of attractiveness   319

Subject Index   373

indicator of fitness   177, 226
indicator of good genes   269–271
indicator of good health   268
indirect mate choice   174
individual recognition   99
individualistic taste   180
industrial exemption   211
industrialization   227
infant mortality   215, 218, 220
inferotemporal cortex   73
information reduction   282
innate perceptual process   74
intellectual beauty   29
intentionality   47
intergroup competition   250
inter-male competition   271
internal activation patterns   50
internal representation   48, 50–51, 53–55, 57, 62–64
intrasexual competition   222

Jugendstil   43

landscape paintings   161
landscape perception   21
landscape preferences   27, 70, 82, 117, 158, 279, 281, 284, 287–291
language   57, 62, 183, 202, 266, 342
lateral inhibition   144
legibility   281, 288, 290
leopard (*Panthera pardus*)   81
leopard recognition   111
leopard spots   110
life span   218
lion (*Panthera leo*)   81
long-tailed macaques (*Macaca fasicularis*)   106, 134

magnetoencephalography   108
maladaptive   341
male competition   174, 179, 300
male preferences   179, 187, 189
Margaret Thatcher illusion   90
masking hypothesis   326
mate abuse   346
mate competition   224
mate-copying   227–228
mathematics   263
mating behavior   58

mating market   179
mating preference   61, 173, 176, 181, 184, 239, 241, 266, 329, 331, 357, 360
mating status   305
mating strategy   181
mating system   178
memory   40, 44
mental representation   166
mestruation   305
MHC (major histocompatibility complex)   181, 211, 328, 330, 333, 362–363
mimesis   132
mimicry   182
mind   48
mirror neurons   319
momentary activation mode   47
moral judgement   251
mother-child complex   160
motion display   300
motion-energy-detection   305
mouse   329–330
mouse lemurs (*Microcebus murinus*)   101
mouthing behavior   87
movement handicaps   303
movement perception   310
Muller's ratchet   327
multiple fitness analysis   201, 207–209, 213–214, 221, 230–231, 298
music   32
mustelid (*Perunium*)   80
mutual mate choice   178
mystery   21, 31, 281, 283–284, 289, 291

natural beauty   131, 183
natural classicism   201–202, 205
natural deprivation   75
natural enviroment   284
natural perception   139
natural playgrounds   285
natural signal   330
naturalism   43, 163, 255
naturalistic fallacy   23
nature of beauty   43
neckties   211
neonate feature   207, 229, 231
nepotism   29, 249
nervous system   42
neural networks   40, 48, 307–308, 317

neurophilosophy   45
neurophysical preference   44
neurophysiology   52
neuroscience   42
nil-crocodiles (*Crocodylus niloticus*)   81, 105
nonadaptation   19
nonadaptedness   192
nonadaptive   341
nonverbal behavior   299
novelty   210

objective world   63
objectivity of knowledge   52–53
odours   325
Oldowan tools   261
optical illusion   144–145
orientation preference   58
outer world   64

Pacific gopher snake (*Pituophis melanoleucus catenifer*)   75
Parathropus robustus   81
parental investment   174
pathogen stress   226–227, 249, 266, 296, 328, 362
pattern recognition   70, 72–76, 112, 114, 117,
peacock's tail   10, 59, 173, 240, 252, 266, 269
peak-shift   185
perceiving patterns   47, 135
perception of 3D shape   304
perception of physical appearance   207
perception of physical attractiveness   205, 318
perception of trees   82
perceptual adaptation   157
perceptual attributes   283
perceptual bias   79, 82, 97, 114, 134, 158, 160, 268–269
perceptual codes   160
perceptual differences   287
perceptual guidelines   164
perceptual processing   149
perceptual psychology   133
perceptual representation   149
perceptual system   282
perfume industry   325

perfumes   325, 337
perfumes- diversity   330
perfumes- function   333
perfumes- ingredients   325, 330
perfumes- preferences   325–326, 334
perfumes- synthetic   334, 337
person perception   317
person recognition   304
personal ecology   218, 221
personality   221
PET (positron emission tomography)   103, 107
phenotypic quality   24, 173, 187, 302
phenotypic solutions   344
phobia   78
physical attractiveness   191, 202–204, 206, 208, 219, 225–229, 231, 290, 303, 319, 350, 355, 357, 359, 361
physical beauty   178, 180–181
physical ecology   221
phytophilia   158, 281
pictorial art   131
piercing   115, 244
pigtailed macaques (*Macaca nemestrina*)   99, 111
Plato   132, 137, 139, 166, 202–203
poetry   60–61, 248
pointed forms   97
polygyny   221
population biology   59
pornography   255
power   250
prägnanztendenz   144
predator recognition   73, 75, 105, 111, 115
predatory risk   79, 85, 176, 182, 265
preparedness   77–78
prestige   245, 247, 251
primate locomotion   300
proboscis monkeys (*Nasalis larvatus*)   105
prototypical templates   149
proximate term   131
psycho-evolutionary theory   279
psychological adapatation   26, 342, 344–345, 347, 351
psychological dynamics   219
psychological landscape   175
puff adder (*Bitis arietans*)   108
pupillary dilation   86, 94, 100

recognition of compatriots   208
red queen hypothesis   328
red sea dragon   344–345
redundant signal hypothesis   298
reflex patterns   47
regularities   133, 141
regularity recognition   136
reliability of paternity   345
religion   221
religious art   110
religious rites   210
reproductive fitness   246
reproductive strategies   179
reproductive success   179, 295, 341–342
reptilian scale patterns   105, 111
reptilian skin products   110
reticulated python (*Python reticulatus*)   106
rhesus macaques (*Macaca mulatta*)   105, 111, 157
role-playing game   97
romantic grooming   218
Rubin's cup   31
runaway process   176

saber-toothed cat (*Megantereon*)   81
savanna hypothesis   83, 280, 287
saw-scaled viper (*Echis pyramidum*)   108
scars   202, 225, 244
scene primitives   72
schematization   148, 151
sculptural art   161
selection   58, 253, 266, 348, 351, 358
selection- direct selection   13, 341–344, 350, 359
selection- frequency dependent selection   240
selection- habitat selection   279–280
selection- indirect selection   342, 358–359
selection- landscape selection   284
selection- mate selection   229, 271, 279, 319
selection- natural selection   70, 72, 74, 77–79, 114, 117, 176, 189, 243, 255, 281, 295
selection- perfume selection   330
selection- runaway selection   176
selection- sexual selection   11, 69, 173–174, 184, 190, 192, 279, 295, 352, 361

selection- sexually antogonistic selection   348
selective sensitivity   145–147, 153
selective sperm transport   350
self-adornment   202, 210
self-deception   350
self-modification   209
self-perception   219
self-presentation   205, 218–219, 230, 241, 245–246
sensory activation   49
sensory bias   176, 178, 185, 333
sensory cognition   133
sensory experience   51
sensory imprinting   40
sensory input   41, 47, 57
sensory processing   135, 166
sensory stimuli   40, 47–48
sex appeal   176
sex difference   290, 348
sex ratio   214–215, 217–218, 220–221, 223–225, 230–231
sex recognition   310
sex-specific   160, 347
sexual attractiveness   192, 325–326, 337, 349
sexual beauty signal   240
sexual behavior   58
sexual competition   247–249
sexual dimorphism   82, 88, 157, 173, 178, 301, 303
sexual dinichism   290
sexual jealousy   346, 358
sexual maturity feature   207
sexual monopolizing   346
sexual preference   353
sexual reproduction   326–327, 337
sexual signals   178
sexual vigilance   346
sharp forms   115
short-sabered cat (*Felis obscura*)   80–81
showing-off   250
sign stimuli   69
signal   42, 46, 155, 175, 177–178, 182–184, 189–191, 230, 240–242, 245, 247–249, 252–253, 255–256, 268, 281, 298, 330, 333
signal evolution   242, 248, 252
signal filtering function   52
signal identification   48

Subject Index

signal perception 48
signal propagation 50, 52
signal receiver 175, 270, 298
signal selection 241
signal sender 175, 182, 270
signal theory 175, 182
signal transfer 41
signaler 242, 252–254, 256
sign-formation 136
skin texture 297–298
skirt-lenghts 214–215, 219
skyscrapers 250
slenderness 216–217, 220, 225, 230
snake phobia 107
snake scale 109–110
snake-species recognition 110
SNNS (stuttgart neural network simulator) 308
social awareness 268
social competition 60, 223, 241, 250
social construction 203, 205
social display 117
social groups 182
social identity 201
social information 228
social learning 103
social monitoring 268, 270
social perception 204, 216
social releasers 155, 157
social scenario 30
social significance 345
social status 229
social stereotypes 206
social structure 221
sociobiology 59–60
socioeconomic level 220, 357
socio-sexual signalism 153
sooty mangabeys (*Cercocebus atys*) 111
Spanish dagger (*Yugga gloriosa*) 91
species recognition 71, 98–99
species-specific 15, 18, 157–158, 160, 280, 290
sperm competition 184
sports 60, 210, 225
spotted hyena (*Crocuta crocuta*) 81
spotted textures 110
squirrel monkeys (*Saimiri sciureus*) 107
standardization 135
status cues 29

sticklebacks (*Gasterosteus aculeatus*) 182, 240, 334
stimulation 210
stock market value 215
stone tools 261
stored models 135
storing information 135
striped bug (*Graphosoma lineatum*) 155
stylization 149
super-sign formation 136, 137, 142
susceptibility to dummies 155
symbol 131
symbolic abstraction 166
symbolic representation 110
symbolism 202
symbols of beautification 202
symmetry 18, 26, 43, 102, 133, 141, 173, 183, 186–187, 190, 201–202, 240, 254, 261, 264, 266–270, 295–298, 302, 319, 350, 352, 354, 356, 360–362
synaptic plasticity 76
system of aesthetics 44

tabula rasa 17, 25
tail symmetry 173, 183
taste of beauty 174
tatoos 202, 204, 225, 231, 244
TDNN (time-delay neural network) 308
teen pregnancy 218
template 156, 147, 167
tessellated graphics 105
testosterone 184–186, 188, 241, 290, 297, 300–301, 353, 355, 360–361, 363
texture 283, 304
the pill 308–310, 329
theoretical physics 63
theory of leisure class 243
theory of mind 318–319, 265
T-lymphocytes 329
trade off 228
tradtional cultures 137
trees 83
turquoise-browed motmots (*Emomota superciliso*) 106
twin studies 76, 78

ultimate term 131
umbrella thorns (*Acacia tortilis*) 83, 116
universal artistic code 154

visible signs of ovulation   190
visual codes   167
visual communication   157
visual cortex   73
visual experience   45
visual information   145, 164
visual orientation   42–43, 52
visual perception   166
visual preference   44
visual representation   143
vitality   244
voice   185

voluptuousness   220
vomeronasal neurons   330

waist-to-hip ratio   188–189, 217
walking style   305
wasp waist   244
water   90
water perception   86, 88
weather   28
within-culture variation   218
wood rat (*Neotoma albigula*)   75

Printing (Computer to Plate): Saladruck Berlin
Binding: Stürtz AG, Würzburg